AF781504

Handbook of Livestock Management

Handbook of Livestock Management

Edited by Ashlie Archer

SYRAWOOD
PUBLISHING HOUSE

New York

Published by Syrawood Publishing House,
750 Third Avenue, 9th Floor,
New York, NY 10017, USA
www.syrawoodpublishinghouse.com

Handbook of Livestock Management
Edited by Ashlie Archer

© 2020 Syrawood Publishing House

International Standard Book Number: 978-1-68286-852-2 (Hardback)

This book contains information obtained from authentic and highly regarded sources. Copyright for all individual chapters remain with the respective authors as indicated. All chapters are published with permission under the Creative Commons Attribution License or equivalent. A wide variety of references are listed. Permission and sources are indicated; for detailed attributions, please refer to the permissions page and list of contributors. Reasonable efforts have been made to publish reliable data and information, but the authors, editors and publisher cannot assume any responsibility for the validity of all materials or the consequences of their use.

The publisher's policy is to use permanent paper from mills that operate a sustainable forestry policy. Furthermore, the publisher ensures that the text paper and cover boards used have met acceptable environmental accreditation standards.

Trademark Notice: Registered trademark of products or corporate names are used only for explanation and identification without intent to infringe.

Cataloging-in-Publication Data

Handbook of livestock management / edited by Ashlie Archer.
p. cm.
Includes bibliographical references and index.
ISBN 978-1-68286-852-2
1. Livestock. 2. Livestock--Management. 3. Food animals. 4. Animal culture. I. Archer, Ashlie.
SF140.L65 H36 2020
636--dc23

Printed in China.

TABLE OF CONTENTS

PREFACE

Over the recent decade, advancements and applications have progressed exponentially. This has led to the increased interest in this field and projects are being conducted to enhance knowledge. The main objective of this book is to present some of the critical challenges and provide insights into possible solutions. This book will answer the varied questions that arise in the field and also provide an increased scope for furthering studies.

Livestock refers to domesticated animals that are raised in an agricultural setting to produce commodities like eggs, fur, meat, milk, wool and leather. Livestock animals can be domestic, semidomestic or captive wild animals. Livestock production is a major economic and cultural contributor in many rural communities. Proper feeding, good husbandry and hygiene are important for ensuring the good health of animals. The routine use of vaccinations and antibiotics promote growth in animals and render immunity against deadly diseases. Intensive animal farming is an upcoming aspect of modern agriculture, where livestock are reared in factory farms. This has potentially raised the yield but has also raised concerns on its negative impact on environment and animal health and welfare. This book includes some of the vital pieces of work being conducted across the world, on various topics related to livestock management. It strives to provide a fair idea about this discipline and to help develop a better understanding of the latest trends in this field. It will prove to be immensely beneficial to students and researchers in this area of study.

I hope that this book, with its visionary approach, will be a valuable addition and will promote interest among readers. Each of the authors has provided their extraordinary competence in their specific fields by providing different perspectives as they come from diverse nations and regions. I thank them for their contributions.

Editor

1

Animal agriculture and the importance of agnostic governance of biotechnology

Alison L. Van Eenennaam* and Amy E. Young

Abstract

Human population growth and rising income levels in developing countries are increasing demand for animal protein. One of the key enablers of the associated increase in global animal protein production has been biotechnology, defined as, "any technological application that uses biological systems, living organisms or derivatives thereof to make or modify products or processes for specific use." Biotechnologies have directly benefitted the three core scientific disciplines of animal science—genetics, nutrition, and health. Significant potential remains to use biotechnologies to improve animal health. Globally, more than 20 % of animal protein is lost as a result of disease. A number of diseases have been targeted by using recombinant DNA (rDNA) techniques in the breeding process to develop disease-resistant food animals, although no such animals have yet been approved anywhere in the world. Part of the reason for this is that "modern" biotechnologies involving the use of rDNA are subject to a unique set of governance and regulatory requirements under the Cartagena Protocol on Biosafety and other national regulatory frameworks. The Protocol defines "modern biotechnology" as the application of in vitro nucleic acid techniques, including recombinant deoxyribonucleic acid (DNA) and direct injection of nucleic acid into cells or organelles, or fusion of cells beyond the taxonomic family, that overcome natural physiological reproductive or recombination barriers and that are not techniques used in traditional breeding and selection. In considering the impact of this modern biotechnology trigger for additional governance and regulatory oversight, a case study is presented of the various biotechnological approaches that might be employed to address the important tropical disease problem of African trypanosomiasis. Some approaches involve the use of natural gene drive systems ("selfish" gene elements that skew inheritance in their favor) and irradiation-induced sterile insect technique. Others involve techniques that trigger the modern biotechnology definition and include the use of an rDNA-derived paratransgenesis, a strategy that employs symbiotic microbes to control pathogens in vector populations, and the development of genetically engineered trypanosomiasis-resistant cattle. Despite the fact that all of these approaches are associated with potential harms and potential benefits, only those that involve the use of modern biotechnology such as rDNA techniques are subject to exceptional regulatory requirements. Triggering governance and regulatory oversight based on an arbitrarily-defined subset of techniques rather than on the outcomes or products resulting from the use of those techniques, does nothing to address the potential harms that might be associated with non-governed processes and disadvantages governed technologies with unique regulatory burdens. Even-handed evaluation that agnostically weighs the potential benefits and risks of products rather than the techniques used to produce those products is essential to ensure that the biotechnology best suited to addressing a problem can be employed, rather than a potentially less efficient approach that is chosen solely because it avoids the complicated regulatory frameworks that are uniquely triggered by the use of a modern biotechnology.

Keywords: Animal biotechnology, Genetic engineering, GMO, Sterile insect technique, Regulation, Governance

*Correspondence: alvaneenennaam@ucdavis.edu
Department of Animal Science, University of California, 2113 Meyer Hall, One Shields Avenue, Davis, CA 95616, USA

Background

On current trends, the United Nations Food and Agriculture Organization (FAO) projects that there will be a 73 percent increase in meat and egg consumption and a 58 percent increase in dairy consumption [1] over 2011 levels worldwide by the year 2050. Already, global meat production has more than doubled since 1970, increasing from 120 million tonnes to more than 270 million tonnes in 2010. Similarly, dairy production in the same period rose from 400 million tonnes to 690 million tonnes [2]. This change in production resulted both from dramatic improvements in the amount of product per animal due to genetic selection and improved nutrition [3], and increases in the sizes of animal populations. This latter point is perhaps most clearly illustrated by the fact that between 1980 and 2010 the world's population of chickens increased almost threefold, from 7.21 to 19.60 billion head [4]. For other livestock species, the current global populations are estimated to be 1.43 billion cattle, 1.87 billion sheep and goats, and 0.98 billion pigs [5]. One important driver for this dramatic increase in animal numbers and animal protein production, termed "the livestock revolution" [6], is growing demand for milk, meat and eggs. Two main forces are driving this demand for animal protein: human demographics and economic development.

Alongside the global livestock population, the world's human population has grown exponentially, from about 4 billion people in 1975 to more than 7 billion today. By 2050, this number is expected to rise to approximately 9.6 billion [7]. Paralleling the human population growth, the world economy has also been growing dramatically over recent decades, with a 20-fold increase in global gross domestic product between 1970 and 2012 [4].

As incomes rise in developing economies, especially among traditionally undernourished populations, an initial increase occurs for dietary calorie intake of all food products equally, and then transitions to a decrease in per capita consumption of cereals and vegetables coupled with an increase in consumption of sugar, fats and animal products. This transition can lead to improved health because the overall nutritional value of animal protein is higher than that of staple foods such as cereals, roots and tubers. It has been shown that even small amounts of animal products can correct amino acid deficiencies in cereal-based human diets [8]. Consumption of milk and other animal-source foods by undernourished children improves anthropometric indices (weight-for-height, height-for-age and weight-for-age) and cognitive function and reduces the prevalence of biochemical and micronutrient deficiencies, reducing morbidity and mortality [9]. Animal source proteins are also more digestible and metabolized more efficiently than plant proteins, meaning they are more rapidly and readily available for use [10]. In one report, the consumption of meat was associated with a 36 % decrease in the likelihood of stunting among toddlers (12–24 months) living in Guatemala, Democratic Republic of Congo, Zambia, and Pakistan [11]. This shift of food consumption patterns in developing countries towards livestock products is a major factor to be considered in the growth of world agriculture.

The importance of animal disease

It is interesting to note that developing economies actually produce more meat and milk than developed economies. However, this supply is still insufficient to meet demand due to the fact that much of the growth in the world's population and in incomes is also occurring in developing economies. Animal system productivity within developing regions is generally less than in developed countries owing to a combination of factors including inadequate animal nutrition (as a consequence of low feed quality, quantity and lack of nutritionally-balanced diets), lack of access to genetic selection techniques, and higher incidence of livestock disease [3]. Developing economies are faced with a number of human, zoonotic (can be transmitted between animals and humans) and livestock diseases, creating a major impediment to both economic development and food safety [4]. The World Organization for Animal Health (OIE) estimates that worldwide an average of more than 20 % of animal protein is lost as a result of disease; therefore, significant potential exists to reduce this loss thereby decreasing the overall environmental impact per unit of animal protein production through improving the health of global livestock populations.

The widespread occurrence of animal disease in developing economies is one of the major factors negatively impacting livestock productivity in these countries. Generally, these diseases have the greatest impact on resource-poor livestock farmers and effective disease control is essential for poverty alleviation. Due to the devastating effects of disease on local livestock populations, many developing economies, particularly in sub-Saharan Africa, are net importers of livestock products. As a result, imports of livestock products from developed to developing economies are increasing. This is in part due to the fact that exports are confined to countries with officially confirmed and carefully monitored disease-free status and the absence of disease agents that are notifiable to the OIE. The net effect of these trends is that poor producers in some developing economies are unable to participate in supplying the domestic demand in their countries due to the prevalence of endemic animal diseases. Embracing solutions such as those afforded by various biotechnologies to prevent disease will be necessary to break this cycle of the economic effects of animal diseases and enable producers to participate in their local economies. This review will focus on the opportunities that biotechnologies offer to address some of the most pressing disease problems of animal agriculture and the use of these tools in meeting the global demands for animal protein.

The role of biotechnology in animal production

One of the key enablers of the increased global production of animal protein is biotechnology, which is defined by the Convention on Biological Diversity [12], as "any technological application that uses biological systems, living organisms or derivatives thereof to make or modify products or processes for specific use". The three core disciplines of animal science—genetics, nutrition, and health—have all benefited from a variety of biotechnological developments. Included under this broad heading are some relatively common and non-controversial biotechnologies such as artificial insemination (AI), use of genetic markers in breeding programs, feed additives, molecular diagnostics and vaccines. Others, such as cloning and the use of recombinant DNA (rDNA) technology to make transgenic or genetically engineered (GE) animals, recombinant vaccines and GE insects, have been associated with considerable controversy and often face both political and public opposition.

To date, the vast majority of biotechnologies have been produced and implemented in developed countries. However, various examples illustrate the potential of these approaches to improve health and productivity in livestock production systems and help reduce poverty and hunger in developing countries. The global elimination of Rinderpest (cattle plague) was achieved through a large-scale vaccination and surveillance program [13]. The sterile insect technique (SIT) has been used to suppress and locally eradicate or prevent the invasion of New World and Old World screwworms in North America (US), Netherlands, Central America and Libya [14]. The 2013 U.N. Food and Agriculture Organization (FAO) report, "Biotechnologies at Work for Smallholders: Case Studies from Developing Countries in Crops, Livestock and Fish," details 19 case studies in which agricultural biotechnologies were used to serve the needs of smallholders in developing countries. Among these was the application of SIT in Zanzibar to eradicate tsetse flies, the vector of trypanosomiasis. A common component of these successful programs has been global collaboration and investment by international organizations such as as the FAO, OIE, International Atomic Energy Agency (IAEA) and the World Health Organization (WHO), along with national political commitments.

These examples only provide a glimpse of the tremendous growth that has taken place in the fields of molecular genetics and genomics since the 1980s, and a range of DNA-based technologies are likely to play an increasingly important role in making future animal production more efficient, economical and sustainable. Some emerging rDNA-based technologies have the potential to address long-standing animal health issues that have proven intractable using traditional approaches. These technologies include GE animals that have been extensively reviewed elsewhere [15]. However, they are being deliberated in an often contentious global setting that includes the voices of many players, including the various bodies that are involved with international governance of animals and animal products, biosafety protocols, multinational animal activist and environmental corporations, and trade partners. Biotechnologies are increasingly being evaluated based not only on their technical effectiveness and feasibility, but also on their broader ethical and social implications, and this can lead to disagreements among interested parties depending upon their values and visions for a desirable future. It is important to employ the most suitable innovations, and in cases where that means rDNA technologies, it is becoming increasingly evident that evaluations need to be conducted in a manner that is agnostic to the technology used to achieve the result and focus on the unique risks and potential benefits that are associated with the end product or outcome.

Reducing livestock susceptibility to infectious diseases would seem to be a goal that aligns with most definitions of sustainability. Such a goal would improve animal welfare, decrease production losses and the use of therapeutic interventions such as antibiotics, and improve production efficiency. In fact, the goal of many traditional breeding programs in both plants and animals is to decrease host susceptibility to pests and disease. This target has been the ambitious goal of animal genetic engineers [16, 17] since the first GE livestock were generated almost 30 years ago [18]. In general, disease-resistant GE farm animals can be generated by either the introduction of resistance genes into the genome of the host (gain-of-function strategy) or the targeting of endogenous or exogenous susceptibility genes (loss or exchange-of-function strategy) [19]. There are many extant examples where researchers have successfully modified the genomes of livestock to render them less susceptible to disease (Table 1). Despite these technical successes, there are currently no GE animals approved for food purposes anywhere in the world. Conversely, GE plants, which were first produced in 1983 [20], were approved for food purposes in the 1990s and have been rapidly adopted by farmers globally. Approximately 180 million hectares

Table 1 Examples of successful modification of livestock genomes to render them less susceptible to disease

Disease	Species	References
African swine fever	pigs	[47]
Avian influenza	chickens	[48]
Avian leukosis virus	chickens	[49]
Bacteria resistance	sheep/pigs/fish	[50–53]
Bovine spongiform encephalopathy	cattle, sheep	[54, 55]
Bovine tuberculosis	cattle	[56]
Foot and mouth disease	various	[57]
Grass carp haemorrhage virus (GCHV)	fish	[58]
Influenza A & classical swine fever	pigs	[59]
Mastitis/health of nursing offspring	various	[60–65]
Porcine reproductive and respiratory syndrome virus (PRRSV)	pigs	[66]
Trypanosomiasis	various	[30]
Visna virus resistance	sheep	[67]

of GE crops (13 % of total arable land) were cultivated worldwide by 18 million farmers in 2014. Of these farmers, 94 %, or more than 17 million, were small acreage and resource-poor farmers in developing countries [21].

Regulation of animal biotechnologies

Part of the reason for the slow development of GE animals is that their potential deployment is associated with a daunting range of regulatory oversight and governance requirements from local and national governments, and international organizations. Globally, the Cartagena Protocol on Biosafety to the Convention on Biological Diversity (Protocol) is the preeminent instrument for the regulation of the products of GE technology at the international level. The stated objective of the Protocol relates to "the conservation and sustainable use of biological diversity", which may be adversely affected by the trade (or "transboundary movements") of "living modified organisms". "Biological diversity" (or biodiversity) refers to the number and variety of living organisms on the planet and is defined in terms of genes, species and ecosystems. The Protocol aims to ensure the safe handling, transport and use of living modified organisms (LMOs), both plants and animals, that "may have adverse effects on the conservation and sustainable use of biological diversity, taking also into account potential harms to human health", with specific focus on transboundary movements. The Protocol defines an LMO as any living organism that possesses a novel combination of genetic material obtained through the use of "modern biotechnology", which is defined as: (a) in vitro nucleic acid techniques, including recombinant deoxyribonucleic acid (DNA) and direct injection of nucleic acid into cells or organelles, or (b) fusion of cells beyond the taxonomic family that overcome natural physiological reproductive or recombination barriers and that are not techniques used in traditional breeding and selection. The Cartagena Protocol is therefore limited to the subset of biotechnology encompassed by the term "modern biotechnologies." Likewise, many national regulations are triggered by the use of rDNA technologies.

The United States of America (US) is not a party to the Cartagena Protocol. However, the trigger for the mandatory US Food and Drug Administration (FDA) regulation of GE animals is specified as those animals modified by rDNA techniques, including the entire lineage of animals that contain the modification. All GE animals are captured under these provisions, regardless of their intended use. Thus, although the regulatory evaluation is based on the product (the GE animal), the method used to produce the genetic change (i.e. rDNA versus other breeding methods) is the trigger for regulatory oversight. The fast-growing GE AquAdvantage salmon is the first GE animal to be evaluated by the FDA for food purposes. University researchers first produced the founder line for this strain in 1989 [22]. Despite the fact that the product has undergone nearly two decades of FDA regulatory scrutiny and the company has spent more than $77 million on development and regulatory costs, no decision regarding the application to commercialize this fish had been made as of October 2015. Much of this delay has been associated with political interference in the science-based regulatory review [23]. In contrast, there is no regulatory oversight of the many lines of fast-growing livestock species, including fast-growing Atlantic salmon, that have been produced using conventional selection and breeding techniques.

A case study: African trypanosomiasis

In considering the impact of "modern biotechnology" or "rDNA" triggers for additional governance and regulatory oversight of biotechnological tools to address animal disease, a case study is presented of biotechnological approaches that might be employed to address the major tropical disease African trypanosomiasis. This disease, caused by blood parasites of the genus *Trypanosoma* that are transmitted by tsetse flies, affects human health, livestock production, agricultural production (lack of draught animals and manure), rural socio-economic development, national economies (import and export of animal products), and the environment (insecticide and trypanocide applications). The WHO estimates that there are currently 10,000–45,000 cases of human African trypanosomiasis (HAT), with 60 million people at risk in 36 African countries. Without treatment, infection is almost always fatal.

Losses in livestock production due to African animal trypanosomiasis (AAT), or nagana, are estimated at US $1 billion annually [24]. Direct losses in meat production and milk yield, plus the costs for trypanosomiasis control programs, are estimated at up to US $1.2 billion each year. In many parts of sub-Saharan Africa (Fig. 1), the presence of tsetse flies is considered the major obstacle to the development of more efficient and sustainable livestock production systems and is one of the most important causes of hunger and poverty.

Current methods of controlling trypanosomiasis and the tsetse fly include: (1) SAT (sequential aerosol technique), an aerial application of extremely low volumes of nonresidual insecticides, (2) stationary bait techniques, i.e. use of insecticide-impregnated targets and traps that can be odor-baited, (3) the live bait technique, i.e. application of residual insecticides on livestock, (4) SIT (sterile insect technique), i.e. the release of sterile male insects, and (5) treatment of cattle with trypanocidal drugs [25]. The treatment of livestock with trypanocides and insecticides is associated with high costs, and resistance is developing to these pesticides and trypanocides. Due to the lack of effective vaccines (due to the high rate of antigenic variation by these blood parasites) and affordable drugs, vector control currently remains the most efficient strategy for sustainable disease management.

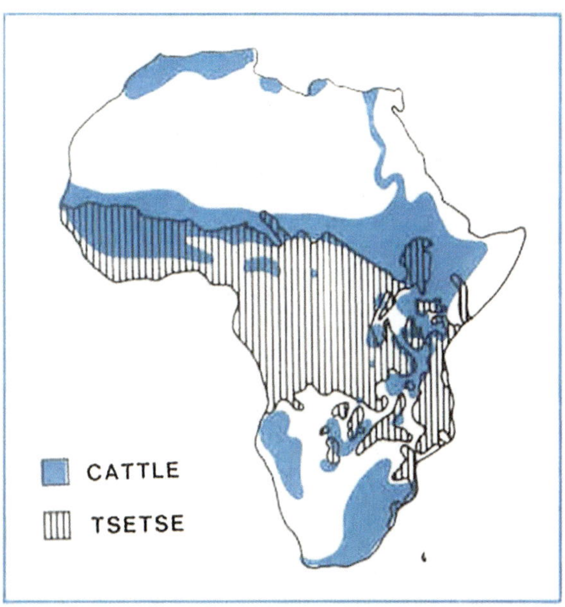

Fig. 1 Major cattle production areas and tsetse-infested zones in Africa. Trypanosomiasis occurs in livestock wherever there are tsetse flies. A lower density of cattle is found in tsetse-infested as compared to tsetse-free areas of Africa [46]. Used with permission (https://ilri.org/InfoServ/Webpub/fulldocs/Ilrad82/Trypano.htm)

An example of this is the successful eradication of the tsetse population on Unguja Island (Zanzibar) using an area wide integrated pest management (AW-IPM) campaign. With a disease prevalence of 19 %, the rural farming communities were unable to maintain livestock as a basis for farming, food production, or income-generation prior to the program. The fly population was initially suppressed using insecticide-based control strategies. SIT was used as an important final component of the campaign to drive the population to extinction [26].

Following eradication, increases in livestock and crop productivity and animal availability for transport and draft purposes [26] resulted in substantial social and economic improvements. A 30 % increase in yearly average income over 1999 levels was recorded in 2002. Demand increased for improved livestock breeds and the number of farmers raising improved cattle breeds increased from two percent in 1985 to 24 percent in 2002. During the same period, milk production nearly tripled and the proportion of farmers selling milk from indigenous cattle increased from 11 to 62 percent. Households with a monthly income of over US $25 increased from 69 to 86 percent; and those with a monthly income of over US $50 increased from 22 to 36 percent.

The project budget was approximately US $3.5 million over 4 years. An additional US $2 million included substantial insectary refurbishment, operational research, and many other activities including the need to establish large tsetse fly populations due to the reduced fitness of the males after irradiation, a facility to irradiate the males, and the weekly expenses involved with releasing sterile males. The successful eradication involved a substantial investment, but costs must be weighed against the many benefits obtained so far and those that are expected in the future, and against the cost of permanently investing in continuous suppression based on insecticides. This example is seen as a success story in the deployment of biotechnology to address an important disease problem.

In spite of these successes using a laboratory colony of the target species, efforts to establish mass rearing of tsetse colonies from the target field population revealed a susceptibility to salivary gland hypertrophy virus (SGHV). Despite initial success, the field colony experienced a steady decline over 2 years and collapsed [27]. The presence of SGHV in the wild tsetse has become a stumbling block for establishing and maintaining colonies for tsetse SIT control. A strategy to manage this tsetse virus will be required to continue this program.

As a result of the successes and limitations of SIT to date, complementary novel biotechnological approaches have recently been proposed and modeled to be sucessful. These involve manipulation of the symbiotic bacteria that have been found in tsetse flies, including the endosymbiont *Sodalis glossinidius*, and the reproductive symbiont

Wolbachia pipientis. *Wolbachia* is an obligatory intracellular and maternally transmitted bacteria that gives the female tsetse flies in which it resides a reproductive advantage over female flies in which it does not, therefore becoming more common over time in the tsetse population. In insects, embryonic lethality occurs when an infected male is crossed with an uninfected female. This lethality rapidly biases inheritance in the population and increases the frequency of *Wolbachia*-infected insects by transmitting to the next generation with near 100 % success, thereby creating a gene drive system for *Wolbachia*. This naturally occurring gene drive system can be used to increase the frequency of other maternally-inherited symbionts. Paratransgenesis, a term used to describe an insect carrying a GE microorganism typically modified to make the insect resistant to the disease-causing organism [28], can benefit from gene drive systems to spread the GE microorganism within an insect population. *Sodalis glossinidius* is ideally suited as a paratransgenic delivery system for anti-trypanosomal components to control parasite development in the fly. A strong gene drive system increases the power of paratransgenesis to replace non-transformed symbionts present in natural vector populations.

Another biotechnological approach that could be used to address trypanosomiasis would be the development of disease-resistant cattle. A consortium of scientists from the City University of New York, the International Livestock Research Institute (ILRI) in Nairobi, Kenya, the Roslin Institute, and Michigan University have been awarded a grant by The National Science Foundation of the USA and the Bill and Melinda Gates Foundation to develop GE cattle that are resistant to African Bovine Trypanosomiasis [29]. They plan to build upon work in transgenic mice where expression of a baboon trypanosome lytic factor resulted in sterile immunity (complete resistance) to trypanosomiasis [30].

In considering these various biotechnologies, radiation-induced SIT and the use of a *Wolbachia*-based invasive gene drive system would not be subject to rDNA regulations or the Cartagena Protocol on Biosafety, whereas paratransgenesis using GE bacteria, which would perhaps achieve the same goal of eliminating African tsetse and trypanosomiasis more effectively, would be subject to both. Gene drive was not an issue that was considered when the terms of the Cartagena Protocol were first negotiated. According to the Protocol, an advance informed agreement (AIA) procedure applies before the first environmental release of GE organisms in another country and grants the importing country the right to request the exporting country to perform a risk assessment at its own expense, part of which is to determine the likelihood of "unintentional transboundary movements". In effect, the *Wolbachia*-based invasive gene drive system carrying the endosymbiont *Sodalis glossinidiu* would be beyond the scope of the Cartagena Protocol. However, if *Wolbachia*-carrying tsetse flies also carried a paratransgenic GE endosymbiont bacteria modified to control parasite development in the fly they would be covered by the Cartagena Protocol. Environmental release would likely be disallowed due to the high probability of the same "unintentional transboundary movements" associated with the *Wolbachia*-based invasive gene drive system, irrespective of the potential disease-control benefits that might be associated with the paratransgenic GE endosymbiont bacteria.

Similarly to paratransgenesis, GE cattle would be subject to the Protocol and, depending upon the country, would also likely fall under additional national regulations that may include both human health and environmental risk assessments, and an evaluation of the social and ethical implications of the resulting phenotype. Although such cattle could be of great benefit to people and livestock in the endemic areas through reduced incidence of trypanosomiasis, to livestock producers through increased profitability, and to the environment through reduced pesticide use, their use could conceivably result in an overall decline in biodiversity due to increased livestock production in areas that were previously unavailable due to trypanosomiasis. It is possible that an increase in cattle numbers resulting from increased survival could lead to detrimental environmental effects such as overgrazing which has been shown to be a contributor to land degradation and desertification. Although it might also be argued that there would be a decreased incidence of non-productive, chronically-ill cattle and that if farmers are confident of stock survival, they might be more willing to invest in improved health and grazing management programs. The eradication of the tsetse fly from Zanzibar was associated with an increase in cattle numbers, an outcome that was considered one of the benefits of the SIT-based eradication program, although the island location of this intervention may have limited the possible impacts of increased cattle numbers on biodiversity. This seemingly inconsistent appraisal of the impacts of biotechnologies developed using rDNA technologies as compared to those developed using non-rDNA approaches is difficult to reconcile from a scientific perspective. The potential harms associated with the products derived from rDNA technologies have proven no higher than those associated with the use of conventional technologies and may be less than those associated with existing production systems; and the benefits from these technologies are potentially high. It would seem that the potential impact of the outcomes of all proposed biotechnologies should be evaluated in the context of achieving a balance between the ethical mandate to control disease on a global scale, respect for the sovereignty of states, and an agnostic evaluation of the potential harms and potential benefits associated with the outcome that is expected to result from a proposed biotechnology in the context of existing production systems, irrespective of the technology that was used to develop it.

Considerations and complications

To some, addressing the problems of global hunger and poverty should be the key and perhaps principle drivers of technology adoption, but considerations such as the protection of biological diversity or animal integrity can complicate the issues. This leads to inherent conflicts when a technological approach leads to the destruction or detriment of an organism (e.g. a pest) that is causing disease or decreased productivity, or an increase in the number of an agricultural species like cattle. While some may see that outcome as helping food security, others may view it as decreasing biodiversity or subjecting a species to an undesirable, unnatural genetic manipulation. While both may agree that the technology is effectively decreasing the prevalence of the vector, disagreements may ensue as to whether this is a desirable outcome based on individual values. These differing worldviews sometimes play out in regulatory evaluations of modern biotechnologies, and although couched as scientifically-based risk evaluations, factors regarding the desirability of the innovation, including ethical and social considerations and not science, often drive regulatory decisions. Regulation is increasingly being used to prevent the deployment of potentially beneficial biotechnologies based on contrasting values and visions for the future.

A 2012 decision in the Philippines regarding the introduction of a GM insect-resistant talong (eggplant) [through the insertion of a gene from *Bacillus thuringiensis* (Bt)] provides an example of this conflict. In considering this technological development, the Court of Appeals in the Philippines ruled against allowing experiments on Bt talong using an argument in support of biodiversity [31]; in this case, the biodiversity was represented by a species of borers that consume talong. The ruling stated "It is a deliberate genetic reconstruction of the eggplant to alter its natural order which is meant to eliminate one feeder (the borer) in order to give undue advantage to another feeder (the humans). The genetic transformation is one designed to make Bt talong toxic to its pests (the targeted organisms). In effect, Bt talong kills its targeted organisms. Consequently, the testing or introduction of Bt talong into the Philippines, by its nature and intent, is a grave and present danger to (and an assault on) the Filipinos' constitutional right to a balanced ecology because, in any book and by any yardstick, it is an ecologically imbalancing event or phenomenon. It is a willful and deliberate tampering of a naturally ordained feed-feeder relationship in our environment. It destroys the balance of our biodiversity. Because it violates the conjunct right of our people to a balanced ecology, the whole constitutional right of our people (as legally and logically construed) is violated". The decision is now on appeal in the Supreme Court of the Philippines.

This decision valued avoidance of a potential threat to biodiversity by targeting an agricultural pest over potential benefits associated with the pest-resistant product or the fact that pests threaten agricultural production systems and food security. All agricultural systems involve using selected crop and animal varieties and altering the species composition of a given ecosystem, presumably to the advantage of food production and human "feeders" and to the disadvantage of pests, weeds, and less productive food species. Frequently, agricultural technologies associated with pest control are controversial. Herbicides, insecticides, fungicides and antibiotics are all employed to control pests that threaten food production, namely weeds, insects, fungi and pathogens. Oftentimes, groups opposed to the use of these technologies in agricultural production systems examine only the potential harms or downsides, such as the potential development of resistance, and do not consider the benefits of their use or the tradeoffs that are associated with precluding their use. It has been estimated that without pesticides 70 % of the world food crop would be lost and, even with pesticide use, 42 % is destroyed by insects and fungal damage [32]. Dispensing with pesticides would require at least 90 % more cropland to maintain present yields [33]. Likewise, the development of disease-resistant varieties of plants and animals is associated with potential harms, benefits and tradeoffs. Ironically, the tradeoffs associated with limiting or precluding the use of rDNA in the process of breeding disease-resistant animals are likely to include diminished animal welfare, decreased animal protein production, per animal thereby necessitating an increased number of animals to produce a given quantity of protein, and an increased use of antibiotics and other pesticides (e.g. insecticides) to help control animal diseases around the world.

Perspective

The rDNA process-based governance of animal biotechnology is being increasingly questioned [34], and it is becoming difficult to justify given our increased understanding of genomes and the unfortunate fact that the current governance situation is resulting in the avoidance or abandonment of potentially useful applications that could be enabled by these technologies. The lack of scientific certainty about any potential adverse effects on biological diversity can be used to delay, or at worst permanently forestall, the adoption of potentially beneficial LMOs produced using modern biotechnologies. The precautionary principle requires no assessment or judgement of the potential benefits of a given modification, nor consideration of the opportunity costs associated with continuing to use existing animal genetic resources. This situation was perhaps most tersely summarized by Dr. Wells of the University of Missouri in Columbia in 2013, "The basic premise for regulation appears to be that any genotype produced by breeding is safe, and that any genotype produced intentionally via rDNA technologies cannot be allowed to go to market" [35].

Further complicating these regulatory issues are newly-available precision gene editing techniques that precisely mimic naturally-occurring evolutionary mutations, a process that can leave no DNA-based signature of modification in the genome to regulate. Recent review papers [36–38] summarize the use of site-specific designer nuclease genome editing techniques to improve livestock for agriculture. These methods are being used to explore a variety of approaches to address issues affecting animal agriculture. These include genetic research on generating offspring of a single sex for animal industries where one sex is favored (e.g. egg and dairy) [39, 40], development of genetic approaches for avoiding painful procedures such as dehorning and castration [40, 41] to improve animal welfare, and gene knockout approaches to preclude the production of endogenous proteins such as food allergens and prion proteins [42–44] to improve food safety. Many of these applications are directed towards improved animal welfare and again would seem to align with many definitions of sustainability. It is currently unclear whether these techniques, which do involve the "application of in vitro nucleic acid techniques", but frequently introduce genetic variation that is indistinguishable from naturally-occurring alleles induced by spontaneous mutation, will be subject to the modern biotechnology regulatory governance, especially given there is often no rDNA sequence(s) present in the product.

Similarly, disease vectors are also being targeted with these techniques. Mosquitoes are known to transmit a number of pathogens that cause diseases such as yellow fever, dengue and chikungunya. Together, these infections are responsible for more than 50,000 human deaths each year. A recent report details the use of genome engineering via the CRISPR-Cas9 system to generate stable germline mutations in the mosquito *Aedes aegypti*. The authors suggested that the techniques outlined in their paper could be useful not only for addressing the disease-transmitting abilities of *A. aegypti* through a variety of mutant alleles, but also for modifying other non-model organisms [45].

Triggering governance and regulatory oversight based on the use of modern biotechnologies in the development process, rather than on the outcomes or products resulting from the use of those technologies, is ineffective for addressing those risks that might be associated with non-governed processes and disadvantages governed technologies with unique regulatory burdens. If the risks and potential benefits resulting from a particular novel phenotype (e.g. disease-resistance) are the same, there would seem to be no rationale for regulating and increasing the development and implementation costs of only those produced using specific techniques (e.g. rDNA) in the breeding program. Perhaps more importantly, in some cases the risks and ecological impacts associated with unregulated processes or existing production systems may in fact be greater than those associated with using systems that employ modern biotechnological approaches.

Even-handed, technology-agnostic evaluation of animal biotechnologies is essential to ensure that one strategy is not chosen over another solely because it avoids onerous regulations. Regulation and governance to ensure the safety and responsible use of new technologies is necessary. However, such efforts should be proportional to the unique risks posed by the product being evaluated, and should not be triggered on the technique or biotechnology that was used to produce those products. Currently, identical products produced using different biotechnologies are subject to very different levels of scrutiny. It is perhaps this lack of logical consistency that most frustrates scientists working to develop new agricultural technologies. It is time to carefully evaluate and weigh both any unique risks and potential benefits posed by novel traits developed in animal breeding programs, irrespective of the breeding method that was used to introduce those traits. Overly burdensome regulation of modern biotechnologies is forestalling the development of solutions to real-world agricultural problems.

Conclusion

The stated objective of the Cartegena Protocol is "the conservation and sustainable use of biological diversity". However, as it is written it only covers transboundary movements of "living modified organisms" produced using specific modern biotechnologies, and considers only the potential risk associated with such developments. It ignores the fact that novel organisms created by traditional breeding and selection techniques may pose equivalent or greater risk, and also the opportunity costs associated with the delayed adoption of beneficial products that have been shown to be safe. And because the trigger is defined as a specific subset of biotechnologies, the definition may become outdated as technologies to make subtle changes to the genome evolve. Good public policy requires that regulation be clearly targeted to address an identified problem. Modern biotechnologies have not proven to be uniquely risky, and it is the attributes of the resulting organism that present both potential risks and benefits, regardless of the technology that was used to develop that organism.

The concern with the current governance of animal biotechnologies is that it actually results in an illogical incentive to pursue those technical approaches that are exempt from regulatory oversight, irrespective of the relative technical feasibility, likelihood of success, speed, potential harms, and potential benefits. A less effective, non-regulated approach may be preferred to a more precise, effective, and efficient regulated approach. In a world facing burgeoning demands for animal-source protein, creating arbitrary regulatory hurdles for potentially beneficial uses of modern animal biotechnologies based solely on their use of in vitro nucleic acid techniques cannot be justified scientifically, and is hindering the development of more productive animal agricultural systems globally.

Abbreviations

AW-IPM: area wide integrated pest management; CI: cytoplasmic incompatibility; FAO: United Nations Food and Agriculture Organization; FDA: US Food and Drug Administration; GE: genetically engineered; HAT: human African trypanosomiasis; IAEA: International Atomic Energy Agency; LMOs: living modified organisms; OIE: World Organization for Animal Health; Protocol: Cartagena protocol on biosafety to the convention on biological diversity; rDNA: recombinant DNA; SGHV: salivary gland hypertrophy virus; SIT: sterile insect technique; US: United States of America; WHO: World Health Organization.

Authors' contributions

AVE initiated this review and wrote and developed the content of this manuscript. AEY contributed to the literature review and assisted with writing and editing. Both authors read and approved the final manuscript.

Authors' information

AVE is an Animal Genomics and Biotechnology Extension Specialist in the Department of Animal Science at the University of California, Davis. AEY is a Staff Research Associate and manages the Van Eenennaam Laboratory at the University of California, Davis.

Acknowledgements

The authors would like to acknowledge support from the W. K. Kellogg endowment and the California Agricultural Experiment Station of the University of California, Davis, for the preparation of this paper.

Competing interests

The authors declare that they have no competing interests.

References

1. Food and Agriculture Organization of the United Nations (FAO). World Livestock 2011—Livestock in food security. Rome: FAO; 2011.
2. Guyomard H, Manceron S, Peyraud J-L. Trade in feed grains, animals, and animal products: current trends, future prospects, and main issues. Anim Front. 2013;3(1):14–8.
3. Capper JL, Bauman DE. The role of productivity in improving the environmental sustainability of ruminant production systems. Annu Rev Anim Biosci. 2013;1:469–89.
4. Food and Agriculture Organization of the United Nations (FAO). World Livestock 2013—changing disease landscapes. Rome. 2013. http://www.fao.org/docrep/019/i3440e/i3440e.pdf. Accessed 26 May 2015.
5. Robinson TP, Wint GR, Conchedda G, Van Boeckel TP, Ercoli V, Palamara E, et al. Mapping the global distribution of livestock. PLoS One. 2014;9(5):e96084.
6. Delgado CL, Rosegrant MW, Steinfeld H, Ehui SK, Courbois C. Livestock to 2020: the next food revolution. Intl Food Policy Res Inst, Food, Agriculture, and the Environment Discussion Paper 28. 1999; 28:1–74. http://www.animalbiotechnology.org/livestock%20revolution.pdf. Accessed 26 May 2015.
7. United Nations, Department of Economic and Social Affairs, Population Division. World Population Prospects: The 2012 revision, key findings and advance tables. Working Paper No. ESA/WP.227. 2013. http://esa.un.org/wpp/documentation/pdf/wpp2012_%20key%20findings.pdf. Accessed 13 Oct 2015.
8. Miller BDD, Welch RM. Food system strategies for preventing micronutrient malnutrition. Food Policy. 2013;42:115–28.
9. Dror DK, Allen LH. The importance of milk and other animal-source foods for children in low-income countries. Food Nutr Bull. 2011;32(3):227–43.
10. McNeill S, Van Elswyk ME. Red meat in global nutrition. Meat Sci. 2012;92(3):166–73.
11. Krebs NF, Mazariegos M, Tshefu A, Bose C, Sami N, Chomba E, et al. Meat consumption is associated with less stunting among toddlers in four diverse low-income settings. Food Nutr Bull. 2011;32(3):185–91.
12. Convention on Biological Diversity Handbook (3rd Edition). 1992. https://www.cbd.int/doc/handbook/cbd-hb-01-en.pdf. Accessed 29 Sep 2015.
13. Roeder PL. Rinderpest: the end of cattle plague. Prev Vet Med. 2011;102(2):98–106.
14. Vargas-Terán M, Hofmann HC, Tweddle NE. Impact of screwworm eradication programmes using the sterile insect technique. In: Dyck VA, Hendrichs J, Robinson AS, editors. Sterile Insect Technique. Springer: Netherlands; 2005. p. 629–50.
15. Forabosco F, Lohmus M, Rydhmer L, Sundstrom LF. Genetically modified farm animals and fish in agriculture: a review. Livest Sci. 2013;153(1–3):1–9.
16. Müller M, Brem G. Transgenic approaches to the increase of disease resistance in farm animals. Rev Sci Tech. 1998;17(1):365–78.
17. Whitelaw CB, Sang HM. Disease-resistant genetically modified animals. Rev Sci Tech. 2005;24(1):275–83.
18. Hammer RE, Pursel VG, Rexroad CE Jr, Wall RJ, Bolt DJ, Ebert KM, et al. Production of transgenic rabbits, sheep and pigs by microinjection. Nature. 1985;315(6021):680–3.
19. Lassnig C, Müller M. Disease-resistant transgenic animals. In: Christou P, Savin R, Costa-Pierce B, Misztal I, Whitelaw CB, editors. Sustainable food production. New York: Springer; 2013. p. 747–60.
20. Bevan MW, Flavell RB, Chilton M-D. A chimaeric antibiotic resistance gene as a selectable marker for plant cell transformation. Nature. 1983;304(5922):184–7.
21. James C. Global status of commercialized biotech/GM Crops: 2014. ISAAA Brief No 49. ISAAA: Ithaca, NY.
22. Du SJ, Gong Z, Fletcher GL, Shears MA, King MJ, Idler DR, et al. Growth enhancement in transgenic atlantic salmon by the use of an "all fish" chimeric growth hormone gene construct. Nat Biotechnol. 1992;10(2):176–81.
23. Van Eenennaam AL, Muir WM. Transgenic salmon: a final leap to the grocery shelf? Nat Biotechnol. 2011;29(8):706–10.
24. Kristjanson PM, Swallow BM, Rowlands GJ, Kruska RL, de Leeuw PN. Measuring the costs of African animal trypanosomosis, the potential benefits of control and returns to research. Agr Syst. 1999;59(1):79–98.
25. Vreysen MJB, Seck MT, Sall B, Bouyer J. Tsetse flies: their biology and control using area-wide integrated pest management approaches. J Invertebr Pathol. 2013;112(Suppl 1):S15–25.

26. Vreysen MJB, Saleh K, Mramba F, Parker A, Feldmann U, Dyck VA, et al. Sterile insects to enhance agricultural development: the case of sustainable tsetse eradication on Unguja Island, Zanzibar, using an area-wide integrated pest management approach. PLoS Neglect Trop D. 2014;8(5):e2857.
27. Kariithi HM, Ahmadi M, Parker AG, Franz G, Ros VI, Haq I, et al. Prevalence and genetic variation of salivary gland hypertrophy virus in wild populations of the tsetse fly Glossina pallidipes from southern and eastern Africa. J Invertebr Pathol. 2013;112(Suppl):S123–32.
28. De Vooght L, Caljon G, De Ridder K, Van Den Abbeele J. Delivery of a functional anti-trypanosome nanobody in different tsetse fly tissues via a bacterial symbiont, *Sodalis glossinidius*. Microb Cell Fact. 2014;13:156.
29. Willyard C. Putting sleeping sickness to bed. Nat Med. 2011;17(1):14–7.
30. Thomson R, Molina-Portela P, Mott H, Carrington M, Raper J. Hydrodynamic gene delivery of baboon trypanosome lytic factor eliminates both animal and human-infective African trypanosomes. Proc NatlAcad Sci USA. 2009;106(46):19509–14.
31. Republic of the Philippines, Court of Appeals, Greenpeace Southeast Asia vs. Environmental Management Bureau of the Department of Environment and Natural Resources, CA-G.R. SP NO. 00013. http://www.ensser.org/fileadmin/files/NoticeOfDecision-CA-G.R.SP-No.00013.pdf. Accessed 30 Sep 2015.
32. Pimentel D, Wilson C, McCullum C, Huang R, Dwen P, Flack J, et al. Economic and environmental benefits of biodiversity. Bioscience. 1997;47(11):747–57.
33. Trewavas AJ. The population/biodiversity paradox. Agricultural efficiency to save wilderness. Plant Physiol. 2001;125(1):174–9.
34. Bruce A, Castle D, Gibbs C, Tait J, Whitelaw CB. Novel GM animal technologies and their governance. Transgenic Res. 2013;22(4):681–95.
35. Wells KD. Natural genotypes via genetic engineering. Proc Natl Acad Sci USA. 2013;110(41):16295–6.
36. Bosch P, Forcato DO, Alustiza FE, Alessio AP, Fili AE, Olmos Nicotra MF, et al. Exogenous enzymes upgrade transgenesis and genetic engineering of farm animals. Cell Mol Life Sci. 2015;72(10):1907–29.
37. Laible G, Wei J, Wagner S. Improving livestock for agriculture—technological progress from random transgenesis to precision genome editing heralds a new era. Biotechnol J. 2015;10(1):109–20.
38. Rajendran SR, Yau YY, Pandey D, Kumar A. CRISPR-Cas9 based genome engineering: opportunities in agri-food-nutrition and healthcare. Omics. 2015;19(5):261–75.
39. Fan Z, Lee S, Park H, Lucibello K, Meng Q, Polejaeva I, et al. Genetic inactivation of the SRY gene in Argali wild and Romney domestic sheep with CRISPR/Cas systems for producing sex-reversed female animals. Reprod Fert Develop. 2013;26(1):167.
40. Tan W, Carlson DF, Lancto CA, Garbe JR, Webster DA, Hackett PB, et al. Efficient nonmeiotic allele introgression in livestock using custom endonucleases. Proc Natl Acad Sci USA. 2013;110(41):16526–31.
41. Tan WS, Carlson DF, Walton MW, Fahrenkrug SC, Hackett PB. Precision editing of large animal genomes. Adv Genet. 2012;80:37–97.
42. Yao J, Huang J, Hai T, Wang X, Qin G, Zhang H et al. Efficient bi-allelic gene knockout and site-specific knock-in mediated by TALENs in pigs. Sci Rep. 2014;4:6926.
43. Ni W, Qiao J, Hu S, Zhao X, Regouski M, Yang M, et al. Efficient gene knockout in goats using CRISPR/Cas9 system. PLoS One. 2014;9(9):e106718.
44. Proudfoot C, Carlson DF, Huddart R, Long CR, Pryor JH, King TJ, et al. Genome edited sheep and cattle. Transgenic Res. 2015;24(1):147–53.
45. Kistler Kathryn E, Vosshall Leslie B, Matthews Benjamin J. Genome engineering with CRISPR-Cas9 in the mosquito aedes aegypti. Cell Rep. 2015;11(1):51–60.
46. International Livestock Centre for Africa. Livestock productivity and trypanotolerance. Nairobi, Kenya. 1981. pp. 59–63.
47. Lillico SG, Proudfoot C, Carlson DF, Stverakova D, Neil C, Blain C et al. Live pigs produced from genome edited zygotes. Sci Rep. 2013;3:2847.
48. Lyall J, Irvine RM, Sherman A, McKinley TJ, Núñez A, Purdie A, et al. Suppression of avian influenza transmission in genetically modified chickens. Science. 2011;331(6014):223–6.
49. Salter DW, Crittenden LB. Artificial insertion of a dominant gene for resistance to avian leukosis virus into the germ line of the chicken. Theoret Appl Genetics. 1989;77(4):457–61.
50. Lo D, Pursel V, Linton PJ, Sandgren E, Behringer R, Rexroad C, et al. Expression of mouse IgA by transgenic mice, pigs and sheep. Eur J Immunol. 1991;21(4):1001–6.
51. Dunham RA, Warr GW, Nichols A, Duncan PL, Argue B, Middleton D, et al. Enhanced bacterial disease resistance of transgenic channel catfish Ictalurus punctatus possessing cecropin genes. Mar Biotechnol. 2002;4(3):338–44.
52. Weifeng M, Yaping W, Wenbo W, Bo W, Jianxin F, Zuoyan Z. Enhanced resistance to *Aeromonas hydrophila* infection and enhanced phagocytic activities in human lactoferrin-transgenic grass carp (*Ctenopharyngodon idellus*). Aquaculture. 2004;242(1–4):93–103.
53. Fletcher GL, Hobbs RS, Evans RP, Shears MA, Hahn AL, Hew CL. Lysozyme transgenic Atlantic salmon (*Salmo salar* L.). Aquac Res. 2011;42(3):427–40.
54. Richt JA, Kasinathan P, Hamir AN, Castilla J, Sathiyaseelan T, Vargas F, et al. Production of cattle lacking prion protein. Nat Biotech. 2007;25(1):132–8.
55. Denning C, Burl S, Ainslie A, Bracken J, Dinnyes A, Fletcher J, et al. Deletion of the alpha(1,3)galactosyl transferase (GGTA1) gene and the prion protein (PrP) gene in sheep. Nature Biotech. 2001;19(6):559–62.
56. Wu H, Wang Y, Zhang Y, Yang M, Lv J, Liu J, et al. TALE nickase-mediated SP110 knockin endows cattle with increased resistance to tuberculosis. Proc Natl Acad Sci USA. 2015;112(13):E1530–9.
57. Jiao Y, Gong X, Du J, Liu M, Guo X, Chen L, et al. Transgenically mediated shRNAs targeting conserved regions of foot-and-mouth disease virus provide heritable resistance in porcine cell lines and suckling mice. Vet Res. 2013;44(1):47.
58. Zhong J, Wang Y, Zhu Z. Introduction of the human lactoferrin gene into grass carp (*Ctenopharyngodon idellus*) to increase resistance against GCH virus. Aquaculture. 2002;214(1–4):93–101.
59. Yan Q, Yang H, Yang D, Zhao B, Ouyang Z, Liu Z, et al. Production of transgenic pigs over-expressing the antiviral gene Mx1. Cell Regeneration. 2014;3(1):11.
60. Maga EA, Cullor JS, Smith W, Anderson GB, Murray JD. Human lysozyme expressed in the mammary gland of transgenic dairy goats can inhibit the growth of bacteria that cause mastitis and the cold-spoilage of milk. Foodborne Pathog Dis. 2006;3(4):384–92.
61. Wall RJ, Powell AM, Paape MJ, Kerr DE, Bannerman DD, Pursel VG, et al. Genetically enhanced cows resist intramammary Staphylococcus aureus infection. Nat Biotech. 2005;23(4):445–51.
62. Zhang J, Li L, Cai Y, Xu X, Chen J, Wu Y, et al. Expression of active recombinant human lactoferrin in the milk of transgenic goats. Protein Express Purif. 2008;57(2):127–35.
63. Bleck GT, White BR, Miller DJ, Wheeler MB. Production of bovine alpha-lactalbumin in the milk of transgenic pigs. J Anim Sci. 1998;76(12):3072–8.
64. Wheeler MB, Bleck GT, Donovan SM. Transgenic alteration of sow milk to improve piglet growth and health. Reproduction Suppl. 2001;58:313–24.
65. Yang B, Wang J, Tang B, Liu Y, Guo C, Yang P, et al. Characterization of bioactive recombinant human lysozyme expressed in milk of cloned transgenic cattle. PLoS One. 2011;6(3):e17593.
66. Li L, Li Q, Bao Y, Li J, Chen Z, Yu X, et al. RNAi-based inhibition of porcine reproductive and respiratory syndrome virus replication in transgenic pigs. J Biotechnol. 2014;171:17–24.
67. Clements JE, Wall RJ, Narayan O, Hauer D, Schoborg R, Sheffer D, et al. Development of transgenic sheep that express the visna virus envelope gene. Virology. 1994;200(2):370–80.

2

Reconnecting people, place, and nature: examining alternative food networks in Newfoundland's fisheries

Chloé Poitevin DesRivières[1*], Ratana Chuenpagdee[2] and Charles Mather[2]

Abstract

Background: Fisheries have played a significant role in the economic development of Newfoundland and Labrador and continue to contribute to coastal livelihoods and local food security. However, the access to and availability of locally harvested fish and seafood in the province is notably lacking, particularly since federal and provincial policies have prioritized the development of export markets in the fisheries over local sales. In this paper, we examine market actors and civil society organizations in St. John's (the provincial capital) and Petty Harbour (a nearby fishing community) that have embarked on initiatives to better include fish and seafood in the local food system. The success of these initiatives depends in part on the degree of connectivity between consumers and harvesters, as well as between people, the culture of fishing, and the marine environment.

Results: Reconnections are the foundations of alternative food networks (AFNs), which aim to develop more environmentally and socially sustainable and localized ways of producing, distributing, retailing, and consuming food. This paper explores the social, cultural, and ecological values of the fisheries as means to forging reconnections. The research reveals that values pertaining to traditional food culture, environmental stewardship and conservation, localization, and social cohesion were significant motivating factors for stakeholders in seeking out reconnection in their food system.

Conclusions: While AFNs in Newfoundland's fisheries are currently underdeveloped, reconnections are nonetheless taking place along the fisheries supply chain and can contribute to enhancing the accessibility and availability of locally harvested fish. The study also finds that AFNs must be context appropriate and reflect local values, as there are significant opportunities in building on Newfoundland's strong fishing culture and heritage.

Keywords: Alternative food networks, Fisheries, Values, Reconnection

Background and rationale

Alternative food networks (AFNs) have arisen within the food movement as practical and localized responses to growing concerns over the wide-ranging socio-economic, cultural, and environmental impacts of conventional food production. In addressing the shortcomings of conventional food systems, AFNs seek to create shortened and localized supply chains in order to enhance consumer awareness about where their food comes from and develop alternate means of producing, distributing, and retailing food [1]. AFNs were shaped by concerns over agricultural food production; however, they are equally relevant to the diverse and complex challenges faced in fisheries-based food systems. Much like industrial agriculture, commercial fisheries are part of a highly globalized, technology-driven, and resource-intensive food system. The advent of highly efficient catch methods has resulted in the overexploitation and degradation of marine resources, particularly in the light of a rising global demand for fish and seafood [2, 3].

Lengthening supply chains are of equal concern in the fisheries, in which consumers and fish harvesters have

*Correspondence: chloe.poitevin@carleton.ca
[1] Department of Geography and Environmental Studies, Carleton University, Ottawa, ON, Canada
Full list of author information is available at the end of the article

become increasingly distanced both socially and geographically [4]. The notion of disconnection highlighted by AFN literature can provide insight into ways to localize and simplify fisheries food systems. While local and sustainable food movements have seldom included fisheries, certain approaches derived from agricultural contexts have been adapted to enhance consumers' access to locally sourced and sustainably caught fish [4, 5]. With the emergence of AFNs that include fish and seafood, there remain significant questions with regard to how these fisheries AFNs operate and are structured, especially given that the institutional, socio-cultural, and economic contexts of fisheries are vastly different from those of agricultural food systems [6].

This research examines ways in which fisheries AFNs may present sustainable and appropriate alternatives to conventional fisheries harvest, marketing, and consumption practices in two locations in Newfoundland: the St. John's metropolitan area and the nearby fishing community of Petty Harbour–Maddox Cove (henceforth Petty Harbour). In the following section, we present the AFN concept of 'reconnection' and assess the extent to which it exists in Newfoundland. Next, we describe the study locations and research methods, followed by the main findings. In this analysis, care ethics and the interactive governance perspective inform an understanding of the ways in which the values and ethical motivations that guide decision-making processes enable reconnections in AFNs. We conclude with a discussion about opportunities and limitations in developing fisheries AFNs in contexts similar to Newfoundland and elsewhere.

Reconnecting through alternative food networks

Critical understandings of AFNs evoke a broad definition and appearance of 'alternative foods' to reflect the diverse contexts in which alternatives take place, rather than simply presenting 'alternative' in a dichotomous opposition to conventional food production [1, 7–9]. Conventional food systems have largely been framed as 'bad', particularly since associated food production processes generate significant ecological issues, including habitat degradation and loss of biodiversity; and socio-cultural and economic issues, such as the homogenization of food cultures and the concentration of power in the hands of a few corporate entities [1, 10]. Despite the many failings of global conventional food systems, it is problematic to assume that alternative food strategies and networks are intrinsically sustainable. The variability in alternative food practices has given considerable weight to the significance of the term 'alternative', particularly since many originally 'alternative' approaches have been absorbed into mainstream and industrial practices [11].

One potential answer to the question of what makes AFNs 'truly' alternative can be found in the concept of 'reconnection'. Reconnection has become a central notion in many leading AFN discourses and actions, given that the connection between the different actors throughout the supply chain is one of the key drivers behind direct marketing schemes and other alternative food initiatives [7, 12]. In accounting for the complex social and natural systems and governing institutions that shape food systems, reconnection as part of AFNs is best viewed as a process that produces varied relations and outcomes [13, 14]. Therefore, reconnection can be economically motivated due to the fact that producers seek to enhance localized market opportunities and add value to their products, although they can also be motivated by a desire to create an ethical and ecologically sustainable food system [14].

These multifaceted rationales for reconnection in food systems are linked to conceptions of moral economies through 'care ethics', which refers to personal concerns and care for ecosystems, culture, local economy, sustainability, and local communities [13]. In this concept, the economy is understood as performative in that it is a part of and produced by social relations, rather than a wholly separate entity [15, 16]. A moral economy is an attempt to rework and challenge the ways in which relations of economic exchange are perceived as enacted by rational actors in the pursuit of profit maximization [15, 17]. Alternative economic possibilities, such as AFNs, include ethical considerations as part of economic relations, thus offering a means to examine interpersonal and economic relations in food systems and fisheries as guided by values and ethics, rather than presumed individualistic self-interest [15, 18]. In this sense, economic choices and preferences are understood as the combined result of behaviours that are learnt through social and economic institutions and transmitted by culture and society by means of personal relationships [17, 19].

In AFNs, food choices are based both on self-interest, including enjoyment, health, and nutrition, and on outward cares and concerns, such as care for others, the local community, and the natural environment [13]. Although these cares are central to the AFN model, consumers are nonetheless constrained by economic factors and are not always able to make food choices based on individual tastes and moral preferences [20]. The concept of reconnection allows for an examination of the values that affect people's food choices, which leads to a better understanding of how AFNs emerge and what motivates people to participate in alternative practices in their food system.

While the motivations for seeking alternatives in fisheries-based food systems bear similarities to those in agriculture, there is a need to understand the differences in the social, biological, and governing systems that create a new and unique set of challenges for this food system. In this light, the idea of 'reflexive localism' is especially important for defining locality in a way that takes context into account and is receptive to change [1]. Reflexive localism is process oriented, in the sense that localization is not an end goal but a continually evolving set of relations between different actors and at different scales. Notions of place making and sense of place are useful in understanding the ways in which people relate to landscapes and places, and how these may impact localization processes. In particular, place is created and recreated through different food practices, cultures, and physical geographies [21].

Landscapes, particularly physical geographic realities such as soil composition, weather, and topography, shape food cultures and localization processes through the biophysical limits imposed on how and what foods can be produced or caught [21]. It is equally important to consider temporal scales, since 'local' foods conceptually include the adoption of a seasonal diet, which can limit the choices and availability of fresh produce throughout the year depending on climate [22]. Both the geographic and temporal scales of fisheries are different from agriculture, and many regions are simply not located in areas that can support commercial fishing [12]. While biological factors, such as seasonal migrations and spawning, play a role in determining the availability of fish and seafood, fisheries management regulations determine when and how much fish can be caught.

AFNs and fisheries in Newfoundland

Fisheries in Newfoundland and Labrador hold an immense cultural and historical significance and are highly complex in terms of actors, institutions, and ecology. The province's cultural identity and historical economic development are intrinsically linked with the fisheries, particularly with respect to Atlantic cod [23]. Newfoundland's fishery has gone through dramatic changes and challenges in the last century, specifically with the collapse of the commercial cod fishery in 1992, following nearly 50 years of intensive overfishing [24]. While the subsequent moratorium on the commercial Atlantic cod fishery significantly impacted coastal communities and livelihoods, fishing continues to be an important source of revenue and employment in the province. Snow crab and shrimp have replaced cod as the primary commercial and export species, currently accounting for 83% of capture fisheries' landed value, and have become more lucrative than the cod fishery was prior to the collapse [25, 26]. There remains a restricted commercial quota for Atlantic cod and a limited recreational cod fishery [25, 26].

More recently, the sustainability and durability of the Northern shrimp fishery has been brought into question, as shrimp stocks have been in decline since the 2000s due to increased catches and unfavourable environmental conditions [27]. There is also a push to re-establish the cod fishery, with some evidence pointing to the recovery of Atlantic cod stocks [26]. Other significant challenges to the fishery include a shrinking workforce due to ageing and outmigration, which is exacerbated by the seasonal, economically prohibitive, and often-insecure nature of fishing employment that discourages new entrants [24, 25].

The development of a resilient and sustainable local food system is vital to ensuring food security, particularly in rural communities in Newfoundland, due to the geographic isolation of the island, which presents significant challenges to residents' access to nutritious, affordable, and culturally appropriate foods. In fact, the majority of fresh fruits and vegetables consumed within the province are imported from mainland Canada or internationally, which can lead to food shortages as food distribution networks are vulnerable to interruptions such as natural events that prevent shipping [28, 29]. The access to and availability of fresh, nutritious foods is further challenged by the remoteness of many rural coastal communities, where residents often face lengthy travel times to the nearest full-service grocery store [29].

Although AFNs in the fisheries are developing in many coastal communities in Canada, they have been particularly slow to emerge in Newfoundland and Labrador compared to other coastal provinces. Efforts to introduce fisheries AFNs in Canada mostly come in the form of community-supported fisheries (CSFs), which aim to create a guaranteed local market for fishers and a source of fresh and sustainable fish for consumers. Notable CSFs in Canada include Off the Hook in Halifax, Nova Scotia and Skipper Otto's in Vancouver, British Columbia. While AFNs in Canadian fisheries are still underdeveloped, the number of AFNs in Newfoundland's agricultural systems has increased, with a growing network of farmers' markets and direct marketing schemes for agricultural products. In addition, there is a rich history of self-provisioning practices, including fishing, gardening, hunting, and berry picking, that have contributed significantly to food security [28, 30]. These practices, and other types of informal labour, hold culturally and economically significance for many in Newfoundland, particularly people residing in remote rural areas, such as the many small fishing communities in the province [31]. The seasonal nature of fisheries economies and the strong social ties in these communities, as well as their geographic isolation, create a strong role for informal economic networks as a key means to ensure livelihoods and food security for many residents [28, 31].

Significant barriers exist in the form of provincial and federal legislation, which restrict the development of alternate and informal markets [32–34]. In particular, both the federal and provincial fisheries governing bodies, the Department of Fisheries and Oceans (DFO) and Department of Fisheries and Aquaculture (DFA), respectively, have neglected the development of local markets for fish and seafood, instead focusing heavily on export-oriented production [25]. Equally, the access to and availability of locally harvested fish was further limited by competing interests in the fisheries, primarily between fishers and fish processors and workers [33]. The direct sales of fish-by-fish harvesters were prohibited in Newfoundland and Labrador until late 2015, justified on the grounds of maintaining the viability of the fish-processing industry and to ensure food safety [33]. Despite strict provincial regulations, a handful of alternative food efforts emerged to increase the availability of locally sourced fish and seafood.

Fisheries regulations and policies are also impacted by the Fish, Food and Allied Workers Union (FFAW-Unifor), an important lobbying group in the province that is active in developing local market opportunities, which the union see as an important means to support inshore fish harvesters. The FFAW-Unifor are also committed to developing external markets for Newfoundland fish and seafood in an effort to support the interests of their diverse membership, which includes large-scale fishing operations and fish plant workers in addition to small-scale inshore fish harvesters.

The geographic focus of the study is on the St. John's metropolitan area and the community of Petty Harbour, located both on the Avalon Peninsula and on the east coast of the island. These cases offer an opportunity to examine emerging alternatives, in terms of how they arise, organize, and operate.

St. John's is the capital of Newfoundland and Labrador and is the province's most densely populated area, with the metropolitan region hosting a population of approximately 200,000 [35]. St. John's serves as the province's economic and administrative centre, with the tourism and the offshore petroleum industries generating significant revenue within the city. The local and sustainable food movement has become increasingly prominent in St. John's, with a number of agriculture-oriented AFNs developing in the region, including direct farm sales, community-supported agriculture (CSA), and farmers markets [36]. In terms of fisheries, there is a growing network of food businesses and other actors looking to include and promote local and sustainable fish, and many of these groups have gained public attention through their efforts [37]. Many food businesses have integrated environmental and ethical values and goals into their mandates [37].

Petty Harbour is an important fishing and tourism hub located 15 km south of St. John's. Fish harvesters in this area have a unique long-standing commitment to sustainability and stewardship practices and are actively involved in local fisheries governance [38]. An example of this mentality is the establishment of the Petty Harbour Fisherman's Co-operative, which operates as a fishing enterprise that is socially just and sustainable while reinvesting in the local community. Moreover, efforts to reconnect people with the local environment and their fishing heritage have been strengthened by the creation of the Petty Harbour Mini-Aquarium, a non-profit organization aimed at raising awareness about local marine habitats and animals, and the Island Rooms, a fisheries and Newfoundland culture education program for children and youth.

Methods

The study employed semi-structured interviews that covered a range of topics relating to the fisheries and food systems sustainability and governance in Newfoundland. This method allows the possibility to ask probing and follow-up questions and enables respondents to provide perspectives, experiences, and reflections in their own words [39, 40]. Selected topics were drawn from the interactive governance framework, focusing particularly on the interactions within and between the different components of the food system and the fisheries [34, 41]. Interactive governance frames values as the foundation of decision-making processes, and stresses that understanding how people come to make choices is essential to understanding and improving governance processes [41].

The interview questions used in the study aimed at discerning the underlying values that govern the decision-making processes of actors participating in alternative food initiatives in Newfoundland's fisheries. These values were drawn from a selection of topics previously identified in fisheries governance literature by Song et al. [34]. In their study, 24 value types were identified and categorized into four value orientations, relating to perspectives on fisheries governance and decision-making processes. These value orientations make explicit the 'ethics of care' performed in fisheries, as well as AFNs. They include 'better world' values (e.g. altruistic, common good), 'good life' (personal well-being), 'personal virtues' (inner personal qualities), and 'outward aspirations' (relationships with others and objects) [34]. In addition, the interviews comprised questions in which participants outlined their roles and relationships in the fishery and/or food system, as well as their involvement in enhancing alternative markets, spaces, and knowledge for fish and seafood.

Respondents were selected using purposive sampling method in order to cover a range of actors engaged in alternative food practices. A total of 13 interviews were completed in the St. John's and Petty Harbour areas, each lasting between 15 and 45 min. Participants included restaurant owners, chefs, retailers, fish harvesters, and local food promotion and conservation education organizations. Interview data were analysed thematically using NVivo 9 software (QSR International), guided partly by the categories outlined by Song et al. [34].

The interviews were complemented by an analysis of secondary data, including academic and grey literature, in particular news articles, podcasts, and documentaries, in order to provide a social, cultural, economic, and historical context to the interview data. Federal and provincial policy documents were examined to assess the roles of governing actors and institutions in Newfoundland's fisheries and identify potential barriers and opportunities for emerging AFNs.

Results

The identification of values underpinning alternative food practices aids in understanding the types of reconnections taking place, and the underlying motivations that lead people to seek these connections. Three primary value types were identified in the interviews: food system localization, social cohesion, and education and conservation.

Food system localization

The interview respondents interpreted 'local' in different ways and associated different benefits and values to the prospect of localizing food systems. However, these values pertained predominantly to developing market opportunities. The scale of 'local' in terms of Newfoundland's fisheries is considered larger than that of land-based food production. For instance, fish and seafood products originating from the whole island were referred to as 'local', while 'local' for agricultural products was generally seen as encompassing only the Avalon Peninsula. As well, fish originating from Atlantic Canada was viewed as relatively local, as they come from the same waters.

Restaurants and retailers interviewed for the study indicated reasons for choosing locally harvested fish and seafood around varied notions of quality. Most replied that local fish was preferable as the product 'hasn't travelled', thus perceived as fresh and possessing a small overall environmental footprint. Better taste, nutrition, and higher product quality were also mentioned. For instance, one restaurant owner viewed Newfoundland waters to be more 'pure and clean', resulting in better tasting fish and seafood. Certain restaurant owners also expressed a preference for purchasing fish and seafood from small-scale fishing vessels as these were viewed as engaging in more sustainable harvest methods and better product handling practices. Another motive for purchasing locally caught fish that many respondents noted was the desire to support the local economy, particularly rural fishing communities.

Despite a preference for local products, interview respondents expressed varying degrees of difficulty accessing locally harvested fish and seafood, either for their business or for personal consumption. The seasonality of fish and seafood was one reason that affected local availability of certain species, due to both biological limits and fisheries management regimes. The restaurant owners and chefs that have made local sourcing a part of their business plan change their menus according to seasonal variability, and turn to in-house methods of preservation for local fish, meats, and produce to overcome limited product availability, especially in the winter. Some respondents did note, however, that for some species, the fishing seasons enforced by DFO are too restrictive and do not align with the province's tourism season (which generally runs from June until September). This mismatch applies particularly to halibut and lobster, which means that restaurant owners either have to resort to frozen product, in the case of halibut, or imports from the mainland, as with lobster.

The export-oriented fisheries policy in Newfoundland was another reason frequently listed as a cause of limited access and availability to locally harvested fish. Nonetheless, export was perceived as a necessity, with one fisher based out of Petty Harbour, stressing that export was necessary to make a living. The fishers interviewed noted that the species with the highest local demand is cod, although quotas and seasons are very restricted, resulting in low profitability. The main species harvested by the Petty Harbour Fisherman's Co-op, particularly crab and capelin, are of high value on the export market, while respondents feel that the local demand simply is not there.

The one species that all of the restaurants and retailers interviewed reported no issues accessing locally was cod. Rather, it is the access to other species, such as whelks, squid, crab, and octopus, which are caught locally and destined for foreign markets or for use as bait, that pose a significant challenge to local consumption. The restaurants in St. John's were limited only to what was available through their suppliers, since they could not purchase fish directly from harvesters as per provincial regulations at the time of the interviews. Product availability was further restricted by minimum processing requirements, which are implemented by DFA to support the fish-processing industry, which dictate in what form fish and seafood can be sold. In fact, many restaurants found it difficult to access whole, unprocessed fish. In the words of one restaurant owner and chef: 'A lot of our work is in the sourcing of the product than [the] preparations. Half the work is actually finding it'.

Social cohesion

The most important value identified throughout the interviews was that of social cohesion, which refers to the processes of social connection, interpersonal relationships, and community values [34]. One respondent noted that the considerable disconnection between consumers and fish harvesters in Newfoundland has impeded the flow of information about how fish gets from ocean to plate. The motivations of respondents to seek personal relationships are focused on increasing the knowledge about the product itself and the processes of harvesting fish and seafood. There is limited, if any, information available to retailers, chefs, and consumers about where, how, and by whom fish is caught, thus impeding traceability. An opportunity for creating social connections, albeit at a distance, is the seafood traceability project headed by the FFAW-Unifor on the west coast of the island. This initiative, adapted from Ecotrust Canada's ThisFish project, allows consumers to connect with fish harvesters on an online platform by means of a traceable tag on fish and seafood products. The group of restaurants in St. John's seeking to access the traceable fish see this approach as an opportunity to engage consumers with their food and the fishery and feel that, to a limited degree, the project may help in creating social connections between fish harvesters, restaurants, and consumers.

With an increasing network of alternatives in the agricultural sector in Newfoundland, restaurateurs and consumers are becoming able to build personal relationships with local food producers. Restaurants in St. John's are able to purchase directly from both local farmers and licensed hunters, and owners expressed a desire to connect in the same ways with fish harvesters: 'We can develop relationships with all kinds of farmers, all kinds of producers of everything, and I have a personal relationship with them, but I can't have a personal relationship with my fisherman,' said a restaurant owner.

Another restaurant owner echoed the importance of personal relationships in accessing local products and stressed that proximity was key in facilitating these networks. The restaurant owner relayed their experience living and working in a rural fishing community in Newfoundland and noted that the proximity to an active fishery allowed them to develop networks and personal relationships with harvesters, which subsequently enhanced their access to locally caught fish. Many respondents viewed this type of relationship with fish harvesters as being very difficult to build from within the urban St. John's area.

While formalized food networks for fish and seafood have, up until very recently, been difficult to form between harvesters, retailers, and consumers, there are nonetheless established informal food networks, particularly in rural communities. One respondent in Petty Harbour noted that she had no trouble accessing local fish in the community due to personal relationships with fish harvesters: 'when the commercial fishery starts up, then that's not a problem for me personally because I have a brother in law who's a commercial fisherman. But if you're not, if you're outside the harbour, you're going to have some problems...'. In Petty Harbour, an informal network did seem to take place with regard to cod. One fisher explained: 'Fishermen do it, fishermen take chances and do it for a few extra dollars. That's the way it is in all the communities'. Employees of the Petty Harbour Mini-Aquarium observed an informal trade in the community, and their location beside the co-operative has positioned the organization as a link between local consumers and tourists that are interested in finding fresh, local seafood.

Education and conservation of culture and nature

Given that the fisheries in Newfoundland are fundamentally intertwined with place, culture, history, and community, it is no surprise that alternative initiatives are seeking to integrate traditional values and practices. Community actors and restaurants have stepped into try to educate people about food traditions, skills, and the natural environment, in order to create a demand for and interest in locally harvested fish and seafood.

A number of restaurants in St. John's have developed menus that feature not only local products, but also traditional Newfoundland dishes. As well, many have sought to diversify consumer tastes for seafood to include a wider variety of species. 'Fish' is synonymous with cod in Newfoundland, and consumer demand for other species that are caught locally is generally low. While the inadequate access to a diversity of fish species creates barriers, another issue lies in the lack of food skills; people often have limited capabilities and knowledge in preparing fish and seafood other than cod, which restricts what they will purchase. A seafood retailer in St. John's pointed out that this issue did affect what he sold to local consumers, noting that people living in rural coastal areas have more experience and knowledge with regard to preparing a larger variety of fish and seafood.

Seeking to increase food skills among consumers can also help to increase food security and solve food-related health issues in the province. Food First NL seeks to emphasize the importance that traditional food skills, such as berry picking, preserving, gardening, and fishing, have played in the past and the ways they can contribute to ensuring food security and self-sufficiency. Equally, Food First NL advocates for the development of food skills and increased access to and availability of local foods, including fish, as means to help mitigate the high prevalence of diet-related health problems in Newfoundland.

The need to build self-sufficiency in Newfoundland's food system was another concerned raised by interview respondents. Young people, in particular, lack knowledge about fishing heritage. A fisher in Petty Harbour asserted that, when he was growing up, fishermen were seen as the heroes in the community, and 'it was easy to fall into the fishery trap'. The Island Rooms program looks at using food and fisheries as a way to engage youth in the food system and with nature by developing food skills through fishing workshops and by teaching other forms of self-provisioning. With the workforce declining in Newfoundland's fisheries, it is hoped that the project will introduce young people to a possible career in fishing.

A number of interview respondents stressed that there is a disconnection between different food systems actors and the natural environment that must be addressed in Newfoundland, which can be tied to a lack of knowledge about the ecological impacts of food production and fisheries. The Island Rooms and the Mini-Aquarium are possible avenues to reconnect people to the marine environment through education initiatives related to ecosystem conservation. The Petty Harbour Mini-Aquarium in particular provides an opportunity for people to connect with the local marine environment by learning about, seeing, and handling different aquatic animals, including species that are commercially harvested in Newfoundland such as cod, lobster, and flatfish. By bringing the 'ocean up to eye level', the Mini-Aquarium not only contributes to public education, but also provides fish harvesters with a new means to understand, experience, and appreciate the marine ecosystem on which their livelihoods depend. For instance, the members of the Petty Harbour Fisherman's co-op had never seen a codfish swim and were unaware of some of the characteristics and natural behaviours of the species they fish. In turn, they were also able to share their knowledge and experience of the marine environment with the employees of the aquarium, and there are plans to create a program that would allow visitors to the aquarium to engage with the local fishers.

Discussion

Developing skills, tastes, and knowledge

The access to locally sourced fish in Newfoundland is a significant challenge, though there are many potential endeavours that, together, may help in developing fisheries AFNs in the province. Opportunities to develop these alternatives lie in strengthening relations between food systems actors, as well as the connections to nature, place, and Newfoundland culture. The re-emergence of traditional Newfoundland cuisine provides a unique prospect to reconnect with place, culture, and fisheries. Thus, consumers are able to gain an appreciation for and a better knowledge of food that is available locally, which in turn can help create a demand for locally source seafood. The interviews illustrated the integral role that fisheries play in place-making and localization processes, as respondents emphasized the significance of fish within food culture, heritage, and identity in Newfoundland. Cod in particular is a reflection of food preferences and tastes for Newfoundlanders, as well as a manifestation of sense of place, culture, and identity that is established by family, social norms and traditions, economic realities, and landscapes [13, 19, 21]. The efforts of both market and civil society actors help to situate food and fish in particular places and ecologies in Newfoundland by reconnecting individuals with the local marine environment, their fishing heritage, and the local food system.

Market actors play a role in enhancing reconnections in the local food system by engaging with local food cultures and attempting to improve product availability. More specifically, the ways in which restaurateurs in St. John's craft menus that are reflective of seasonality and locally available products can make consumers more aware about the fish and seafood being harvested in Newfoundland. Through this practice, restaurants help to re-instil a pride of place and culture locally and with visitors. Despite the key role that restaurants and chefs can play in promoting locally sourced food and sustainable food systems, their adherence to 'local' is often oriented towards the characteristic of the food itself rather that the processes and conditions through which food is produced [42]. In doing so, they are engaging with weaker conceptions of 'alterity' that neglect broader socio-economic issues. To some extent, this is true of the restaurateurs interviewed, as quality and taste were mentioned as important reasons for preferring locally harvested fish. Nevertheless, most chefs and restaurant owners also expressed a desire to engage in the local food system as a means to support fish harvesters and rural communities, as well as to celebrate local food cultures and traditions.

Encouraging the diversification of the fish species that are consumed locally is a fundamental part of developing alternatives in the fisheries and enhancing local economies. However, as previously noted, food skills impact food choices by limiting the foods that individuals are able to prepare and can prevent people from accessing healthy foods and participating in food cultures [43]. In order for fisheries-based AFNs to take root and be sustained in Newfoundland, consumers not only need to be interested in and able to access local fish, but also must have the capacities to prepare it.

The education initiatives put forward by the Island Rooms, Food First NL, and the Mini-Aquarium, in addition to the efforts of St. John's restaurateurs to develop tastes for locally sourced fish, are a means of diversifying local consumer food preferences and enhancing food skills. O'Hara and Stagl [19] assert that food preferences are dynamic and continuously adapting, and it is through education and learning opportunities that aim to increase food knowledge and skills that consumer preferences and values can be altered. This reality is evidenced through the endeavours of the Island Rooms and Food First NL, specifically in their attempts to revitalize self-provisioning skills in order to generate a local food system that is truly sustainable. More specifically, one of the main aims of the Island Rooms is to enable young people to engage in subsistence fishing throughout the province in hopes of creating a more culturally appropriate and secure food system. Subsistence food production and fishing enhance food security in Newfoundland, particularly in rural communities, since people at all income levels are able to access food outside formal markets [28]. Equally, self-provisioning skills allow people to reduce dependency on conventional food supply chains, which lessens the impacts of unforeseen shocks to food supply in Newfoundland, including weather and transportation issues [28].

Challenges to developing fisheries AFNs

Despite the considerable opportunities and efforts to develop AFNs in Newfoundland's fisheries, the complex political economic structures of the fisheries and food system generate significant challenges. Specifically, fisheries policies at the provincial and federal levels remain highly problematic, even with the recent change in direct sale regulations. Andrée et al. [9] found that in regions where export-oriented food production policies are implemented, the development of AFNs is arduous, since governments prioritize the growth of international markets over stimulating local opportunities. This is the case in Newfoundland's fisheries, where governments at both federal and provincial levels have sought to develop external markets to the detriment of the local one.

The implications of the new regulations for direct fish sales on local markets remain unclear, as many details have yet to be worked out, such as the structure of reporting and monitoring systems. The benefits of this regulatory change for fish harvesters and buyers is equally uncertain, given that the majority of fish harvesters still rely on export markets for most of their catch due to the small local market. This may lead to a more 'hybridized' version of AFNs, in which food producers continue to depend on some conventional aspects of the food system in order for their business to subsist, while waiting for the local demand to expand [9, 44]. As with the fisheries in Newfoundland, this hybridity can often be the result of policy environments that encourage and incentivize all producers, including those engaging in alternative practices and niche markets, to participate in the conventional supply chain [9]. Direct marketing strategies therefore offer a means to enhance the strength, or 'alternativeness' of AFNs in these environments [45].

Prior to the shift in provincial regulation, the social connections between harvesters and consumers were maintained through informal food networks that have enabled the direct distribution of fish and seafood. Self-provisioning practices, including the food fishery, are important alternatives to conventional market practices, as well as meaningful facets of Newfoundland heritage. However, as noted, these informal food networks in the fisheries are geographically and socially limited to those near active fisheries and to those who know fish harvesters. AFN literature is limited in its ability to address these informal practices, despite the fact that many of these are, by current definitions, AFNs. Conceptually, AFNs are positioned as a response to conventional food production practices, while informal networks such as those taking place in Newfoundland's fisheries existed before conventional practices were developed. People in many Newfoundland communities have historically bought fish directly from harvesters on the wharf. As one fish harvester put it: 'I mean it was done for year and years and years before all this [provincial legislation] came in.'

AFNs that operate as part of informal economies, such as bartering and self-provisioning practices, are culturally significant to Newfoundlanders and will likely remain an important means to access fish, particularly in rural and remote coastal communities. However, these informal practices are not accessible to all, and AFNs that operate within the formal economy also need to be developed. In many instances, formalized alternative food economies are required to enable consumer–producer connections, which can take place in many different ways, such as through direct marketing, local retailers, and restaurants. Newfoundland food systems actors can look elsewhere in Canada, where direct fish sales have legally been allowed for longer periods of time, to draw inspiration in developing fisheries AFNs. Community-supported fisheries (CSFs) are cited as a means for people in Newfoundland, particularly in urban areas, to access fresh fish and seafood while enhancing small-scale fisheries and coastal community livelihoods [46]. The CSF model has the capacity to be adapted to a variety of contexts and to meet diverse socio-economic and ecological needs, as long as ethical and environmental concerns are integrated and enacted adequately [6, 18]. Community and economy are intertwined and integral to the CSF model, thus opening up possibilities for alternative means to managing fisheries resources in a way that builds towards a just and sustainable food system [6, 18].

In the light of the legalized direct sales of fish and the increased opportunity to develop strategies such as CSFs, careful consideration must be given to which groups are targeted by AFNs and how these strategies can be made to include a diverse demographic. Many have criticized cases of the practical implementation of AFNs as catering to higher-income and more privileged groups, based on the fact that strategies such as CSA and farmers' markets can prove costly to the consumer [11]. In these cases, central values to AFNs such as social justice, democracy, and inclusivity are neglected, especially as only certain groups have access and are able to participate in alternatives [11]. Many AFNs operate under the assumption that consumers are willing to pay a premium to access sustainably produced and local foods [20]. Conversely, profitability remains a challenge for many alternative food strategies in agricultural contexts, particularly with the goal of attaining a fair price for consumers while ensuring producers' livelihoods [20, 47]. These challenges are likely to pose limitations on fisheries AFNs in balancing the need to create an inclusive and affordable model while guaranteeing fish harvesters an adequate income.

Many of the efforts towards creating fisheries AFNs in St. John's tend to cater to higher-income populations, particularly due to the fact that most of the restaurants that are most active in supporting local fish are higher end and quite expensive. Equally, the reliance on market-driven mechanisms to create more sustainable, localized, and just food systems places the responsibility squarely on the shoulders of consumers, who are expected to vote with their dollar [6, 48]. These strategies restrict all but those with the economic means from participating in their local food system and having the ability to make food choices based on their values and ethics. These more exclusive AFN strategies risk mirroring the problematic and unequal political economic structures that underpin the conventional food system [6, 11, 20]. In order for systemic changes to take place in the food system, the access to and availability of locally harvested fish needs to be tied to broader food security and food justice goals [6]. The efforts of civil society organizations, such as Food First NL and the Island Rooms, to highlight the importance of local fish and food in building towards community food security in Newfoundland offer a means to better engage with issues of accessibility and inclusivity in AFNs.

Conclusions

The question of how alternatives can be developed in Newfoundland's fisheries is best answered by looking at the reconnections taking place within the food system, especially in terms of re-engaging consumers with their food through culture and traditions. The emphasis on relationships and reconnections in AFNs allows for the recognition of the multiple actors involved in building alternative markets and how they relate to one another. This study found that, while fisheries hold important socio-cultural values in Newfoundland, they are primarily valued by government policy for their economic contributions, which has created a significant barrier to building AFNs. Conversely, recognizing these alternate values is key to strengthening the local access and availability of fish, especially those that relate to and build on Newfoundland food cultures and traditions. In fact, the values identified demonstrate that choosing locally harvested fish and seafood is not simply related to personal ideals such as quality, health, and nutrition, but to 'care ethics' related to outward aspirations such as environmental sustainability and enhancing coastal communities.

Furthermore, there is a need for improved access to local fish and seafood in diverse contexts and for many different people in Newfoundland. Along with AFNs, informal food practices enhance the food system in Newfoundland and can provide opportunities to develop alternative practices in the fisheries. More work needs to be done on informal networks in the fisheries, particularly with regard to how these networks and traditional practices in rural coastal communities may be impacted by the new possibility for legal direct sales.

This research contributes to an understanding of how AFNs can emerge and develop in unfavourable policy environments that impede the development of localized markets for fish and seafood. Moreover, while export-oriented fisheries policies are barriers to developing AFNs, this study finds that reconnections are already taking place in the broader food system, and specifically within the fisheries, that aim to create better access to and availability of locally caught fish and seafood in Newfoundland. These reconnections offer a starting point on which to build fisheries AFNs by strengthening existing food systems and economic relations that include ethical and moral values. In other words, there is significant value in 'starting where you are', as it draws attention to the existing community assets and efforts in building towards sustainable and just food systems [49, 50]. These alternate ways of valuing fisheries are reflective of the diverse possibilities in Newfoundland that already exist and that can help to strengthen and create AFNs in the fisheries. In summation, there must be careful consideration of the types of AFNs that are best suited to support community livelihoods, cultural ties, and ecological sustainability in order to create a just and secure food systems.

Abbreviations

AFN: alternative food network; DFO: Department of Fisheries and Oceans Canada; DFA: Department of Fisheries and Aquaculture (Provincial government of Newfoundland and Labrador); FFAW-Unifor: Fish, Food and Allied Workers Union; SJFM: St. John's Farmer's Market.

Authors' information

Chloé Poitevin DesRivières is a Ph.D. candidate at Carleton University in Geography with a specialization in Political Economy. Dr. Ratana Chuenpagdee is a professor in the Department of Geography at Memorial University. Dr. Charles Mather is a professor in the Department of Geography at Memorial University.

Authors' contributions

CPD conceived of the study, participated in its design, performed the primary data collection and analysis, and drafted the manuscript. RC and CM participated in the study design and data analysis and contributed to the writing of the manuscript. All authors read and approved the final manuscript.

Author details

[1] Department of Geography and Environmental Studies, Carleton University, Ottawa, ON, Canada. [2] Department of Geography, Memorial University of Newfoundland, St. John's, NL, Canada.

Acknowledgements

This research is part of the Master thesis of the first author and contributes to Too Big to Ignore: Global Partnership for Small-Scale Fisheries Research. The authors would like to express their gratitude to all those who participated in the interviews. We are grateful to Brennan Lowery for his helpful feedback. The authors also thank the reviewers for their useful comments on the manuscript.

Competing interests

The authors declare that they have no competing interests.

Funding

As part of the Too Big to Ignore project, this study was funded by the Social Sciences and Humanities Research Council of Canada (SSHRC; Grant No. 895 -2011-1011).

References

1. Harris EM. Eat local? Constructions of place in alternative food politics. Geogr Compass. 2010;4(4):355–69. doi:10.1111/j.1749-8198.2009.00298.x.
2. Ponte S. Greener than thou: the political economy of fish ecolabeling and its local manifestations in South Africa. World Dev. 2008;36(1):159–75. doi:10.1016/j.worlddev.2007.02.014.
3. Chuenpagdee R, Morgan LE, Maxwell SM, Norse EA, Pauly D. Assessing collateral impacts of fishing methods in US waters. Front Ecol Environ. 2003;1(10):517–24.
4. Lowitt K, Nagy M, Nelson C, Bavington D. Where's the fish? Altern J. 2013;39(4):1–6 **(Food and Drink)**.
5. Loring PA, Gerlach SC, Harrison HL. Seafood as local food: food security and locally caught seafood on Alaska's Kenai Peninsula. J Agric Food Syst Community Dev. 2013;3(3):13–30.
6. Olson J, Clay PM, Pinto da Silva P. Putting the seafood in sustainable food systems. Mar Policy. 2014;43:104–11. doi:10.1016/j.marpol.2013.05.001.
7. Winter M. Embeddedness, the new food economy and defensive localism. J Rural Stud. 2003;19(1):23–32. doi:10.1016/S0743-0167(02)00053-0.
8. Holloway L, Kneafsey M, Cox R, Venn L, Dowler E, Helena T. Beyond the 'alternative'-'conventional' divide? Thinking differently about food production-consumption relationships. In: Maye D, Holloway L, Kneafsey M, editors. Alternative food geographies: representation and practice. Bingley: Emerald; 2008. p. 77–93.
9. Andrée P, Dibden J, Higgins V, Cocklin C. Competitive productivism and Australia's emerging "alternative" agri-food networks: producing for farmers' markets in Victoria and beyond. Aust Geogr. 2015;2010(41):307–22. doi:10.1080/00049182.2010.498038.
10. Howard PH. Concentration and power in the food system. London: Bloomsbury; 2016.
11. Guthman J. Bringing good food to others: investigating the subjects of alternative food practice. Cult Geogr. 2008;15:431–47. doi:10.1177/1474474008094315.
12. Campbell LM, Boucquey N, Stoll J, Coppola H, Smith MD. From vegetable box to seafood cooler: applying the community-supported agriculture model to fisheries. Soc Nat Resour. 2014;27(1):88–106. doi:10.1080/08941920.2013.842276.
13. Dowler E, Kneafsey M, Cox R, Holloway L. "Doing food differently": reconnecting biological and social relationships through care for food. Editor Board Sociol Rev. 2009;57(s2):200–21.
14. Kneafsey M, Cox R, Holloway L, Dowler E, Venn L, Tuomainen H. Reconnecting consumers, producers and food: exploring alternatives. Oxford: Berg Publishers; 2008.
15. Lind C. Rumours of a moral economy. Halifax: Fernwood Publishing; 2010.
16. Gibson-Graham JK. Diverse economies: performative practices for "other worlds". Prog Hum Geogr. 2008;32(5):613–32. doi:10.1177/0309132508090821.
17. Sayer A. Moral economy as critique. New Polit Econ. 2007;12(2):261–70. doi:10.1080/13563460701303008.
18. Snyder R, St. Martin K. A fishery for the future: the midcoast fishermen's association and the work of economic being-in-common. In: Roelvink G, St. Martin K, Gibson-Graham JK, editors. Making other worlds possible: performing diverse economies. Minneapolis: University of Minnesota Press; 2015. p. 26–52.
19. O'Hara SU, Stagl S. Endogenous preferences and sustainable development. J Socio Econ. 2002;31:511–27.
20. Allen P. Realizing justice in local food systems. Camb J Reg Econ Soc. 2010;3(2):295–308. doi:10.1093/cjres/rsq015.
21. DeLind LB. Of bodies, place, and culture: re-situating local food. J Agric Environ Ethics. 2006;19(2):121–46. doi:10.1007/s10806-005-1803-z.
22. Parkins W, Craig G. Culture and the politics of alternative food networks. Food Cult Soc An Int J Multidiscip Res. 2009;12:77–103. doi:10.2752/155280109X368679.
23. Kurlansky M. Cod: a biography of the fish that changed the world. Toronto: Random House Canada; 1997.
24. Schrank WE. The Newfoundland fishery: ten years after the moratorium. Mar Policy. 2005;29(5):407–20. doi:10.1016/j.marpol.2004.06.005.
25. Department of Fisheries and Aquaculture (DFA). Annual Report 2013-14. 2014. http://www.fishaq.gov.nl.ca/publications/pdf/annual_report_2013_14.pdf.
26. Mather C. From cod to shellfish and back again? The new resource geography and Newfoundland's fish economy. Appl Geogr. 2013;45:402–9. doi:10.1016/j.apgeog.2013.06.009.

27. Foley P. The political economy of marine stewardship council certification: processors and access in Newfoundland and Labrador's inshore shrimp industry. Agrar Change. 2012;12(2):436–57.
28. Lowitt K. Examining fisheries contributions to community food security: findings from a household seafood consumption survey on the west coast of Newfoundland. J Hunger Environ Nutr. 2013;8(2):221–41. doi:10.1080/19320248.2013.786668.
29. Food First NL. Everybody eats: a discussion paper on food security in Newfoundland and Labrador. 2015. http://static1.squarespace.com/static/54d9128be4b0de7874ec9a82/t/56572239e4b0b807738c5ad0/1448550969272/Everybody+Eats_NL+Discussion+Paper+2015.pdf.
30. Teitelbaum S, Beckley T. Harvested, hunted and home grown: the prevalence of self-provisioning in rural Canada. J Rural Community Dev. 2006;1:114–30.
31. Felt LF, Sinclair PR. "Everyone does it": unpaid work in a rural peripheral region. Work Employ Soc. 1992;6(1):43–64. doi:10.1177/095001709261003.
32. Murphy I, Neis B. Navigating the legislative requirements for fisheries—tourism initiatives in Newfoundland and Labrador. 2012. http://www.curra.ca/documents/TCR_Fisheries Tourism_Regulations_Report.pdf.
33. Eric Dunne Consulting Initiatives. Report of the review of the regulations and policy for direct fish sales in Newfoundland. 2010: p. 1–51. http://www.fishaq.gov.nl.ca/publications/pdf/Direct_Fish_Sales_NL_2010.pdf.
34. Song AM, Chuenpagdee R. A principle-based analysis of multilevel policy areas on inshore fisheries in Newfoundland and Labrador, Canada. In: Jentoft S, Chuenpagdee R, editors. Interactive governance for small-scale fisheries. MARE Publication Series. Vol. 5. Cham: Springer; 2015. p. 435–65. doi:10.1007/978-3-319-17034-3.
35. Department of Finance- Newfoundland & Labrador Statistics Agency. Population estimates. 2016. http://www.stats.gov.nl.ca/statistics/population/PDF/Population_Estimates_CDCMA.pdf. Accessed 26 Jan 2017.
36. Root Cellars Rock! Local Links. 2014. http://rootcellarsrock.ca/local-links/. Accessed 14 April 2014.
37. Gollner AL. Jeremy Charles: Canada's real top chef. Globe Mail. 2014. http://www.theglobeandmail.com/life/jeremy-charles-canadas-real-top-chef/article20495580/. Accessed 1 May 2015.
38. Protected Areas Association of Newfoundland and Labrador. Ancient rights: The protected fishing area of Petty Harbour-Maddox Cove. 1996. http://collections.mun.ca/PDFs/ich_avalon/AncientRightsBooklet.pdf.
39. Cope M. Interviewing. In: Warf B, editor. Encyclopedia of human geography. Thousand Oaks: Sage Publications; 2006. p. 262–63. doi:10.4135/9781412952422.n163.
40. Mason J. Semi-structured interview. In: Lewis-Beck MS, Bryman A, Liao TM, editors. The SAGE encyclopedia of social science research methods. Vol. 3; Thousand Oaks: Sage Publications; 2004. p. 1020–22. doi:10.4135/9781412950589.
41. Kooiman J, Jentoft S. Meta-governance: values, norms and principles, and the making of hard choices. Public Adm. 2009;87(4):818–36. doi:10.1111/j.1467-9299.2009.01780.x.
42. Inwood SM, Sharp JS, Moore RH, Stinner DH. Restaurants, chefs and local foods: insights drawn from application of a diffusion of innovation framework. Agric Hum Values. 2008;26(3):177–91. doi:10.1007/s10460-008-9165-6.
43. Stead M, Caraher M, Wrieden W, Longbottom P, Valentine K, Anderson A. Confident, fearful and hopeless cooks: findings from the development of a food-skills initiative. Br Food J. 2004;106(4):274–87. doi:10.1108/00070700410529546.
44. Watts DCH, Ilbery B, Maye D. Making reconnections in agro-food geography: alternative systems of food provision. Prog Hum Geogr. 2005;29(1):22–40. doi:10.1191/0309132505ph526oa.
45. Mount P. Growing local food: scale and local food systems governance. Agric Hum Values. 2011;29(1):107–21. doi:10.1007/s10460-011-9331-0.
46. Neis B, Ommer R. Moving forward: building economically, socially and ecologically resilient fisheries and coastal communities. 2014. http://www.curra.ca/documents/Revised_CURRA.
47. Galt RE. The moral economy is a double-edged sword: explaining farmers' earnings and self-exploitation in community-supported agriculture. Econ Geogr. 2013;89(4):341–65.
48. Konefal J. Environmental movements, market-based approaches, and neoliberalization: a case study of the sustainable seafood movement. Organ Environ. 2012;26(3):336–52. doi:10.1177/1086026612467982.
49. Gibson-Graham JK. Postcapitalist politics. Minneapolis: University of Minnesota Press; 2006. doi:10.1017/CBO9781107415324.004.

3

Evaluation of dry matter yield and nutritive value of Sugar graze and Jumbo plus at different spacing in the *yala* season in the dry zone of Sri Lanka

Sagana Pushparajah and Jeyalingawathani Sinniah*

Abstract

Background: A field experiment to evaluate the dry matter yield and nutritive value of Sugar graze and Jumbo plus under irrigation was conducted at three different plant spacings (15 x 30, 15 x 45, 15 x 60 cm) on a red–yellow latosol in the dry zone of Sri Lanka from March to June 2016. The design was a factorial randomised complete block design with three replications. Morphological parameters measured were plant height, leaf length, leaf width, leaf area and number of leaves per plant, inter-node elongation and stem girth. Chemical composition of forages, viz. crude protein, crude fibre, neutral detergent fibre, acid detergent fibre and dry matter yield, was also determined.

Results: There was no difference in the chemical composition of Sugar graze and Jumbo plus 60 days after planting. Sugar graze had significantly higher values ($p < 0.05$) for stem girth, leaf length, leaf width and leaf area 60 days after planting. Sugar graze produced significantly higher dry matter yield than Jumbo plus. The narrowest spacing 15 × 30 cm gave the highest yield than the other 2 spacings.

Conclusions: It could be concluded that Sugar graze with 15 × 30 cm spacing will give more dry matter yield than Jumbo plus with similar chemical composition in the yala season of dry zone of Sri Lanka.

Keywords: Crude fibre, Crude protein, Neutral detergent fibre, Acid detergent fibre

Background

The performance of dairy animals depends on the consistent availability of quality fodder in adequate amount. Therefore, the critical limitation on profitable animal production in developing countries is the inadequacy of quality forage [1]. In many developing countries, because of ever growing human need for food, only limited cultivated land can be allocated to fodder production.

Most tropical grasses, like Guinea, *Brachiaria* and Napier, have high fibre content and low sugar and protein content. This affects the intake and digestibility of the grass, and most of the indigestible portions are passed as excreta. Furthermore, the critical lack of protein in many of these grasses directly limits the ability of the cow to produce milk.

In contrast, grasses like Sugar graze (sorghum), maize and improved Napier variants (CO-3 and CO-4) are considered high-quality cattle feed because of the nutritive value they bring. Among these grasses, sorghum and maize are high-energy sources when compared with the improved Napier varieties. While they are high in nutritive value, maize and improved Napier variants (CO-3 and CO-4) require a considerable amount of water for cultivation. In addition, maize can only be cut once while sorghum can be cut 4–5 times with one planting. Considering its nutritive value, production capacity and profitability, sorghum brings more benefit to farmers [2].

In arid and semi-arid areas of the world, water is the principal limiting factor of agricultural production primarily due to low and/or uneven distributions of annual

*Correspondence: jeyalingawathani@gmail.com
Department of Animal Science, Faculty of Agriculture, University of Jaffna, Ariviyal Nagar, Kilinochchi, Sri Lanka

rainfall [3, 4]. Pearl millet (*Pennisetum miliaceum* L.) and sorghum (*Sorghum bicolor* L.) are important forages in several arid and semi-arid regions of the world and are well adapted to environments with limited rainfall, high temperatures and low soil fertility [5, 6]. It has been shown that sorghum and millet were more drought resistant and have higher water-use efficiency compared with corn and are able to produce acceptable forage yields when exposed to drought [7, 8].

Sorghum (*Sorghum bicolor (L) Moench*) is an annual crop grown in the summer season for fodder production and is considered the fifth major cereal in terms of area and production in the world [9]. Recently, some varieties of this crop evolved as the hybrids by crossing between sorghum (*Sorghum bicolor (L) Moench*) and Sudan grass (*Sorghum bicolor subsp. drummondii*) that are known for their quick growth, multi-cuts and nutritious fodder with higher total digestive nutrients (TDN) and crude protein (CP) values exceeding 53-60% and 9–15%, respectively [10]. Fodder yield in sorghum is mainly affected by row spacing and seed rate [11]. To expand the utility of sorghum and corn as forage crops, breeders have focused on traits likely to affect its yield and forage quality [12].

Until recently little research has been done in forage productivity of sorghum especially on Sugar graze and Jumbo plus hybrid varieties in Sri Lanka. Hence, the present investigation was undertaken

- To assess the forage productivity and nutritive value of two multi-cut Sorghum hybrid varieties which were recently introduced to the dry zone of Sri Lanka viz. Sugar graze and Jumbo plus.
- To study the growth parameters of Sugar graze and Jumbo plus
- To study the performance of Sugar graze and Jumbo plus in terms of dry matter yield and chemical composition
- To study the effect of spacing on the performance of Sugar graze and Jumbo plus.

Materials and methods

Location

The study was carried out at the livestock farm of the Department of Animal Science, Faculty of Agriculture, Ariviyal Nagar, Kilinochchi. This falls in the Northern Province and dry zone of Sri Lanka. The longitude of the location is 80°39′E, and latitude is 09°38′N. The Northern Province tends to be hot and dry in the dry season (Feb. to Sept.), and moderately cool and wet in the wet season (October to January). The soil type on the location is red–yellow latosol.

Planting material

The planting materials used in this study were 2 cultivars of fodder sorghum, viz. Sugar graze and Jumbo plus.

Experimental design

Treatments were arranged in factorial completely randomised block design, with 3 replicates. Treatments included 2 cultivars, viz. Sugar Graze and Jumbo Plus with 3 spacings (15 x 30, 15 x 45, 15 x 60 cm). Plot size was arranged in such a way to have 64 plants per plot with respective spacing. Basal dressing was done at the rate of 50 kg urea, 25 kg triple superphosphate and 1 2.5 kg muriate of potash/acre. Seeds were sown at the rate of 2 per hill in March 2016. Seedlings were thinned to a single plant per hill 2 weeks after sowing. One month after planting top dressing was done at the rate of 50 kg urea and 12.5 kg Muriate of Potash/acre. Irrigation was done once in 4 days. At the initial stage, manual weeding was done to reduce competition from weeds.

Data collection

Growth parameters

Sixty days after planting the growth parameters, viz. plant height, leaf area, leaf length, leaf width, stem girth, number of leaves per plant, were measured for each cultivar at different spacing.

Fresh and dry matter determination

Harvesting was done 60 days after planting. Forage was cut 15 cm above ground, and soon after harvest fresh matter yield per plot was taken. Ten plants from each plot were sampled to determine dry matter weight and to prepare samples for chemical analysis. For dry matter determination, 2 kg of chopped (1–2 cm long) samples were air-dried for 2 days and oven-dried at 80 °C till constant weight was reached.

Chemical composition

After dry matter determination, from each sample, 100 g of ground sub-samples were taken for chemical analysis and packed in polyethene bags till analysis. Crude fibres (CF), acid detergent fibre (ADF) and neutral detergent fibre (NDF) were determined in the laboratory on dry matter basis by method of [13]. Crude protein was determined on dry matter basis as per standard procedure [14].

Data analysis

Data were subjected to analysis of variance, and mean separation was done with Duncan's multiple range test by SAS 9.1.3 software.

Results

Weather parameters

The air temperature and precipitation (30 days average values) that prevailed at the experimental site during the growing period between March and May 2016 are presented in Table 1. Though unexpected heavy rainfall was experienced in May, the harvest of the current study was completed before the commencement of heavy rainfall.

Vegetative growth parameters

In terms of plant morphology, the 2 cultivars generally responded differently to variation in plant spacing (Table 2). Except plant height and stem diameter, all other parameters did not differ significantly at different spacing for both cultivars, for which plant height significantly decreased with increasing spacing. In general, the thickest stems were produced at the wider spacing and the thinnest at the narrower spacing. Plant height, inter-node elongation and number of leaves per plant were unaffected by cultivars. In contrast, there were significant differences between cultivars for leaf length, leaf width, leaf area and stem diameter. For all 4 parameters, Sugar graze had significantly higher value than Jumbo plus (Table 3).

Dry matter yield

Sugar graze produced higher dry matter yields than Jumbo plus (Table 4). Dry matter yields were inversely related to plant spacing with yield decreasing progressively as plant spacing increased and the differences were significant (Table 5).

Table 1 Monthly weather parameters during the study period. *Source*: Meteorology Department, Vavuniya, Department of Irrigation, Iranamadu (2016)

Parameter	Mean temperature (°C)	Total rainfall (mm)
March	29.4	1.6
April	31.4	16.2
May	32.1	478.1

Table 3 Effects of variety on growth parameters 60 days after planting

Growth parameters	Sugar graze (mean ± SD)	Jumbo plus (mean ± SD)
Height (cm)	205.68 ± 17.34^a	197.48 ± 22.52^a
Leaf length (cm)	94.04 ± 1.41^a	88.98 ± 2.68^b
Leaf width (cm)	9.21 ± 0.88^a	6.36 ± 0.46^b
Leaf area (cm^2)	865.39 ± 73.49^a	565.51 ± 30.25^b
Stem girth (cm)	10.19 ± 0.32^a	8.14 ± 0.55^b
Inter-node elongation (cm)	21.32 ± 0.55^a	19.43 ± 0.81^a
Number of leaves per plant	14.4 ± 0.43^a	14.13 ± 0.44^a

^{a,b} Within a row, means without a common superscript differ ($P < 0.05$)

Nutritive value

Table 6 shows the proximate composition of Sugar graze and Jumbo plus, 60 days after planting. Proximate composition was not influenced by cultivar and spacing.

Discussion

Plant height

Plant height as a growth parameter is a result of elongation of the stem internodes, which is influenced by the environment as suggested by Hozumi et al. [15]. In the current study, taller plants were observed with narrow spacing, which is in agreement with reports in the literature that narrower spacing will give taller plants as a result of competition for sunlight [16]. Taller plants led to narrower stem diameter as well.

The plant height observed in the current study for Sugar graze with 15 cm × 60 cm spacing is in agreement with the value (188.833 ± 1.922 cm) reported by Pahuja et al. [17] in India. The plant height at the spacing 15 cm × 30 cm was similar to the value (218 cm)

Table 2 Effects of row spacing on growth parameters and yields of Sugar graze and Jumbo plus 60 days after planting

Growth parameters	Sugar graze			Jumbo plus		
	15 x 30 cm	15 x 45 cm	15 x 60 cm	15 x 30 cm	15 x 45 cm	15 x 60 cm
Height (cm)	222.93 ± 6.45^a	203.26 ± 13.9^b	190.86 ± 13.57^c	221.0 ± 13.01^a	196.93 ± 10.34^b	174.53 ± 11.53^c
Leaf length (cm)	95.26 ± 1.13^a	93.73 ± 1.20^a	94.13 ± 0.65^a	89.73 ± 2.41^a	86.73 ± 2.44^a	90.70 ± 1.55^a
Leaf width (cm)	8.66 ± 0.07^a	8.80 ± 0.72^a	8.60 ± 0.21^a	6.03 ± 0.04^a	6.92 ± 0.14^a	6.14 ± 0.37^a
Leaf area (cm^2)	824.9 ± 11.77^a	861.04 ± 8.27^a	810.24 ± 25.61^a	540.96 ± 10.93^a	559.00 ± 15.90^a	556.57 ± 24.11^a
Stem girth (cm)	9.93 ± 0.46^b	10.44 ± 0.083^a	10.2 ± 0.1^a	7.42 ± 0.27^b	8.27 ± 0.04^a	8.69 ± 0.06^a
Inter-node elongation (cm)	21.3 ± 0.75^a	21.1 ± 0.66^a	21.5 ± 0.36^a	19.83 ± 0.51^a	19.93 ± 0.30^a	18.52 ± 0.69^a
Number of leaves/plant	14.66 ± 0.23^a	14.13 ± 0.57^a	14.6 ± 0.34^a	13.936 ± 0.30^a	13.86 ± 0.30^a	14.6 ± 0.34^a

^{a,b,c} Within a row, means without a common superscript differ ($P < 0.05$)

Table 4 Effects of variety on dry matter yield 60 days after planting

Variety	Dry matter yield (t/ha)
Sugar graze	13.074 ± 2.17[a]
Jumbo plus	9.998 ± 1.89[b]

[a,b] Within a column, means without a common superscript differ ($P < 0.05$)

Table 5 Effect of row spacing on dry matter yield 60 days after planting

Spacing	Dry matter yield (t/ha)
15 cm × 30 cm	14.03[a]
15 cm × 45 cm	10.82[b]
15 cm × 60 cm	9.76[c]

[a,b,c] Within a column, means without a common superscript differ ($P < 0.05$)

reported by Epasinghe et al. [18] in the wet zone of Sri Lanka for Sugar graze, 60 days after planting where the spacing was not mentioned. Similarly [19] in wet zone of Sri Lanka reported lower value (213.5 cm) for Sugar graze and higher value (236.5 cm) for Jumbo plus where too the spacing was not mentioned. The plant height observed under spacing of 15 cm × 60 cm is similar to the value (191.1 cm) reported by Bandara et al. [20] in the wet zone of Sri Lanka for Sugar graze.

Stem girth

As far as the stem girth is concerned, the value observed for both varieties was higher than the range (0.01–0.05 m) reported for sweet sorghum in Portugal [21]. The current values were also higher than the mean stem girth (5.38 ± 0.15 cm) reported by Pahuja et al. [17] in a trial to evaluate 17 forage sorghum hybrids and 6 check varieties in India. The differences may be due to climatic condition, soil fertility and stage of maturity of the plants. In the current study, stem girth was affected by cultivars which is in agreement with the findings of Yousef et al. [22] and Ayub et al. [23] who reported a significant variation in stem girth among different cultivars of sorghum.

Leaf length, leaf width and leaf area

Leaf development has been described extensively for fodders, as growth is mostly reflected in large increase in leaf length as plants grow to maturity, accompanied by relatively small increase in width and thickness [24]. Large leave lengths are also important for the survival of individual plants within a sward [25]. In the current study, leaf length, leaf width and leaf area were unaffected by spacing within cultivar. The higher values for leaf length, leaf width and leaf area of Sugar graze than Jumbo plus are attributed to the genetic effect of these cultivars. Leaf length of Sugar graze was similar to the value 95 ± 2.0 cm reported by Pahuja et al. [17] in India for the first cut at 50% flowering and a spacing of 15 × 45 cm, whereas leaf width was higher than that recorded by the same authors (6 ± 0.58 cm).

The higher mean leaf area in Sugar graze than Jumbo plus in the current study may be attributable to genetic makeup of the cultivars [26–31] also observed variation in leaf area among various cultivars and varieties of forage sorghum.

Dry matter yield

Plant spacing has a marked impact on the efficiency of use of land, light, water and nutrients. By optimising plant spacing, highest yield potential can be achieved from the smallest possible area [32]. Significantly higher dry matter yield of Sugar graze is due to its genotype, and the growth parameters contributed for the higher yield were the higher values of leaf length, leaf width and stem girth than for the Jumbo plus. Erandathi et al. [19] and Bandara et al. [20] in the wet zone of Sri Lanka reported 6.6 and 6.01 t/ha dry matter yield, respectively, for Sugar graze where the harvesting was done at 50% flowering stage. Erandathi et al. [19] also reported the dry matter yield of 5.1 t/ha for Jumbo plus. Current values of dry matter yield of both were higher than the above values recorded in the wet zone of Sri Lanka. Similar value of the dry matter yield of current study was reported for Sugar graze in the wet zone of Sri Lanka [18] where the harvesting was done 60 days after planting as in the current study. Chaudhry et al. [26] recorded yields of 10,400–13,100 kgDM/ha for advanced lines of forage sorghums and [33] recorded 13,262 kgDM/ha at 15 cm row spacing 78 days after sowing for Sugar graze. The differences in the dry matter yield of cultivars of the present study with the literature are attributed to differences in the spacing, soil fertility, and environmental conditions.

Table 6 Proximate composition of Sugar graze and Jumbo plus 60 days after planting

Variety	DM% (mean ± SD)	CF% (mean ± SD)	NDF% (mean ± SD)	ADF (mean ± SD)	CP (mean ± SD)
Sugar graze	17.91 ± 0.95[a]	29.22 ± 3.59[a]	69.360 ± 5.33[a]	47.08 ± 5.57[a]	8.77 ± 1.35[a]
Jumbo plus	17.26 ± 1.16[a]	31.09 ± 3.33[a]	70.578 ± 4.06[a]	45.54 ± 2.9[a]	7.52 ± 0.33[a]

[a] Within a column, means without a common superscript differ ($P < 0.05$)

Significant reduction in dry matter yield with increasing plant spacing may be due to decreasing plant density with increasing spacing. In a study conducted at Kemptville Research Station, Canada [34], planting density had a significant effect on fresh and dry matter production and two higher densities had significantly higher production than the 2 lower densities for 3 sweet sorghum hybrids including Sugar graze. In maize, total forage mass was greater at higher plant densities than at lower plant densities, but forage mass was not different among plant populations that exceeded 58,130 plants/ha [35]. Forage dry matter yield declined linearly from 10.9 to 8.1 Mg/ha as within-row spacing increased from 50 to 60 cm in Sorghum [36].

The dry matter percentage of Jumbo plus was higher than the values (15.9%) reported by Tauqir et al. [37] in Pakistan and [18] in Sri Lanka who reported a dry matter percentage of 12.66% for Sugar graze 45 days after planting. The low dry matter percentage of the latter may be due to 15 days early harvesting than the current study.

The crude fibre percentage of this study is in agreement with the finding of Epasinghe et al. [18] in Sri Lanka who reported CF% of 32.23 for Sugar graze 60 days after planting. Higher value of 36.8% was reported by Bandara et al. [20] in the wet zone of Sri Lanka at 50% flowering stage for Sugar graze, whereas lower value (25.12%) was reported by Bozorgvar et al. [12] for Sugar graze in Iran. The differences may be due to stage of maturity and soil fertility.

The NDF% of this study was similar to the finding of Epasinghe et al. [18] (2012) in Sri Lanka where the NDF% was 67.55 for Sugar graze 60 days after planting in the wet zone of Sri Lanka. Similar values NDF for Sugar graze and Jumbo plus were reported in Turkey [38]; the values were 67.97 and 74.89% for Sugar graze and Jumbo plus, respectively. But the result for Sugar graze was higher than the finding of Davison [39] in Manawatu, New Zealand, as he stated that NDF% of Sugar graze was 65.2, 78 days after sowing. For Jumbo plus cultivar, observed value was lower than related study [35] in Pakistan (75.2%). The differences may be due to stage of maturity at harvesting, soil fertility and other environmental factors.

The ADF% of this study was higher than the finding in Sri Lanka [18] reporting the ADF% for Sugar graze as 40.12, 60 days after planting. The current result for Sugar graze was higher than the finding in Manawatu, New Zealand [39], stating that ADF % of Sugar graze was 36.2 after 78 days of sowing. For Jumbo plus cultivar, observed value (39.7) was higher than related study in Pakistan [2]. But for Jumbo plus cultivar, observed value (48.32%) was in agreement with related study in Turkey [34], but the value (41.86) reported for Sugar graze was lower than the current study.

The CP% of this study was lower than the finding in Sri Lanka [18] reporting that CP% for Sugar graze was 17.2, 60 days after planting; also the result for Sugar graze was lower than the finding in Manawatu, New Zealand [37], stating that CP% of Sugar graze was 10.2 after 78 days of sowing. A lower value (7.45) than the current study was reported in Iran [12].

For Jumbo plus, observed CP value was lower than (11.05) the related study in Pakistan [37]. But for Jumbo plus and Sugar graze, observed values were in agreement with related study in Turkey [38] where the values reported were 7.2 and 6.82% at milking stage, respectively. The differences may be attributed to differences in soil fertility, stage of maturity and other environmental factors.

From the above discussion, Sugar graze showed significantly higher values for the growth parameters stem girth, leaf length, leaf width and leaf area than Jumbo plus. Sugar graze showed significantly higher dry matter yield than Jumbo plus. Dry matter yield was inversely related to spacing, and the narrowest spacing gave the highest dry matter yield. Nutritive value was unaffected by cultivar and spacing within cultivar.

Conclusions

Therefore, it can be concluded that Sugar graze with 15 × 30 cm could be used to get maximum yield in the dry season with irrigation in the dry zone of Sri Lanka.

Authors' contributions

SP carried out research, collected data and analysed the data. JS conceived of the study, participated in its design and coordination and helped in the statistical analysis and drafting the manuscript. Both authors read and approved the final manuscript.

Competing interests

The authors declare that they have no competing interests.

Funding

No funding was obtained.

References

1. Sarwar M, Khan MA, Iqbal Z. Feed resources for livestock in Pakistan. Int J Agric Biol. 2002;2002(4):186–92.
2. Sivayoganathan B. The SOLID approach to producing year-round cattle feed. USAID; 2017. p. 68.
3. Keshavarz Afshar R, Chaichi MR, Asareh MH, Hashemi M, Liaghat A. Interactive effect of deficit irrigation and soil organic amendments on seed yield and flavonolignan production of milk thistle (*Silybum marianum* L. Gaertn.). Ind Crops Prod. 2014;58:166–72.
4. Jahanzad E, Jorat M, Moghadam H, Sadeghpour A, Chaich MR, Dashtaki M. Response of a new and a commonly grown forage sorghum cultivar to limited irrigation and planting density. Agric Water Manag. 2013;117:62–9.
5. Keshavarz Afsha R, Ansari Jovini M, Chaichi MR, Hashemi M. Grain sorghum response to Arbuscular mycorrhiza and phosphorous fertilizer under deficit irrigation. Agron J. 2014;106(4):1212–8.
6. Amer S, Hassanat F, Berthiaum R, Seguin P, Mustafa AF. Effects of water soluble carbohydrate content on ensiling characteristics, chemical composition and in vitro gas production of forage millet and forage sorghum silages. Anim Feed Sci Technol. 2012;177:23–9.
7. Jaster EH, Fisher CM, Miller DA. Nutritive value of oatlage. barley/pea, pea, oat/pea, pearl millet, and sorghum as silage grown under a double cropping forage system for dairy heifers. J Dairy Sci. 1985;68:2914–21.
8. Singh BR, Singh DP. Agronomic and physiological responses of sorghum, maize and pearl millet to irrigation. Field Crop Res. 1995;42:57–67.
9. CGIAR. Sorghum. CGIAR. 2013. http://www.cgiar.org/ourresearch/crop-factsheets/sorghum/. Accessed 10 Feb 2016.
10. Newman Y, Erickson J, Vermerris W, Wright D. Forage sorghum (*Sorghum bicolor*): overview and management. No.SS-AGR-333. IFAS Extension. University of Florida, USA; 2013.
11. Scott S, Gordon B, Taylor R, Duncan S, Fjell D. Narrow-row grain sorghum production in Kansas. Manhattan: Kansas State University; 1999. p. 1–4.
12. Bozorgvar N, Khademosharich MM, Neamatollahi E, Jahansuz MR. Determine the best varieties of forage sorghum. Adv Environ Biol. 2013;7(6):1105–12.
13. Van Soest PJ, Robertson JB, Lewis BA. Methods for dietary fiber, neutral detergent fiber and non-starch polysaccharides in relation to animal nutrition. J Dairy Sci. 1991;74:3583–97.
14. AOAC. Official methods of analysis. 18th ed. Washington: Association of Official Analytical Chemists; 2005.
15. Hozumi E, Weston RH, Hesketh J. Factors limiting the intake of feed by sheep: studies with wheat hay. Aust J Agric Res. 1965;18:983–1002.
16. Lamana MCL. Effect of spacing between plants on growth and forage yield of two maize (*Zea mays* L.) cultivars. Masters Dissertation. University of Khartoum; 2003.
17. Pahuja S, Arya S, Kumari S, Panchta R. Evaluation of forage sorghum hybrids [*Sorghum bicolor (l.) Moench*]. Forage Res 2014;40(3):159–162. [cited 25 Jan 2016]. http://forageresearch.in/wpcontent/uploads/2014/12/159-162.pdf.
18. Epasinghe TM, Jayawardena VP, Premalal GGC. Comparison of growth, yield and nutritive value of maize, multi-cut fodder sorghum and hybrid Napier (var. Co3) grown in wet zone of Sri Lanka. In: Proceedings of 22nd Annual Students Research Session, Department of Animal Science; 2012. [cited 12 Feb 2016]. http://agri.pdn.ac.lk/dept/pdf/Proceedings%20-%202012%20As.pdf.
19. Erandathi SAT, Premaratne S, Premalal GGC. Yield, chemical composition and adverse factors of different varieties of sorghum (*Sorghum bicolour* (L.) *Moench*) and hybrid var.CO-3 (*Pennisetum americanum* × *Pennisetum Purpureum*) in the mid country of Sri Lanka. In: Proceedings of 23rd annual students research session, Department of Animal Science, Faculty of Agriculture, University of Peradeniya; 2013. p. 17–18.
20. Bandara PGG, Premalal GGC, Nayananjalie WAD. Assessment of nutritive value of sorghum (*Sorghum bicolor (L.) Moench*) fodder in Sri Lanka. Int J Livest Res. 2016;6(8):44–9.
21. Jose S, Tomas M, Bohumil H, Tatiana I. Sweet sorghum (*Sorghum bicolour (L.) Moench*) bio energy value—importance for Portugal. Agric Trop Subtrop. 2012;45(1):12–9.
22. Yousef E, Carmi A, Nikbachat M, Zenou A, Umiel N, Miron J. Characteristics of tall versus short-type varieties of forage sorghum grown under two irrigation levels, for summer and subsequent fall harvests, and digestibility by sheep of their silages. Anim Feed Sci Technol. 2009;152:1–11.
23. Ayub M, Tanveer A, Mahmud K, Ali A, Azam M. Effect of nitrogen and phosphorus on the fodder yield quality of two sorghum cultivars (*Sorghum bicolor* L.). Pak J Biol Sci. 1999;2:247–50.
24. Skinner RH, Nelson CJ. Role of leaf appearance rate and the coleoptiles tiller in regulating tiller production. Crop Sci. 1994;34:71–5.
25. Barre P, Turner LB, Gutierrez AJE. Leaf length variation in perennial forage grasses (Review). Agriculture. 2015;5:682–96.
26. Chaudhry GN, Riaz M, Ahamad G. Comparison of advanced lines of *Sorghum bicolor* (L.) Moench for green fodder and dry matter yield and morpho-economic parameters. J Agric Res. 2006;44:191–6.
27. Musa M, Gondal MR, Ali A, Ashraf M. Effect of cutting at different plant growth stages and heights above ground levels on green fodder and dry matter yields of Sadabahar. J Agric Res. 1993;31:409–13.
28. Naeem M, Chauhan MSM, Khan AH, Salahaddin S. Evaluation of different varieties of sorghum for green fodder yield potential. Asian J Plant Sci. 2002;1:142–3.
29. Mehmud K, Ahmad I, Ayub M. Effect of nitrogen and phosphorus on the fodder yield and quality of two sorghum (*Sorghum bicolor* L.) cultivars. Int J Agric Biol. 2003;5:61–3.
30. Chohan MSM, Naeem M, Khan AH, Salah-ud-Din S. Performance of newly developed forage varieties of sorghum (*Sorghum bicolor* (L.) Moench). J Plant Sci. 2003;2:48–50.
31. Chohan MSM, Naeem M, Khan AH, Kainth RA. Performance of pearl millet (*Pennisetum americanum* L.) varieties for forage yield. J Agric Res. 2006;44:23–7.
32. Oseni TO, Fawusi MO. Influence of nursery spacing and plant arrangement on growth and leaf nutrient content of three citrus root stock seedlings. Trop Agric. 1986;64(1):41.
33. Silungwe D, Millner JP, McGill CR. Evaluation of forage yield and quality of sorghum, Sudan-grass and pearl millet cultivars in Manawatu. Masters Thesis. Massey University, New Zealand; 2011.
34. Ashraf T, Alessandro S. Effects of planting density and genetics on production of sweet sorghum. Poster presented at the Northeast Agricultural and Biological Engineering Conference (NABEC), Kemptville, 27–30 July 2014.
35. Cuomo GJ, Redfearn DD, Blouin DC. Plant density effects on tropical corn forage mass, morphology, and nutritive value. Agron J. 1998;90:93–6.
36. Caravetta GJ, Cherney JH, Johnson KD. Within-row spacing influences on diverse sorghum genotypes: II. Dry matter yield and forage quality. Agron J. 1990;82(2):210–5.
37. Tauqir NA, Sarwar M, Jabbar MA, Mahmood S. Nutritive value of Jumbo grass (*Sorghum bicolour Sorghum sudanefe*) silage in lactating Nili-Ravi buffaloes. Cellul Pak Vet J. 2009;29(1):5–10.
38. Uzun F, Ugur S, Sulak M. Yield, Nutritional and chemical properties of some sorghum × sudan grass hybrids (*Sorghum bicolour (L.) Moench* × *Sorghum Sudanese Stapf.*). J Anim Vet Adv. 2009;8(8):1602–8.
39. Davison S, Plant density effects on tropical corn forage mass, morphology, and nutritive value. M.Sc Thesis, Massey University, Palmerston North, New Zealand; 2011.

4

Phenotypic characterization of sheep breeds in Gamogofa zone

Fsahatsion Hailemariam*, Dawit Gebremicheal and Hagos Hadgu

Abstract

Background: Information on the morphometric character of indigenous sheep population in their habitat is pre-requisite to set up the breed improvement under the farmer level in Ethiopia. Still, there is a long way to characterize, identify and document the existing indigenous sheep type. The study was conducted in Gamogofa zone of Southern Ethiopian with the objective of describing the morphological characterization of the indigenous sheep breed.

Methods: Purposive sampling techniques were employed to select target farmers. Qualitative, body weight and body measurement were taken from 736 sheep. All the collected data were analyzed using Generalized Linear Model procedure of the statistical analysis system (release 9.1, 2003).

Results: Plain black (34%) and red (19.6%) are the dominant color types with long thin tail (88.2%) in both ewe and ram sheep. Almost 99.3% of ewe's and 100% of rams had no wattle. The majority (90.4%) of the sheep had a straight tail form at the tip, straight head profiles (88.7%) and also horizontal ear forms (51.2). Agroecology, sex, and age had a significant effect on body weight and linear body measurements. The mean body weight, whiter height, body length, chest girth, shoulder width, cannon bone width, tail length, and tail circumference for ram and ewe sheep were 28.1 ± 0.37 and 24.0 ± 0.18, 58.6 ± 0.32 and 56 ± 0.05, 65.5 ± 0.35 and 62.3 ± 0.17, 75.9 ± 0.47 and 71.0 ± 0.21, 16.95 ± 0.18 and 15.24 ± 0.09, 7.45 ± 0.06 and 6.5 ± 0.03, 27.44 ± 0.3 and 21.04 ± 0.26, 17.11 ± 0.42 and 9.56 ± 0.14 cm, respectively. Body weight and linear body measurements were higher in ram and increased with age. The correlation coefficient between body weight and most of the linear body measurement at different sex and age group were positive. The highest correlation coefficient was obtained between body weight and chest girth for both sex and age groups.

Conclusion: It was concluded that native sheep had a possible for the versatile role to generate income for livestock keepers. Therefore, genetic improvment program should aim at farmers need to cope with trait preference and existing traditional herding and breeding practice.

Keywords: Agroecology, Qualitative character, Sex, Age, Linear measurement, Correlation

Background

About 29.33 million heads of sheep population are found in Ethiopia [1]. The sheep's are widely dispersed and adapted across the different agro-ecologies of the country, out of which about 72.77% are female and about 27.23% are males. Majority of the sheep are found in the highlands, while one-fourth of them are reared in the lowlands [2]. Morphological and molecular both characterizations of sheep breeds in the country are traditionally recognized by ethnic and/or geographic nomenclatures. Based on this, the sheep breeds have been classified into 14 traditional populations in 9 breeds within 6 major breed groups [3]. However, some parts of the country viz. Gambella regions and the northern part of Tigray bordering Eritrea were not considered in his study implying the possibilities of the existence of additional populations in the country.

Generally, the current state of knowledge on the characterization of farm animal generic resources in Ethiopia shows that there is still lack of information about

*Correspondence: fhailema@gmail.com
Department of Animal Science, College of Agriculture, Aksum University Shire Campus, Tigray, Ethiopia

the production characters of local breeds managed in their native production system [4] although the country is widely known to possess a large population of livestock with enormous diversity. Indigenous sheep genetic resources have developed specific adaptations to survive and produce under adverse condition of climatic stresses, poor quality feed, seasonal feed and water shortage, endemic disease and parasite challenge that make them suitable for their use in the traditional, low-external-input production system [5]. Therefore, selection and breeding based on zoometric/morphometrical measurement, fast growth rate, good body size and conformation [6] could result in improvement in live weight of indigenous sheep and goat for meat production [7, 8]. The productivity of sheep as in the case of most of the ruminants is markedly low due to several genetic and environmental factors [9].

The information on body measurement is the basis for the establishment of further advanced characterization, conservation, breeding and selection strategies for indigenous sheep breed [10, 11] and used to assess the type and function and the value of the animal as potential breeding stocks [12]. Body measurements are well-thought-out as qualitative development indicator which reflects the conformational fluctuations occurring during the life span of animals [13].

Gamogofa zone is geographically located in Southern, Nation, Nationalities and Peoples of Regional State (SNNPRS) of Ethiopia. Even though the study area is rich in livestock resources including small ruminants, still there is a long way to go to characterize, identify and document the existing indigenous sheep type/s of the region. The overall objective of this study, therefore, was to describe the morphological of the indigenous sheep population of Gamogofa Zone based on their morphological and quantitative trait.

Methods

The study was carried out within the period October 2015 to May 2016 in four 'Woredas.' The sampling frame was established using purposive sampling to selected the four 'Woredas' based on agroecology, sheep population distribution, availability of transportation and their contribution to farmers. From each selected 'Woredas,' two rural 'Kebeles' were selected based on flock size, sheep population and agroecology area. The sample size was determined according to [14], and based on this formula, the calculated sample size is 384 but, for the higher accuracy, the total number of sampling size for this study was increased to 736. Therefore, qualitative and quantitative measurements were made on 736 (181 males and 555 non-pregnant females in the proportion of 25 males/75 females) mature sheep were selected randomly (184 per district, 92 per rural 'Kebele'). Quantitative traits including body length, pelvic width, chest width, tail length, tail circumference, head length, head width, cannon bone length and width, horn length, ear length and scrotum circumstance were measured using measuring tape, and height at wither and shoulder width were measured using self-devised scale (which is locally made from wooden material), while the body weight was measured using suspended spring balance of 50 kg (weighing scale). The age of the animal was estimated by recall and dentition methods. Adult sheep were classified in to five groups; no pair of permanent incisors (0PPI), one pair of permanent incisors (1PPI), two pair of permanenet incisors (2PPI), three pair of permanent incisors (3PPI) and four pair permanent incisors (4PPI) to represent age of <15, 15.5–20, 22.5–27, 28–36 and above 39 months, respectively.

Statistical analysis

Quantitative characters (body weight and linear body measurements) were analyzed using the Generalized Linear Model (GLM) procedures of the Statistical Analysis System (SAS, release 9.1, 2003). For adult animals, agroecology, sex and age group were fitted as fixed independent variables, while body weight and linear body measurements (except scrotum circumference) were fitted as dependent variables. Scrotum circumference was analyzed for each type by fitting agroecology and age group as a fixed factor. Mean separation was under taken when it was significant to reveal the difference between mean using Duncan multiple range test. The model to analyze body weight and linear body measurements was:

$$Y_{ijk} = \mu + E_i + A_j + S_k + (\mathrm{AS})_{jk} + e_{ijk}$$

where Y_{ijk}, the observed k (body weight or linear body measurement) in the ith agroecology, jth age group and kth sex; μ, overall mean; E_i, the effect of ith agroecology ($i=$'Dega,' 'Wenya-dega'); A_j, the effect of jth age group ($j=1$, 2, 3 and 4); S_k, the effect of kth sex ($k=$intact ram and ewe); $(\mathrm{AS})_{ij}$, interaction effect of ith age group with jth of sex; e_{ijk}, random residual error.

Body weight (BW) and other linear body measurements including chest girth (CG), body length (BL), wither height (WH), pelvic width (PW), ear length (EL), horn length (HL), tail length (TL), tail circumference (TC), head length (HL), head width (HW), shoulder width (SW), cannon bone length (CBL), cannon bone circumference (CBC) and Scrotum circumference (SC) were conducted for rams, whereas scrotum circumference (SC) was excluded from the analysis of ewes sheep.

Results

Qualitative character

The physical body characterization on sheep production is presented in Table 1. Most of the sheep in the study

Table 1 Qualitative character of sheep

Trait	Class level	Ewe *N* (%)	Ram *N* (%)	Overall *N* (%)	χ^2
Coat color pattern	Plain	388 (69.9)	110 (60.8)	498 (67.7)	13.13**
	Patchy	96 (17.3)	40 (22.1)	136 (18.5)	2.09
	Spotted	71 (12.8)	31 (17.1)	102 (13.9)	2.15
Coat color type	White	61 (11)	18 (9.9)	79 (10.7)	0.156
	Black	191 (34.4)	59 (32.6)	250 (34)	1.201
	Red	101 (18.2)	43 (23.8)	144 (19.6)	0.465
	Brown	43 (7.7)	23 (12.7)	66 (9)	4.11*
	Gray	15 (2.7)	3 (1.7)	18 (2.4)	0.625
	White dominant	33 (5.9)	8 (4.4)	41 (5.6)	2.604
	Black dominant	80 (14.4)	7 (3.9)	87 (11.8)	11.228**
	Red dominant	17 (3.1)	14 (7.7)	31 (4.2)	12.92
	Brown dominant	9 (1.6)	4 (2.2)	13 (1.8)	1.952
	Gray dominant	5 (0.9)	2 (1.1)	7 (1.0)	0.605
Hair type	Short and smooth	182 (32.8)	52 (28.7)	234 (31.8)	82.81***
	Long and coarse	373 (67.2)	129 (71.3)	311 (68.2)	82.81**
Head profile	Straight	497 (89.5)	156 (96.2)	653 (88.7)	110.47**
	Concave	31 (5.6)	11 (6.1)	42 (5.7)	0.061
	Convex	27 (4.9)	14 (7.7)	41 (5.6)	2.310
Tail type	Long fat tail	–	87 (48.1)	87 (11.8)	–
	Long thin tail	555 (100)	94 (51.9)	649 (88.2)	302.5**
Tail form	Carved at the tip	16 (2.9)	55 (30.4)	71 (9.6)	118.46*
	Straight at the tip	539 (97.1)	126 (69.6)	665 (90.4)	118.46**
Ear form	Erect	91 (16.4)	32 (17.7)	123 (16.7)	1.411
	Horizontal	264 (47.6)	113 (62.4)	377 (51.2)	12.068**
	Semi-pendulous	200 (36)	36 (19.9)	236 (32.1)	16.01**
Horn	Present	356 (64.1)	115 (63.5)	471 (64)	0.022
	Absent	199 (35.9)	66 (36.5)	265 (36)	0.022
Wattle	Present	5 (0.9)		5 (0.7)	1.64
	Absent	550 (99.1)	181 (100)	731 (99.3)	2.03
% of total		555 (75.4)	181 (24.4)	736 (100)	

N number of observation

*Significant at $P<0.05$; **significant at $P<0.01$

area were thin-tailed, and the most common type of tail in both rams and ewes was long thin-tailed (88.2%) having different coat colors. The major coat colors were plain black (34%), red (19.6%) and black dominant (11.8%). The dominant of black coat color is obvious in the cold region as the black coat helps in absorption of solar radiation thereby assisting in maintaining a body temperature of the animals. Almost all ewes (99.3%) and rams had no wattles. Most of the sheep had a straight tail at the tip (90.4%), while 9.6% had carved at the tip.

Effect of agroecology, sex and age on body weight and linear body measurement

Agroecology seemed to have been the highest contributing ($P<0.01$) factors to the variation in linear body measurement (LBM) and body weight (BW) of sheep (Tables 2, 3). However, the result of Tables 2 and 3 indicates that, there is no significant ($P>0.05$) variation in the different linear body measurement of head width (HW), shoulder width (SW), tail length (TL), and tail circumference (TC) for ewe and head width, head length, shoulder width, ear length, horn length and cannon bone length for ram. The difference in the linear body measurement and the body weight within the same sex with different agroecology might be due to the non-genetic factors (like feed quality and quantity, environmental factors, management level).

The body weight, chest girth, wither height, head length, body length and pelvic width of ewe sheep found in 'Wenya-dega' agroecology were significant ($P<0.01$) higher than those in the 'Dega' agroecology. A similar trend was observed for body weight and linear body measurement of rams found in 'Wenya-dega' agroecology except for head length which is not significant ($P>0.05$). The superiority in the body weight and linear body measurement of 'Wenya-dega' over 'Dega' agroecology could be a result of the difference in nutritional/feed availability, location, management level of the sheep and high population density in 'Dega' agroecology results more grazing/feed resources shifts to crop production/cultivation.

The body weight of mature ram and ewe sheep found in 'Wenya-dega' agroecology was 29.38 and 25.13 kg, respectively. This value is higher than the body weight of mature rams and ewes reared in the Dega areas. As shown in Tables 2 and 3, sex of the sheep had a significant ($P<0.01$) effect on BW, WH, HW, CG, BL, SW, TL, TC, HL, CBL, and CBC. The study indicates that EL and PW were not significantly ($P>0.05$) influenced by the sex. The difference in body weight and LBM of rams and ewes shows that these parameters are sex dependent. Scrotum circumference too increased with age and hence significantly ($P<0.01$) differed from each other. The value of scrotum circumference obtained in the study was 22.75, 26.16 and 31.30 cm for the lower-age (1PPI), middle-age (2PPI) and higher-age groups ($\geq$ 3PPI), respectively.

Body weight, chest girth, body length, wither height and shoulder width of ram sheep at one pair of teeth (1PPI) were 25.49 kg, 72.64 cm, 62.76 cm 56.90 cm and 14.69 cm, respectively. The value of ewe sheep at the same age group was 20.88 kg, 67.42 cm, 59.89 cm,

Table 2 Least square mean and standard error (LSM ± SE) of body weight (kg) and linear body measurement (cm) of sheep population

Effect	WH	HW	HL	BL	PW	CH	SW	BW
Agroecology	**	ns	**	**	**	**	ns	**
'Dega' ewe	54.9 ± 0.17	8.81 ± 0.06	15.6 ± 0.08	61.4 ± 0.21	12.9 ± 0.10	70.0 ± 0.24	15.4 ± 0.10	22.9 ± 0.22
WD ewe	57 ± 0.23	8.89 ± 0.08	16.4 ± 0.06	63.5 ± 0.25	14.6 ± 0.08	71.9 ± 0.34	15.1 ± 0.14	25.1 ± 0.28
	**	ns	ns	**	**	**	ns	**
'Dega' ram	57.5 ± 0.49	9.80 ± 0.13	17.2 ± 0.14	64.4 ± 0.51	12.6 ± 0.21	74.9 ± 0.73	17.07 ± 0.30	26.7 ± 0.63
WD ram	59.8 ± 0.40	10.1 ± 0.12	16.9 ± 0.13	66.6 ± 0.45	14.5 ± 0.13	76.8 ± 0.60	16.82 ± 0.22	29.4 ± 0.43
Sex	**	**	**	**	ns	**	**	**
Ram	58.6 ± 0.32	9.92 ± 0.09^{a}	17.1 ± 0.10	65.5 ± 0.35	13.6 ± 0.1	75.9 ± 0.47	16.95 ± 0.18	28.1 ± 0.37
Ewe	56.0 ± 0.15	8.80 ± 0.05^{b}	16.0 ± 0.05	62.3 ± 0.17	13.8 ± 0.1	71.0 ± 0.21	15.24 ± 0.09	24.0 ± 0.18
Age	**	*	**	**	**	**	**	**
1PPI	54.9 ± 0.23^{a}	8.77 ± 0.07^{a}	15.64 ± 0.09^{a}	60.96 ± 0.35^{a}	12.4 ± 0.08^{a}	69.4 ± 0.38^{a}	14.8 ± 0.12^{a}	23.5 ± 0.27^{a}
2PPI	57.5 ± 0.24^{b}	8.98 ± 0.08^{b}	16.32 ± 0.08^{b}	63 ± 0.29^{b}	13.4 ± 0.10^{b}	74.0 ± 0.41^{b}	15.5 ± 0.16^{b}	25.9 ± 0.31^{b}
≥ 3PPI	59.2 ± 0.22^{c}	9.48 ± 0.08^{b}	16.70 ± 0.07^{b}	65.87 ± 0.24^{c}	15.0 ± 0.07^{c}	76.6 ± 0.33^{c}	18.0 ± 0.13^{c}	28.9 ± 0.27^{c}
Sex by age	**	ns	ns	*	ns	**	**	**
Ewe*1PPI	53.7 ± 0.21^{a}	8.32 ± 0.07	15.1 ± 0.09	59.9 ± 0.27^{a}	12.1 ± 0.09	67.2 ± 0.41^{a}	14.2 ± 0.13^{a}	20.9 ± 0.25^{a}
Ewe*2PPI	55.7 ± 0.23^{b}	8.72 ± 0.09	16.1 ± 0.09	62.9 ± 0.29^{bc}	13.2 ± 0.11	70.2 ± 0.38^{b}	14.9 ± 0.15^{b}	23.6 ± 0.28^{b}
Ewe* ≥ 3PPI	57.4 ± 0.22^{c}	9.22 ± 0.07	17.5 ± 0.07	64.0 ± 0.23c	14.9 ± 0.08	73.4 ± 0.32^{c}	16.0 ± 0.13^{c}	26.0 ± 0.26^{c}
Ram*1PPI	56.9 ± 0.43^{bc}	9.53 ± 0.10	16.5 ± 0.14	62.8 ± 0.79^{b}	12.8 ± 0.16	72.6 ± 0.62^{c}	14.7 ± 0.21^{b}	25.5 ± 0.44^{c}
Ram*2PPI	59.2 ± 0.54^{d}	9.99 ± 0.16	17.1 ± 0.18	65.5 ± 0.51^{cd}	13.6 ± 0.24	76.9 ± 0.69^{d}	17.5 ± 0.31^{c}	28.6 ± 0.59^{d}
Ram* ≥ 3PPI	60.4 ± 0.38^{d}	10.9 ± 0.16	17.8 ± 0.15	68.4 ± 0.54^{e}	15.4 ± 0.21	80.0 ± 0.58^{e}	18.8 ± 0.29^{e}	31.6 ± 0.55^{e}

WD 'Wenya-dega', *WH* wither height, *HW* head width, *HL* head length, *BL* body length, *PW* pelvic width, *CG* chest girth, *SW* shoulder width, *BW* body weight, *1PPI* one pair of permanent teeth, *2PPI* two pair of permanent teeth, ≥ *3PPI* ≥ three pair of permanent teeth, *ns* nonsignificant

a, b, c, d means with different superscripts across column are significantly different

**Significant at ($P<0.01$), *significant at ($P<0.05$)

53.70 cm and 14.21 cm, respectively. The measurement for the higher-age group (≥ 3PPI) was 31.62 kg, 80 cm, 68.43 cm, 60.40 cm and 18.78 cm for ram, respectively, and the values of ewe sheep at the same age group were 25.97 kg, 73.44 cm, 64.04 cm, 57.40 cm and 16.03 cm, respectively. CBL, CBW and PW of ram sheep at one pair of teeth were 12.66, 7.12 and 12.78 cm, respectively, and the value of ewe at the same age group was 11.99, 6.26 and 12.14 cm, respectively.

Correlation

The phenotypic correlation coefficients between body weight and linear body measurements within age group and sex are presented in Table 4. The result of the study indicates that WH, HW, BL, CG, SW, CBL, CBW and TC were significantly ($P<0.01$) correlated with BW at all of the age groups and in both of the sexes, while SC was also observed to be significantly ($P<0.01$) correlated with the BW of the rams. The results further indicate that horn length (HOL) was significantly ($P<0.01$) correlated with BW in ram at 1PPI age groups and ewes at 2PPI and ≥ 3PPI age group, while it was also significantly ($P<0.05$) correlated with BW in ewe at the 1PPI age group. The results for the correlations between TL and BW indicate that it is significantly ($P<0.01$) correlated with rams at 1PPI and 2PPI. It is also significantly ($P<0.05$) correlated for rams at ≥ 3PPI age group. However, still, it is negative but significant ($P<0.01$) for ewes at all age groups. The correlation values between TC and BW indicate that they were significantly ($P<0.05$) correlated with the rams at 2PPI age group, while they were significantly ($P<0.01$) correlated for all age groups and both sexes.

Discussion

The difference of linear body measurement and body weight within the same sex with different agroecology might be due to the non-genetic factors and also location. However, the result obtained in the present study was in consonance with the observation of Belete et al. [15], and Kunene et al. [16] indicated that location was the highest contributing factor for the variation of linear body measurement and body weight of animals.

The superiority in the body weight and linear body measurement of 'Wenya-dega' over 'Dega' agroecology could be a result of the difference in nutritional, location and management level of the sheep. The obtained result is in consonance with the result of Oke and Ogbonnaya

Table 3 Least square mean and standard error (LSM ± SE) of linear body measurement (cm) sheep breed

Effect	EL	CBL	CBW	TL	TC	HOL	SC
Agroecology	*	*	*	ns	ns	ns	na
'Dega' ewe	8.62 ± 0.15	12.5 ± 0.07	6.44 ± 0.03	21.4 ± 0.37	9.69 ± 0.19	6.34 ± 0.15	
WD ewe	9.72 ± 0.16	12.8 ± 0.08	6.57 ± 0.05	20.7 ± 0.37	9.42 ± 0.21	7.09 ± 0.16	
	ns	ns	**	**	**	ns	*
'Dega' ram	8.46 ± 0.19	13.27 ± 0.19	7.19 ± 0.09	26.41 ± 0.37	15.07 ± 0.62	20.5 ± 0.86	25.4 ± 0.73
WD ram	8.82 ± 0.14	13.64 ± 0.14	7.72 ± 0.08	28.46 ± 0.41	19.16 ± 0.49	20.3 ± 1.42	28.1 ± 0.59
Sex	ns	**	**	**	**	**	na
Ewe	9.03 + 0.09	12.67 + 0.05	6.50 + 0.03	21.04 ± 0.26	9.56 ± 0.14	6.72 ± 0.11	
Ram	8.68 ± 0.11	13.49 ± 0.11	7.45 ± 0.06	27.44 ± 0.30	17.11 ± 0.42	20.4 ± 0.76	
Age	ns	**	**	ns	*	ns	**
1PPI	8.74 ± 0.12	12.2 ± 0.07[a]	6.58 ± 0.05[a]	23.2 ± 0.41	12 ± 0.31[a]	12.8 ± 0.64	22.8 ± 0.42[a]
2PPI	8.82 ± 0.14	13.8 ± 0.09[b]	6.70 ± 0.06[b]	24.6 ± 0.44	13.7 ± 0.38[b]	13.6 ± 0.38	26.2 ± 0.57[b]
≥ 3PPI	9.18 ± 0.12	14.4 ± 0.08[c]	6.91 ± 0.05[c]	26.3 ± 0.37	14.1 ± 0.31[b]	14.0 ± 0.43	31.3 ± 0.63[c]
Sex by age	ns	**	**	ns	**		
Ewe*1PPI	8.90 ± 0.16	11.9 ± 0.09[a]	6.26 ± 0.04[a]	21 ± 0.53	8.4 ± 0.27[a]		
Ewe*2PPI	8.85 ± 0.17	12.6 ± 0.08[b]	6.43 ± 0.05[a]	21.3 ± 0.48	9.3 ± 0.26[a]		
Ewe* ≥ 3PPI	9.22 ± 0.13	13.1 ± 0.08[bc]	6.68. ± 0.05[b]	22.0 ± 0.38	10.5 ± 0.21[b]		
Ram*1PPI	8.49 ± 0.15	12.7 ± 0.12[b]	7.12 ± 0.07[c]	26.9 ± 0.40	15.3 ± 0.53[c]		
Ram*2PPI	8.72 ± 0.24	13.7 ± 0.21[c]	7.60 ± 0.12[c]	27.3 ± 0.66	17.6 ± 0.84[d]		
Ram* ≥ 3PPI	9.01 ± 0.25	14.9 ± 0.18[d]	8.16 ± 0.11[d]	29.4 ± 0.51	20.1 ± 0.85[e]		

WD 'Wenya-dega', *EL* ear length, *CBL* cannon bone length, *CBW* cannon bone width, *TL* tail length, *TC* tail circumference, *HOL* horn length, *SC* scrotum circumference, *1PPI* one pair of permanent teeth, *2PPI* two pair of permanent teeth, ≥ *3PPI* ≥ three pair of permanent teeth, *na* not applicable, *ns* nonsignificant

a, b, c, d means with different superscripts across column are significantly different

**Significant at ($P < 0.01$), *significant at ($P < 0.05$)

Table 4 Phenotypic correlation between body weight and linear body measurement of sheep breed within sex and age group

Traits	1PPI		2PPI		≥ 3PPI	
	Ram	Ewe	Ram	Ewe	Ram	Ewe
WH	0.762**	0.717**	0.758**	0.634**	0.620**	0.566**
HOL	0.533**	0.228*	0.063	0.314**	0.017	0.305**
EL	0.085	0.093	0.109	0.047	0.053	0.112
HW	0.634**	0.428**	0.527**	0.449**	0.570**	0.569**
HL	0.465**	0.132[ns]	0.241*	0.349**	0.310*	0.429**
BL	0.833**	0.799**	0.716**	0.771**	0.664**	0.707**
PW	0.668**	0.563**	0.683**	0.442**	0.617**	0.439**
CG	0.921**	0.841**	0.826**	0.805**	0.925**	0.832**
SW	0.575**	0.379**	0.526**	0.459**	0.530**	0.509**
CBL	0.674**	0.590**	0.705**	0.454**	0.545**	0.569**
CBW	0.624**	0.632**	0.620**	0.605**	0.600**	0.635**
TL	0.410**	−0.300**	0.400**	− 0.298**	0.337*	− 0.347**
TC	0.483**	0.436**	0.355*	0.405**	0.481**	0.508**
SC	0.715**	–	0.547**	–	0.627**	–

WH wither height, *HOL* horn length, *EL* ear length, *HW* head width, *HL* head length, *BL* body length, *PW* pelvic width, *CG* chest girth, *SW* shoulder width, *CBL* cannon bone length, *CBW* cannon bone width, *TL* tail length, *TC* tail circumference, *SC* scrotum circumference, *1PPI* one pair of permanent teeth, *2PPI* two pair of permanent teeth, ≥ *3PPI* ≥ three pair of permanent teeth, *ns* nonsignificant

**Significant at $P < 0.01$, *significant at $P < 0.05$

[10] and Mengistie et al. [17], and the variation in linear body measurement (LBM) was influenced by the location where an animal was raised and by its age [16]. In contrary to this result, Legaze et al. [18] reported that location effect showed a very low significance in the morphological traits, leaving a more important influence on the flocking effect nested into the location effect.

The body weight of mature ram and ewe sheep found in 'Wenya-dega' agroecology was 29.38 and 25.13 kg, respectively. This result is in line with the body weight of afar and Menz sheep breed [19]. Birhan and Aynealem [20] reported a wide variation in growth performance between districts could be an advantage to improve the genetic potentials of local sheep with a long-term selection effort. Scrotum circumference of rams for 'Wenya-dega' agroecology was wider than for 'Dega' rams. This may be due to existing of the nutritional level of the feed. According to the report of Nugussie et al. [21], feeding high quality and quantity of feed could enhance testicular growth as the indicator of the scrotum circumference.

Body weight of mature ewe flocks found in this study is comparable to the reports from in Bensa district of southern Ethiopia and Sub-Saharan Africa [22, 23]. The effect of sex on body weight, chest girth, pelvic width and wither height obtained in this study is comparable to the value of Bonga, Abergelle, Washera and Gumuz sheep breed [13, 17, 24, 25]. The average value of body length (65.49 cm ram and 62.43 cm ewe) obtained in the study was higher than the average body length of central highland sheep, Afar and Menz sheep breed [13, 19] and Washera sheep of western highland sheep [17]. The average value of chest girth both for the rams (75.86 cm) and ewes (70.98 cm) was also comparable to the average value of Abergelle sheep breed [13].

Body weight and body measurements of body length and chest girth obtained in this study are comparable with the previous report on Washera sheep under traditional management system [17] but higher than the sheep type found in Bale zone [15]. Body weight obtained at the higher-age group of ewes and rams in the in the study is comparable with the on-farm study by Tesfaye [19] for Afar ewes and rams and higher than Menz ewes and rams. Body weight and body measurement of mature sheep in the present study are higher than values reported for Abergelle sheep breed under the traditional management of Northern Tigray [26] but lower than Gumuz sheep breed [25]. Similarly, the chest girth obtained at the lower-age group (1PPI) in the present study was higher than the chest girth for Menz and Horro sheep at 365 days [27].

The head width and shoulder width obtained in this study were lower than Yakassa and Balami sheep breed of Nigeria [11] and longer head length and wider head width for West Africa dwarf sheep [7].

Pelvic width at the oldest age group obtained in this study is comparable with the result of Mengistie et al. [17], and the relatively lower body weight for both ram and ewes recorded in the current study than by Adilo for ewe, Bonga, Horro and Gumuz [24, 28] might be attributed to the difference in the level of management, genetic factor and related to the location effect.

The average value of tail length for mature ewes and rams obtained in the study was lower than the average value of Gumuz and Washera ewe and rams with an on-farm management [25, 29] and longer than Afar ewe and rams [19].

The observations pertaining to the correlations between linear measurements and BW have also been more or less documented [10, 19, 25, 29]. The moderate to high correlation coefficients between body weight and linear body measurements for all age groups suggest that either of these variables or their combination could provide a better evaluation for forecasting live weight of sheep.

Conclusion

Gamogofa sheep population is a multicolored breed and identified as thin-tailed with the straight tail form at the tip. Agroecology, sex and age group had a significant ($P<0.01$) effect on body weight and linear body measurement. The body weight of rams and ewes in 'Wenya-dega' were higher when compared to the body weight of ram and ewes of 'Dega' agroecology, respectively. Generally, positive and significant ($P<0.01$) correlation was obtained between the body weight and most of the linear body measurement. Therefore, it strongly recommended that further genetic analyses should be used to determine the genetic variation between and within these small populations to develop an effective conservation and utilization program.

Authors' contributions

FH is the principal researcher, prepared interview schedule to collect data, analysis and prepared this manuscript. DG and HH have guided FH while undertaking this research from the very beginning of the study. They helped for FH in data entry, analysis, revised manuscript minutely with critical comments and suggestions and drafted the manuscript. All authors read and approved the final manuscript.

Acknowledgements

Authors sincere thanks to the SARI for the reliable budget support and Gamogofa zone Agricultural and Rural Development Office of southern Ethiopia for giving us the chance and allowing us to support for selecting high sheep population density.

Competing interests

The authors declare that they have no competing interests.

Funding
This study was financially supported by SARI.

Significance statements
This study will help the researcher to uncover the critical area of phenotypic characterization in sheep population that many researchers were not able to explore. This study discovers the effect of agroecology, sex and age on body weight and linear body measurements. These value used as reference in the future studies on the genetic characterization and improvement through conservation of the breeds.

References

1. CSA (Ethiopia Central Statistical Agency). Agricultural Sample Survey. The National Statistics. Report on livestock and livestock characteristic in Addis Ababa, Ethiopia. February, 2015; vol. II.
2. Assen E, Aklilu H. Sheep and goat production and utilization in different agro-ecological zones in Tigray, Ethiopia. Liv Res Rur Dev. 2012;24(1):16.
3. Solomon G. Sheep resource of Ethiopia: genetic diversity and breeding strategy. Ph.D. thesis, Wageningen University, the Netherlands. 2008; P. 47–97.
4. Workneh A, Ephrem G, Markos T, Yetnayet M, Rege JEO. Current status of knowledge on the characterization of farm animal genetic resources in Ethiopia. In: Proceeding of the 11th annual conference of Ethiopia Society of Animal Production (ESAP). 28–30 August, Addis Ababa, Ethiopia. 2004; p. 1–21.
5. IBC (Institute of Biodiversity Conservation). The State of Ethiopia's Farm Animal Genetic Resources: Country Report. A Contribution to the First Report on the State of the World's Animal Genetic Resources, Addis Ababa, Ethiopia. 2004.
6. Zewdu E, Aynalem H, Markos T, Sharma AK, Sölkner J, Wurzinger M. Breeding practices of indigenous sheep breeds of smallholders for designing community based-breeding strategies in Ethiopia. In: Proceedings of the 16th annual conference of the Ethiopian Society of Animal Production (ESAP), 8–10 October, Addis Ababa, Ethiopia. 2009; P. 241–6.
7. Sowande OS, Sobola OS. Body measurements of West African dwarf sheep as parameters for estimation of live weight. Trop Anim Health Prod. 2008;40:433.
8. Sowande OS, Oyewale BF, Iyasere OS. Age- and sex-dependent regression models for predicting the live weight of West African Dwarf goat from body measurements. Trop Anim Health Prod. 2010;42:969–75.
9. Markos T. Productivity and health of indigenous sheep breeds and crossbred in the central highland of Ethiopia. Ph.D. dissertation. Swedish University of Agricultural Sciences, Uppsal, Sweden. 2006; P. 89.
10. Oke UK, Ogbonnaya EO. Application of physical body traits in the assessment of breed and performance of WAD sheep in a humid tropical environment. Liv Res Rur Dev. 2011;23(2):24.
11. Ibrahim Yakubu A, Isa A. Multivariate analysis of morpho-structural characteristics in Nigerian indigenous sheep. Ital J Anim Sci. 2011;10:e17.
12. Aamir HM, Babiker SA, Youssif GM, Hassan YA. Phenotypic characterization of Sudanese Kenana cattle. Res J Anim Vet Sci. 2010;5:43–7.
13. Sisay L. Phenotypic characterization of indigenous sheep breeds in the Amhara National Regional State of Ethiopia. Master thesis, Haramaya University, Haramaya, Ethiopia. 2009; P. 87–102.
14. Thrusfield MV. Veterinary epidemiology. 3rd ed. Edinburgh: Black Wellsience Ltd.; 2007. p. 229–50.
15. Belete A, Tadesse A, Eshetu A. Phenotypic characterization of indigenous sheep types in Bale Zone, Oromia Regional State, Ethiopia. J Vet Sci Technol. 2017;8:4.
16. Kunene NW, Nesamvuni EA, Fossey A. Characterization of Zulu (Nguni) sheep using linear body measurements and some environmental factors affecting these measurements. S Afr J Anim Sci. 2007;37(1):11–20.
17. Mengistie T, Girma A, Solomon G, Sisay L, Abebe M, Tibbo M. Traditional management systems and linear body measurements of Washera sheep in the western highlands of the Amhara National Regional State, Ethiopia. Liv Res Rural Dev. 2010;22(9):169.
18. Legaze E, Cervantes I, Pérez-Cabal MA, Fuente LF, Mártinez R, Goyachee F, Gutiérrez JP. Multivariate characterization of morphological traits in Assaf Sheep. Small Rum Res. 2011;100(2–3):122–30.
19. Tesfaye G. Characterization of Menz and Afar indigenous sheep breeds of smallholders and pastoralists for designing community-based breeding strategies in Ethiopia. Master of thesis; Haramaya University, Ethiopia. 2008; P. 52–76.
20. Berhanu B, Aynalem H. Reproductive performance of traditionally managed sheep in the south western part of Ethiopia. Liv Res Rur Dev. 2009;21:154.
21. Nugussie D, Shenkoru T, Tegegne A. Growth rates and testicular characteristics of Ethiopian highland sheep offered chickpea haulm supplemented with incremental levels of *Leucaena leucocephala* leaf hay. Liv Prod Sci. 2000;65:209–17.
22. Hizkel K, Yoseph M, Yosef T. On farm phenotypic characterization of indigenous sheep type in Bensa District, Southern Ethiopia. J Bio Agric Health. 2017;7:224–3208.
23. FAO (Food and Agriculture Organization of the United Nations). Cattle and small ruminant production systems in sub-Saharan Africa—a systematic review. In: Otte MJ, Chilonda P, editors. Livestock information sector analysis and policy branch. Rome: FAO; 2000.
24. Zewdu E. Characterization of Bonga and Horro indigenous sheep breeds of smallholders for designing community-based breeding strategies in Ethiopia. Master thesis, Haramaya University, Haramaya, Ethiopia. 2008; P. 106.
25. Solomon A, Hegde BP, Mengistie T. Growth and physical body characteristics of Gumuz sheep under traditional management systems in Amhara Regional State, Ethiopia. Liv Res Rur Dev. 2011;23(5):117.
26. Tajebe S, Gangwar SK, Kefelegn K. Performance and physical body measurement of Abergell sheep breed under traditional management system of Tigray Regional State, Northern Ethiopia. Int J Sci Nat. 2011;2(2):225–30.
27. Kassahun A. Comparative performance evaluation of Horro and Menz sheep of Ethiopia under grazing and intensive feeding condition. Ph.D. dissertation, Humboldt-University. 2000; P. 129.
28. Solomon G, Komen H, Hanotte O, Van Arendonk JAM. Indigenous sheep resources of Ethiopia: types, production systems, and farmer's preferences. Anim Genet Resour Inf. 2008;43:25–39.
29. Mengistie T, Girma A, Sisay L, Solomon G, Abebe M, Marikos T. Reproductive performance and survival of Washera sheep under traditional management system at Yilmanadensa and Quarit district of the Amhara National Reginal State of Ethiopia. J Anim Vet Adv. 2011;10(9):1158–65.

5

Livestock ownership, animal source foods and child nutritional outcomes in seven rural village clusters in Sub-Saharan Africa

Jack B. Hetherington[1,2], Anke K. Wiethoelter[2,3], Joel Negin[4] and Siobhan M. Mor[2,3*]

Abstract

Background: Sub-Saharan Africa currently has the highest prevalence of malnutrition worldwide. In children under the age of 5 years, malnutrition can have long-term effects on physical and cognitive development, with implications at the national scale. Theoretically, livestock-based interventions are well placed to overcome constraints faced by micronutrient and/or food-based interventions. However, there is limited empirical evidence to support this hypothesis.

Methods: This study utilised agriculture, nutrition and anthropometry data from the Millennium Villages Project to investigate relationships between livestock ownership, animal source foods (ASF) consumption and child nutritional outcomes across seven rural village clusters in Sub-Saharan Africa. Village clusters were located in different agro-ecological zones and included: Bonsaaso, Ghana; Mayange, Rwanda; Mwandama, Malawi; Tiby, Mali; Pampaida, Nigeria; Potou, Senegal; and Ruhiira, Uganda. Data from 1624 households (including 1543 children) were included in the analysis.

Results: Overall, the proportion of children with stunting, underweight or wasting across the seven village clusters was 40, 18 and 5%, respectively. Livestock ownership, ASF consumption and child nutritional outcomes varied between village clusters. Households that owned livestock were generally more likely to consume associated ASF. For example, the proportion of households that consumed milk was higher in households that owned cows compared to those that did not in Pampaida, Mayange and Ruhiira ($P < 0.05$), while poultry meat consumption was generally higher in poultry-keeping households in Mayange and Ruhiira ($P < 0.05$). The relationship between ASF consumption and anthropometric measurements was complex, ranging from positively to negatively associated depending on the food commodity and village cluster. For instance, in Ruhiira, the mean weight-for-age Z score (WAZ) was significantly higher (better) in children from households that consumed eggs in the last 30 days, while in Potou, mean WAZ was significantly lower (worse) in children from households that consumed eggs in the last 30 days ($P < 0.05$).

Conclusions: This study contributes to the growing body of research that investigates the relationships between livestock ownership, ASF consumption and nutritional outcomes in children. Our results reveal complex patterns that vary across agro-ecological zones. More research is needed to assess seasonal variations in these factors, effects of gender roles on intra-household distribution of ASF, as well as effects of zoonotic food-borne diseases on nutritional status of children in these sites.

Keywords: Animal source foods, Child nutrition, Livestock, Sub-Saharan Africa, Millennium Villages Project, One Health

*Correspondence: siobhan.mor@sydney.edu.au
[2] Sydney School of Veterinary Science, The University of Sydney, Camperdown, NSW 2006, Australia
Full list of author information is available at the end of the article

Background

United Nation's projections indicate that the global population will reach 9.7 billion people by 2050, with a further 1.5 billion people by 2100 [1]. To meet the growing demand for food, it has been estimated that agricultural outputs will need to increase by at least 70% [2]. Currently, Sub-Saharan Africa has a high prevalence of malnutrition, with approximately half of the world's malnourished human population inhabiting this region [3]. The situation is expected to worsen in the future due to the expected population growth, with countries such as Nigeria and Uganda expected to double their population by 2050 [1].

Malnutrition in children, particularly in children under 5 years of age, can have lasting effects on their physical growth and cognitive development [4, 5]. Deficiencies in micro- and macro-nutrients [4, 6], recurrent infection [7] and gut dysbiosis [8] can lead to physical manifestations of malnutrition including stunting, underweight or wasting, with extreme cases leading to death. Furthermore, maternal malnutrition has a significant effect on a child's physical and mental development [6, 9]. The implications of malnutrition also extend beyond the individual, having significant implications for national economies, such as forgone economic productivity caused by partial or total incapacitation from work, including premature death [10].

A range of interventions have been implemented globally to address malnutrition, ranging from micronutrient supplementation programmes (where single micronutrients in the form of injections, formulas or tablets are administered to an individual), to food-based interventions (where nutrient deficiencies are targeted through the provision of different food groups). For example, school-based food programmes have focused on delivery of animal source foods (ASF) including milk and meat to children with some positive gains recorded in terms of improvements to cognitive and physical development [11]. ASF are rich in bioavailable vitamin B12, riboflavin, calcium, zinc, iron and amino acids, and therefore, access to ASF can improve dietary quality [12]. Nevertheless, programmes that rely on government-subsidised micronutrient supplementation or school-based food programmes are unlikely to be sustainable in the long term and may not reach the poorest of households [13].

Evidence is emerging that agricultural interventions can play a role in addressing malnutrition in children. For instance, Palmer et al. [14] demonstrated improved pupillary responsiveness through consumption of provitamin A carotenoid-biofortified maize, while Jones and de Brauw [15] showed promising evidence of orange sweet potato in reducing childhood diarrhoea. While these crop-based interventions have proven to have substantive benefits, to date they have mostly targeted replacement of single micronutrients [16].

Livestock production has the potential to help alleviate food and nutrition insecurity as it not only provides a source of nutrient-dense foods, but also a wide range of saleable commodities and hence a source of financial security and insurance for the rural poor [12, 17, 18]. Benefits from livestock to food and nutrition security arise both directly (by improving household diet through increasing access to ASF) and indirectly (by improving income and ability to purchase more diverse foods) [18, 19]. With adequate livestock nutrition, renewable ASF, such as eggs and milk, provide an opportunity for a steady supply of essential micro- and macro-nutrients. Indeed, Drewnowski [20] demonstrated that milk and eggs are one of the lowest cost sources of protein amongst plant and animal source foods.

Despite these positive claims, there is limited empirical evidence to support a relationship between livestock production, ASF consumption and improved nutritional outcomes in children. Due to the highly complex nature of food and nutrition insecurity, a One Health approach that specifically aims to bridge the divide between agricultural production and nutrition research is required. This study adapts methods used by Azzarri et al. [21]—who performed a detailed analysis of livestock ownership, ASF consumption and child nutritional outcomes in Uganda—to examine these relationships across seven rural village clusters in Sub-Saharan Africa as a basis for informing future nutrition security and livestock-based interventions.

Methods

This study utilised data from seven rural village clusters participating in the Millennium Villages Project (MVP). The MVP sought 'to demonstrate the feasibility of practical economic transformation in rural tropical Africa through targeted multisectoral investments' [22]. Project sites were chosen to represent the major agro-ecological zones and farming systems in Sub-Saharan Africa [23]. All sites were classified as 'hunger hot spots' where rates of child undernutrition exceeded 20% at baseline and where the majority of residents were engaged in agriculture [22]. The project included 12 village clusters, across 10 countries in Sub-Saharan Africa, where a cluster included several villages totalling 5000–55,000 people [22]. Sanchez et al. [22] describe village/site selection in more detail.

Data analysed in this study were collected in year 3 of the MVP (2008–2009) and included households from 7 of the 12 village clusters, namely: Bonsaaso, Ghana; Mayange, Rwanda; Mwandama, Malawi; Tiby, Mali; Pampaida, Nigeria; Potou, Senegal; and Ruhiira, Uganda.

These village clusters were established in 2006 and 2007. By the time of data collection (2008–2009), a few interventions had taken place that likely had some effect on nutritional status. A school meal programme focused on provision of beans, leafy vegetables and soya-enriched maize was implemented in a number of the villages [22]. In addition, provision of fertilizer and improved seed varieties led to increased crop yields in a number of sites [24]. These interventions are unlikely to have had an impact on livestock production and/or ASF consumption. The five remaining village clusters participating in the MVP (Koraro, Ethiopia; Mbola, Tanzania; Dertu, Kenya; Sauri, Kenya; and Ikaram, Nigeria) were not included in this analysis due to differences in questionnaire design and/or comparability of data across countries.

Data used in this study were collected through a series of cross-sectional, face-to-face interviews. The survey modules are described in more detail in the supplementary materials of Sanchez et al. [22] and were largely based on Demographic and Health Surveys (DHS), UNICEF's Multiple Indicator Cluster Surveys (MICS) and the World Bank Living Standards Measurement Study (LSMS) surveys. A household was defined as people who shared the same cooking pot. Questionnaires captured information on household demographics, agriculture and maternal and child health. In addition, questionnaires on food security and food frequency were adapted and piloted at each site and administered to men and women in sampled households (described in more detail in Remans et al. [25]). Finally, anthropometric measurements (height/length, weight and mid-upper arm circumference) were taken on children in participating households.

Food insecurity status was measured as the number of months the respondent/respondents reported as having inadequate food in the past 12 months. We calculated the median (interquartile range, IQR) number of months of food insecurity as well as the proportion of households reporting ≥ 1 month of food insecurity. Children were classified as being stunted, underweight or wasted on the basis of their height-for-age Z score (HAZ), weight-for-age Z score (WAZ) and weight-for-height Z score (WHZ), respectively, when their score was two standard deviations (SD) below the reference median according to the World Health Organization (WHO) [26]. The mean HAZ/WAZ/WHZ was calculated as well as the proportion of children 2SD below the reference level.

Livestock ownership assessment was based on major food animal species as categorised by the Food and Agriculture Organization of the United Nations (FAO) [23]. Thus, households were classified according to whether they owned one or more of the following: poultry (chicken, ducks and/or other birds), pigs, small ruminants (sheep and/or goats) and cattle (cows, heifers, bulls and/or oxen). Other animals, such as camels, donkeys and horses, were not included as they did not represent a major livestock species across the seven countries.

Food consumption patterns were assessed via a questionnaire, in which an adult member of the household was asked to identify how frequently they consumed specific food items in the last 30 days (none, once per month, 2–3 times per month, once per week, 2–3 times per week, 4–6 times per week, once per day, twice or more per day). For purposes of this analysis, data were re-coded into a binary variable denoting whether or not a particular ASF had been consumed (as main or part of a meal) in the last 30 days. We focused on food commodities produced by the major food animals above, namely: poultry meat, eggs, pork, sheep or goat meat, beef (including veal) and milk and milk products (including those served as a drink or included in sour milk and/or yoghurt). Meat commodities were considered to comprise flesh meat; organ meat (or offal) was not included.

Given diversity of livestock ownership and ASF consumption patterns across the seven sites, we elected to evaluate each site separately. Fisher's Chi-squared (non-parametric) and t tests (parametric) were performed to examine relationships between livestock ownership, ASF consumption and child nutritional status. To adjust for clustering at the household level, mean HAZ/WAZ/WHZ was calculated for each household. These summary measures were then used in subsequent analyses comparing mean Z scores across households according to the livestock ownership status and ASF consumption patterns. Assessment of ASF consumption patterns in relation to livestock ownership was limited to foods derived from each livestock group (e.g. cattle ownership and beef consumption). Statistical analysis was performed in R version 3.2.4 (The R Foundation) and SPSS version 22 (IBM, Chicago, IL).

Results

Household demographics and food security status

Household demographics are shown in Table 1. A total of 1624 households were included in the analysis. Across the seven countries, 18% of households had female heads, ranging from less than 1% (Tiby) to one-third (Mwandama). Approximately 69% of the respondents did not have any level of formal education. The median household size was 6 people (range 1–60). Of the households surveyed, 47% reported that they had experienced at least 1 month of food insecurity during the last 12 months. Anthropometric measurements were available for 1543 children belonging to 967 households. The median number of children per household was 1 (range 1–7). The

Table 1 Household demographic characteristics and food security status by village cluster

Characteristic	Millennium Villages Cluster						
	Bonsaaso, Ghana	Pampaida, Nigeria	Potou, Senegal	Tiby, Mali	Mayange, Rwanda	Mwandama, Malawi	Ruhiira, Uganda
No. of households	169	267	274	213	234	245	222
No. of children	145	325	293	210	142	174	254
Household characteristics							
Household size, median (IQR)	6 (3)	7 (4)	9 (7)	13 (12)	5 (2)	5 (3)	5 (3)
Female head of household, no. (%)	40 (24)	2 (1)	35 (13)	1 (0)	58 (25)	84 (34)	65 (29)
Household head education level, no. (%)							
No formal education	73 (43)	212 (79)	270 (99)	210 (99)	148 (63)	151 (62)	163 (73)
Primary school	31 (18)	45 (17)	3 (1)	1 (0)	63 (27)	45 (18)	41 (18)
Lower secondary school	58 (34)	7 (3)	0	1 (0)	8 (3)	40 (16)	7 (3)
Upper secondary school or higher	7 (4)	3 (1)	1 (0)	1 (0)	15 (6)	9 (4)	11 (5)
Food insecurity							
≥1 month of food insecurity, no. (%)	115 (68)	157 (59)	140 (51)	69 (32)	77 (33)	102 (42)	109 (49)
No. months of food insecurity, median (range)	2 (0–12)	1 (0–6)	1 (0–8)	0 (0–6)	0 (0–12)	0 (0–7)	0 (0–12)
Child anthropometry							
Mean *Z* score (SD)							
Height-for-age	−1.40 (1.6)	−1.66 (2.12)	−1.02 (1.83)	−1.49 (1.81)	−1.25 (1.58)	−1.70 (1.47)	−1.81 (1.68)
Weight-for-age	−1.13 (1.24)	−1.01 (1.81)	−0.78 (1.30)	−1.08 (1.32)	−0.73 (1.48)	−0.76 (1.14)	−0.80 (1.26)
Weight-for-height	−0.28 (1.10)	−0.24 (1.27)	−0.40 (1.45)	−0.26 (1.25)	−0.09 (1.21)	0.34 (1.08)	0.42 (1.01)
Malnourished children, no. (%)							
Stunted	54 (38)	129 (47)	75 (29)	80 (40)	45 (32)	72 (42)	124 (50)
Underweight	27 (19)	56 (26)	46 (17)	51 (25)	22 (15)	17 (10)	36 (14)
Wasted	3 (2)	17 (8)	33 (11)	11 (5)	6 (4)	2 (1)	1 (0)

No. number, *IQR* interquartile range, *SD* standard deviation

proportion of children with stunting, underweight or wasting across the seven village clusters was 40, 18 and 5%, respectively.

Livestock ownership patterns

Livestock ownership patterns were highly variable between village clusters (Table 2). Overall, 77% (range 44–97%) of households owned one or more livestock group. The most commonly owned groups in descending order were poultry (57% of households; range 40–81%), small ruminants (55%; range 14–85%), cattle (30%; range 0–91%) and pigs (5%; range 0–21%). In general, households in this study were smallholders. More than half of households in Bonsaaso and Mwandama owned a single livestock group only (poultry; data not shown). In other sites, mixed livestock farming was practised by a majority of households.

Consumption of animal source foods

Animal source foods consumption patterns varied considerably across village clusters (Table 3). Overall, 61% (range 31–84%) of households reportedly consumed sheep or goat meat as either part of a meal or as the main meal in the past 30 days, 59% (range 19–86%) consumed beef, 59% (range 27–99%) consumed milk/milk products, 58% (range 24–77%) consumed poultry meat, 57% (range

Table 2 Livestock ownership patterns by village cluster

Livestock group	Unit	Millennium Villages Cluster						
		Bonsaaso, Ghana	Pampaida, Nigeria	Potou, Senegal	Tiby, Mali	Mayange, Rwanda	Mwandama, Malawi	Ruhiira, Uganda
Cattle	No. (%)	0	142 (53)	52 (19)	193 (91)	63 (27)	0	40 (18)
	Median (range)	NA	2 (1–60)	4 (1–105)	4 (1–54)	2 (1–6)	NA	3 (1–15)
Small ruminants	No. (%)	24 (14)	162 (61)	232 (85)	171 (80)	92 (39)	58 (24)	147 (66)
	Median (range)	5.5 (1–12)	5 (1–160)	8 (1–160)	5 (1–45)	2.5 (1–12)	3 (1–10)	4 (1–23)
Poultry	No. (%)	68 (40)	155 (58)	164 (60)	173 (81)	61 (26)	175 (71)	133 (60)
	Median (range)	10 (1–50)	9 (1–200)	6 (1–50)	9 (1–340)	4 (1–30)	6 (1–53)	4 (1–19)
Pigs	No. (%)	0	0	0	0	6 (3)	27 (11)	46 (21)
	Median (range)	NA	NA	NA	NA	1 (1–3)	2 (1–7)	1 (1–9)
Any livestock[a]	No. (%)	74 (44)	215 (81)	244 (89)	206 (97)	132 (56)	190 (78)	187 (84)

Values depict the number (%) of households that own each livestock group and median (range) herd/flock size in those households that own each livestock group

NA not applicable (livestock not owned)

[a] Any livestock includes poultry, pigs, small ruminants and/or cattle

Table 3 Animal source food consumption patterns, by village cluster

Animal source food	Millennium Villages Cluster						
	Bonsaaso, Ghana	Pampaida, Nigeria	Potou, Senegal	Tiby, Mali	Mayange, Rwanda	Mwandama, Malawi	Ruhiira, Uganda
Beef	56 (33)	230 (86)	203 (74)	144 (68)	135 (58)	46 (19)	152 (69)
Milk/milk products	56 (33)	179 (67)	270 (99)	193 (91)	115 (49)	66 (27)	85 (38)
Sheep/goat meat	52 (31)	196 (73)	168 (61)	178 (84)	77 (33)	189 (77)	131 (59)
Poultry meat	95 (56)	204 (76)	190 (69)	128 (60)	57 (24)	188 (77)	77 (35)
Eggs	146 (86)	195 (73)	231 (84)	24 (11)	55 (24)	205 (84)	71 (32)
Pork	26 (15)	NA	NA	NA	21 (9)	116 (47)	78 (35)

Values depict the number (%) of households consuming each animal source food (ASF) in the last 30 days

NA not applicable (ASF not included in questionnaire)

11–86%) consumed eggs, and 28% (range 0–47%) consumed pork.

Consumption of animal source foods by livestock ownership status

Relationships between ASF consumption and livestock ownership status are shown in Table 4. In general, the proportion of households that consumed a particular ASF in the last 30 days tended to be higher in households that owned the livestock group from which the food was sourced compared to households that did not own that livestock group. This was particularly true in Mayange and Ruhiira, where significant positive associations (P value <0.05) existed between ASF consumption and livestock ownership across several commodity–livestock pairings. Of all ASF commodities, the relationship between cattle ownership and dairy consumption was the most consistent across village clusters, with a greater proportion of households that owned cattle reporting that they had consumed milk/milk products in the last 30 days compared to households that did not own cattle. This association was statistically significant in Mayange, Pampaida and Ruhiira (P value <0.05). Similarly, compared to households that did not own poultry, a greater proportion of poultry-keeping households reported eating poultry meat in the last 30 days in four villages clusters, a finding that was statistically significant for Mayange (P value <0.05) and Ruhiira (P value <0.001). Although egg consumption was generally higher in such households, poultry ownership was only significantly associated with higher rates of egg consumption in Mwandama (P value <0.05).

Livestock ownership and child nutritional status

The results of the univariate analysis of anthropometric measurements and livestock ownership across the 7

Table 4 Animal source food consumption, by livestock ownership status and village cluster

Animal source food (ASF)	Livestock group	Livestock ownership status (Yes/No)													
		Millennium Villages Cluster													
		Bonsaaso, Ghana		Pampaida, Nigeria		Potou, Senegal		Tiby, Mali		Mayange, Rwanda		Mwan-dama, Malawi		Ruhiira, Uganda	
		Yes	No	Yes	No	Yes	No	Yes	No	Yes	No	Yes	No	Yes	No
Beef	Cattle	NA	33	86	86	75	74	68	60	68	54*	NA	19	68	68
Milk/milk products	Cattle	NA	33	73	60**	100	98	92	80*	63	44**	NA	27	53	35**
Sheep/goat meat	Small ruminants	46	28*	71	77	64	48*	82	88	32	34	86	74*	61	56
Poultry meat	Poultry	56	56	73	81	66	74	61	55	34	21**	79	70	45	19***
Eggs	Poultry	87	86	72	75	85	84	13	5	31	20	88	73**	36	26
Pork	Pigs	NA	15	NA	NA	NA	NA	NA	NA	33	8*	44	48	50	31**

Values depict the proportion of households that consumed a particular animal source food (ASF) in the last 30 days, stratified by livestock ownership status. Only those ASF that were derived from each livestock group were compared. Proportions that differed significantly (Chi-square test) are indicated with asterisks

NA not applicable (livestock group not owned/ASF not consumed or not included in questionnaire)

* $P < 0.1$; ** $P < 0.05$; *** $P < 0.001$

village clusters are shown in Fig. 1. Trends varied between villages and across different livestock groups, with few clear patterns. An exception was Pampaida where mean HAZ was significantly higher (better) in children from households that owned cattle or 'any livestock' ($P < 0.05$). In contrast, households that owned 'any livestock' in Mayange had significantly lower (worse) HAZ ($P < 0.05$), suggesting that in this village cluster, owning livestock is a risk factor for child stunting.

Animal source foods consumption and child nutritional status

The results of the univariate analysis of anthropometric measurements and ASF consumption across the seven village clusters are shown in Fig. 2. Trends again varied between clusters and across different food commodities; however, some consistent patterns emerged for some village clusters. For example, children from households in Mayange and Ruhiira that consumed ASF in the previous

Z score	Livestock group	Livestock ownership status (Yes/No)													
		Bonsaaso, Ghana		Pampaida, Nigeria		Potou, Senegal		Tiby, Mali		Mayange, Rwanda		Mwandama, Malawi		Ruhiira, Uganda	
		Yes	No	Yes	No	Yes	No	Yes	No	Yes	No	Yes	No	Yes	No
HAZ	Cattle	NA	-1.46	-1.43	-2.03 **	-1.10	-1.11	-1.50	-1.14	-1.22	-1.20	NA	-1.65	-1.89	-1.70
	Small ruminants	-1.99	-1.38 *	-1.53	-1.94	-1.11	-1.10	-1.43	-1.67	-1.42	-1.02	-1.40	-1.72	-1.73	-1.77
	Poultry	-1.45	-1.47	-1.52	-1.95	-1.12	-1.09	-1.48	-1.41	-1.50	-1.08	-1.70	-1.52	-1.81	-1.65
	Pigs	NA	-1.46	NA	-1.70	NA	-1.11	NA	-1.47	-0.19	-1.24	-1.17	-1.69	-1.79	-1.73
	Any livestock[1]	-1.52	-1.41	-1.54	-2.40 **	-1.12	-0.96	-1.47	-1.51	-1.46	-0.80 **	-1.67	-1.61	-1.82	-1.34
WAZ	Cattle	NA	-1.17	-1.15	-1.04	-0.88	-0.77	-1.13	-0.79	-0.85	-0.72	NA	-0.72	-0.76	-0.74
	Small ruminants	-1.29	-1.15	-1.17	-1.00	-0.78	-0.90	-1.10	-1.09	-0.93	-0.60	-0.58	-0.75	-0.70	-0.82
	Poultry	-1.14	-1.19	-1.09	-1.13	-0.80	-0.79	-1.15	-0.85	-0.98	-0.66	-0.74	-0.66	-0.76	-0.73
	Pigs	NA	-1.17	NA	-1.11	NA	-0.80	NA	-1.10	-0.57	-0.76	-0.20	-0.76	-0.65	-0.77
	Any livestock[1]	-1.23	-1.11	-1.06	-1.37	-0.76	-1.14	-1.10	-1.27	-0.83	-0.63	-0.73	-0.68	-0.77	-0.63
WHZ	Cattle	NA	-0.33	-0.22	-0.13	-0.64	-0.28	-0.35	-0.19	-0.13	-0.12	NA	0.37	0.54	0.43
	Small ruminants	-0.81	-0.26 *	-0.31	0.01	-0.32	-0.64	-0.40	-0.07	-0.09	-0.15	0.25	0.40	0.46	0.42
	Poultry	-0.19	-0.44	-0.19	-0.16	-0.43	-0.24	-0.36	-0.22	-0.33	-0.04	0.38	0.34	0.51	0.37
	Pigs	NA	-0.33	NA	-0.18	NA	-0.36	NA	-0.34	-0.56	-0.10	0.60	0.35	0.54	0.43
	Any livestock[1]	-0.31	-0.36	-0.17	-0.22	-0.32	-0.80	-0.34	-0.64	-0.12	-0.12	0.36	0.40	0.47	0.33

Fig. 1 Heat map of child anthropometric measurements by livestock ownership status and village. Values represent the mean *Z* score, adjusted for clustering at the household level. In pairwise comparisons (own livestock/do not own livestock), *orange* represents lower (worse) *Z* scores, while *blue* represents higher (better) *Z* scores. *White* indicates no difference between *Z* scores. * $P < 0.1$; ** $P < 0.05$; *NA* not applicable (livestock not owned); [1]Any livestock includes cattle, small ruminants, poultry and pigs

Z score	Animal source food (ASF)	ASF consumed in the last 30 days (Yes/No)													
		Bonsaaso, Ghana		Pampaida, Nigeria		Potou, Senegal		Tiby, Mali		Mayange, Rwanda		Mwandama, Malawi		Ruhiira, Uganda	
		Yes	No	Yes	No	Yes	No	Yes	No	Yes	No	Yes	No	Yes	No
HAZ	Beef	-1.64	-1.39	-1.61	-2.12	-1.30	-0.64 **	-1.51	-1.37	-1.09	-1.32	-1.72	-1.64	-1.76	-1.70
	Milk/milk products	-1.54	-1.43	-1.57	-1.94	-1.13	0.19	-1.57	-0.58	-0.92	-1.45 *	-1.83	-1.57	-1.53	-1.91
	Sheep/goat meat	-1.54	-1.43	-1.68	-1.74	-1.18	-1.01	-1.46	-1.54	-1.07	-1.25	-1.69	-1.51	-1.66	-1.87
	Poultry meat	-1.40	-1.52	-1.57	-2.11	-1.17	-0.98	-1.51	-1.42	-1.39	-1.15	-1.61	-1.79	-1.71	-1.76
	Eggs	-1.43	-1.63	-1.70	-1.79	-1.24	-0.54 **	-1.37	-1.48	-1.33	-1.17	-1.62	-1.87	-1.53	-1.85
	Pork	-1.80	-1.42	NA	NA	NA	NA	NA	NA	-0.70	-1.25	-1.84	-1.49	-1.85	-1.68
WAZ	Beef	-1.40	-1.07	-1.13	-0.99	-0.82	-0.73	-1.15	-0.99	-0.61	-0.89	-0.77	-0.70	-0.73	-0.77
	Milk/milk products	-1.14	-1.18	-1.09	-1.13	-0.80	-0.45	-1.19	-0.30	-0.57	-0.91	-0.73	-0.71	-0.60	-0.85
	Sheep/goat meat	-1.12	-1.19	-1.23	-0.83	-0.88	-0.68	-1.17	-0.73	-0.26	-0.93 **	-0.75	-0.59	-0.66	-0.88
	Poultry meat	-1.13	-1.21	-1.04	-1.38	-0.86	-0.66	-1.24	-0.93	-0.50	-0.82	-0.65	-0.93	-0.70	-0.77
	Eggs	-1.17	-1.18	-1.26	-0.68 *	-0.89	-0.38 **	-1.21	-1.08	-0.57	-0.79	-0.70	-0.84	-0.49	-0.87 **
	Pork	-1.34	-1.15	NA	NA	NA	NA	NA	NA	-0.41	-0.79	-0.90	-0.56 *	-0.73	-0.75
WHZ	Beef	-0.61	-0.21 **	-0.20	-0.10	-0.31	-0.50	-0.42	-0.17	-0.05	-0.19	0.35	0.38	0.49	0.35
	Milk/milk products	-0.29	-0.35	-0.20	-0.14	-0.36	-0.45	-0.38	0.01	-0.20	-0.05	0.39	0.36	0.50	0.40
	Sheep/goat meat	-0.51	-0.26	-0.27	0.01	-0.37	-0.35	-0.39	-0.07	0.17	-0.22	0.37	0.37	0.49	0.38
	Poultry meat	-0.22	-0.45	-0.10	-0.49	-0.49	-0.10 *	-0.44	-0.21	0.17	-0.19	0.46	0.06 **	0.42	0.46
	Eggs	-0.34	-0.29	-0.20	-0.13	-0.42	-0.11	-0.36	-0.34	0.32	-0.21 *	0.40	0.21	0.57	0.39
	Pork	-0.48	-0.31	NA	NA	NA	NA	NA	NA	0.04	-0.14	0.22	0.51	0.49	0.42

Fig. 2 Heat map of child anthropometric measurements by consumption of animal source foods (ASF) and village. Values represent the mean *Z* score, adjusted for clustering at the household level. In pairwise comparisons (ate ASF/did not eat ASF), *orange* represents lower (worse) *Z* scores, while *blue* represents higher (better) *Z* scores. *White* indicates no difference between *Z* scores. $^{*}P < 0.1$; $^{**}P < 0.05$; *NA* not applicable (ASF not included in questionnaire)

30 days in general displayed higher (better) anthropometric scores compared to households that did not. In contrast, consumption of ASF tended to be associated with lower (worse) anthropometric scores in children in Potou and Tiby. Mean WAZ was significantly higher (better) amongst children from households that consumed sheep/goat meat or eggs in the last 30 days in Mayange and Ruhiira, respectively ($P < 0.05$). Mean WHZ was also significantly higher amongst children from households in Mwandama that consumed poultry meat in the previous 30 days. In contrast, in Bonsaaso, mean WHZ was significantly lower (worse) amongst children from households that consumed beef in the last 30 days. Mean HAZ was also significantly lower in children from households that consumed beef or eggs in the previous 30 days in Potou.

Discussion

This study is one of few to provide a comparative, descriptive analysis of the relationships between livestock ownership, ASF consumption and child nutritional outcomes in Sub-Saharan Africa. We found some evidence that owning livestock was associated with increased household consumption of ASF in some study sites; however, owning livestock and/or consuming ASF was not consistently associated with improved child growth in any of the village clusters. Moreover, the results paint a complex picture with considerable variation across sites, highlighting the need to consider local circumstances in any programme promoting livestock development and/or ASF consumption as a means to alleviate childhood malnutrition.

Farming systems throughout Sub-Saharan Africa are highly variable and dependent on factors such as geography and climate, as well as socio-cultural influences. Across the 15 major farming systems in Sub-Saharan Africa, the contribution of livestock production to rural livelihoods ranges from subsistence or semi-subsistence through to being the primary source of income [23]. Livestock ownership patterns differed considerably across the village clusters, with the proportion of households owning any livestock ranging from less than half in Bonsaaso to nearly all households in Tiby. At least a quarter of households across all seven sites owned poultry, confirming the central role of this livestock species in rural African households [27]. Cattle were important sources of livestock in all but two village clusters: in humid, tropical Bonsaaso, where the presence of tse-tse flies limits the viability of cattle [23], and in Mwandama, where small farm sizes might preclude grazing [28]. Pigs represent an increasingly important sector in Uganda, with 17% of all households keeping pigs at the recent census [29]. In contrast, the lack of pig production in Nigeria, Mali and Senegal can be explained by the fact that these are predominantly Muslim countries [30].

Food consumption patterns were highly variable across the village clusters and likely influenced by many factors beyond whether or not a household owned a particular livestock species (e.g. cultural preferences and different strategies for accessing ASF). A large number of households consumed alternative ASF in addition to those detailed here. For example, 24 and 27% of households in Pampaida and Potou reported eating camel and tripe in the previous 30 days, respectively. These outcomes were not included in our analysis because they did not represent a significant livestock group or ASF across all sites. Food derived from other animal groups undeniably contributes to total food intake, including fish and fish-related products [31, 32], bush meat [33] and insects [34]. We note that nutrition surveys have tended to focus on skeletal meat and food preferences of western cultures. In the future, efforts to truly understand how ASF contribute to food and nutrition security need to consider the whole animal, in particular offal, and other animal groups such as fish and insects.

Although previous studies have demonstrated a positive association between livestock ownership and ASF consumption [21, 35], this is the first study to investigate these relationships across several Sub-Saharan countries. Overall, there were no consistent relationships between livestock ownership and ASF consumption in the preceding 30 days. Our results show that in some village clusters, ASF were significantly more likely to be consumed by households that owned livestock from which the commodity is derived. For instance, we found that cattle, poultry and pig ownership were significantly associated with consumption of milk/milk products, poultry meat and pork, respectively, in Ruhiira, a finding which is consistent with previous studies in Uganda [21]. Similar positive associations were also observed in this study for Mayange, Rwanda. Notably, ASF consumption was amongst the lowest in these village clusters compared to other Millennium Villages. In contrast, poultry, small ruminant and cattle ownership in Tiby were amongst the highest across the 7 village clusters, and the consumption of ASF did not vary significantly between livestock owners and non-owners. This may suggest that there is a threshold effect whereby once a level of livestock ownership is reached in a village, ASF is more readily accessible to all households, including those that do not own a particular livestock species. In other instances, we found consumption of ASF occurred in the absence of a particular livestock species. For example, beef and milk/milk products were consumed with some frequency in Bonsaaso and Mwandama, although no households in these village clusters reported owning cattle. These patterns highlight the complexities of household decision-making regarding food production, trade and consumption [36].

Few studies have investigated the relationship between child nutritional status and livestock ownership and/or ASF consumption in observational studies. Jin and Iannotti [37] found a modest positive association between co-owned/female-owned livestock and WAZ in Kenya, while Mosites et al. [38] found a slight beneficial effect of livestock ownership on stunting prevalence using demographic and health survey data from Ethiopia, Kenya and Uganda. The latter study found no association between ASF consumption (based on 24-h recall) and stunting rates in any country. In this study, we disaggregated ASF into different groups (e.g. poultry meat, pork and eggs), which allowed more detailed analysis of the potential impact of different ASF on child growth. We also captured ASF consumption in the previous 30 days, rather than 24-h recall which does not adequately capture the diversity of a usual diet [39]. Nonetheless, our findings were similar to Mosites et al. [38], in that there was no consistent association between ASF consumption and improved nutritional status in children in any of the village clusters.

There are several limitations of this study. Data analysed in this study are from countries at the very beginning of the nutrition transition [40], with food insecurity and malnutrition still widespread, especially in children [41]. Other research suggests that household diets in Sub-Saharan countries are exposed to large seasonal variety and insecurity around food [42]. It is important to recognise that our study does not account for seasonal variations or extreme weather events (e.g. droughts). Further, it is not representative of national trends as the sample populations were individual villages from the lowest socio-economic bracket and are not representative of rural areas in those countries. Long-term research needs to capture fluctuations in nutritional outcomes, the shifts in livestock ownership and consumption patterns during this transition.

Secondly, interviews were conducted with the household head and we therefore cannot comment on within-household differences in livestock ownership and ASF consumption patterns over the past 30 days. Although 30-day consumption recall has its limitations [43] and is likely to result in bias, we suspect that recall is unlikely to be different between groups (e.g. households that do and do not own livestock), and therefore, any findings would be biased towards the null. Frequency and amount of ASF consumption could plausibly differ, for example, between men, women and children in the household [44, 45]. Given potential intra-household variability of food allocation, particularly between males and females [46], future studies should measure consumption at the individual child level. The importance of women in health and nutritional outcomes of their children is well documented [47–49]. For example, Kariuki et al. [48] demonstrated a positive relationship between female-owned cattle and increased consumption of ASF in Tanzania, Kenya and Mozambique, while Jin and Iannotti [37] found that child HAZ and WAZ scores were higher in households with co owned/female-owned livestock in Kenya. Future studies are required to investigate how gender roles influence the allocation of food within the household and the impacts on child nutritional outcomes.

Finally, given the high degree of variability across countries and the limited significant findings in the univariate analysis, we elected not to perform further multivariate analysis. Any future study aiming to establish an independent effect of livestock ownership on household food and nutrition security will need to control for factors such as household size, wealth and gender of household head. Additionally, data on breastfeeding and complimentary feeding practices, as well as nutritional status of pregnant or lactating mothers, should be collected and considered in the analysis, given their important influence on child development [50]. Notably, time and labour inputs associated with livestock rearing activities might impact these child feeding practices; more empirical evidence is required to investigate these trade-offs. Access to safe food is also critical for ensuring positive nutritional outcomes in children [6]. Concerns have been raised about possible negative impacts of livestock ownership on child health and development, by increasing infection risks and potentially affecting enteric functions of the gut microbiome [51]. Exposure of infants to livestock faeces has been shown to have a significant impact on bacterial exposure leading to diarrhoea [52, 53]. There is some evidence that housing of poultry in close proximity to the household can have a negative effect on child growth [54]. Given constraints imposed by the data, we were unable to assess the potential effect of livestock housing, zoonoses and child growth outcomes in the current study.

Food and nutritional security in Sub-Saharan Africa is a complex issue with multiple entry points for creating sustainable improvements. It is plausible that livestock production and ASF can play a role in alleviating hunger, although robust evidence for this impact pathway is still widely lacking. Any intervention will need to be multifaceted, multidisciplinary and tailored to the specific context of the country to ensure a sustained impact is realised. A One Health approach that combines assessment of nutritional outcomes, diet, agricultural and social practices is required. In particular, further detailed research is needed to (1) evaluate the quantity and utilisation of ASF in relation to number of livestock owned by a household; (2) assess the time dimension of food security using longitudinal studies to account for seasonal patterns and changes over time; (3) understand how gender roles influence intra-household variation of food consumption and how livestock ownership may impact this, with emphasis on the most vulnerable (i.e. women and children); and (4) investigate how the potential negative impact of zoonotic/food-borne diseases in households that own livestock or consume ASF can be mitigated to ensure improved nutritional outcomes through livestock-based interventions.

Abbreviations

ASF: animal source foods; DHS: Demographic and Health Surveys; FAO: Food and Agriculture Organization of the United Nations; HAZ: height-for-age *Z* score; IBM: International Business Machines Corporation; IQR: interquartile range; LSMS: Living Standards Measurement Study; MICS: Multiple Indicator Cluster Surveys; MVP: Millennium Villages Project; NA: not applicable; SD: standard deviation; SPSS: Statistical Package for the Social Sciences; UN: United Nations; UNICEF: United Nations International Children's Emergency Fund; WAZ: weight-for-age *Z* score; WHO: World Health Organization; WHZ: weight-for-height *Z* score.

Authors' contributions

JBH performed the analysis, interpreted the data and wrote the manuscript; AKW performed data cleaning, interpreted the data and provided critical evaluation of manuscript drafts; JN conceived of the study idea, facilitated data access and provided critical evaluation of manuscript drafts; and SMM conceived of the study idea, contributed to the analysis, interpreted the data and provided critical evaluation of manuscript drafts. All authors read and approved the final manuscript.

Author details

[1] Australian Centre for International Agricultural Research, Bruce, ACT 2612, Australia. [2] Sydney School of Veterinary Science, The University of Sydney, Camperdown, NSW 2006, Australia. [3] Charles Perkins Centre, The University of Sydney, Camperdown, NSW 2006, Australia. [4] Sydney School of Public Health, The University of Sydney, Camperdown, NSW 2006, Australia.

Acknowledgements

The authors would like to thank Roseline Remans, Cheryl Palm and Bocary Kaya from Columbia University for providing the data. Their efforts in the Millennium Villages Project made this study possible. Thanks also to the Australian Centre for International Agricultural Research (ACIAR) for supporting the lead author to undertake this study.

Competing interests

The authors declare that they have no competing interests.

Funding

Data used in this study were collected as part of the Millennium Villages Project which was funded by the Lenfest, Blaustein, Sara McCune, Stephen Lewis/ MACAID, and Yara Foundations and the Government of Japan through the United Nations Development Programme's Human Security Trust Fund and was headed by Professor Jeffrey Sachs, Earth Institute, Columbia University, USA. The analysis, interpretation of data and writing of the manuscript did not have direct funding.

References

1. UN. World population prospects revision. New York: United Nations; 2015. p. 2015.
2. FAO. How to feed the world in 2050. Rome: Food and Agriculture Organization of the United Nations; 2009.
3. UNICEF. Levels and trends in child mortality. New York: United Nations International Children's Emergency Fund; 2013.
4. Black MM. Micronutrient deficiencies and cognitive functioning. J Nutr. 2003;133:3927S–31S.
5. Alderman H, Hoddinott J, Kinsey B. Long term consequences of early childhood malnutrition. Oxf Econ Pap. 2006;58:450–74.
6. Müller O, Krawinkel M. Malnutrition and health in developing countries. CMAJ. 2005;173:279–86.
7. Schaible UE, Kaufmann SHE. Malnutrition and infection: complex mechanisms and global impacts. PLoS Med. 2007;4:e115.
8. Kane AV, Dinh DM, Ward HD. Childhood malnutrition and the intestinal microbiome malnutrition and the microbiome. Pediatr Res. 2015;77:256–62.
9. Osrin D, Vaidya A, Shrestha Y, Baniya RB, Manandhar DS, Adhikari RK, Filteau S, Tomkins A, Anthony MDL. Effects of antenatal multiple micronutrient supplementation on birthweight and gestational duration in Nepal: double-blind, randomised controlled trial. Lancet. 2005;365:955–62.
10. FAO. Incorporating nutrition considerations into development policies and programmes: brief for policy-makers and programme planners in developing countries. Rome: Food and Agriculture Organization of the United Nations; 2004.
11. Whaley SE, Sigman M, Neumann C, Bwibo N, Guthrie D, Weiss RE, Alber S, Murphy SP. The impact of dietary intervention on the cognitive development of Kenyan school children. J Nutr. 2003;133:3965S–71S.
12. Murphy SP, Allen LH. Nutritional importance of animal source foods. J Nutr. 2003;133:3932S–5S.
13. Alders RG, Aongolo A, Bagnol B, de Bruyn J, Kimboka S, Kocj R, Li M, Maulaga W, Mcconchie R, Mor S. Using a One Health approach to promote food and nutrition security in Tanzania and Zambia. Planet@ Risk. 2014;2:187–90.
14. Palmer AC, Healy K, Barffour MA, Siamusantu W, Chileshe J, Schulze KJ, West KP, Labrique AB. Provitamin A carotenoid-biofortified maize consumption increases pupillary responsiveness among Zambian children in a randomized controlled trial. J Nutr. 2016;146:2551–8.
15. Jones KM, de Brauw A. Using agriculture to improve child health: promoting orange sweet potatoes reduces diarrhea. World Dev. 2015;74:15–24.
16. Choi CW. Nutritionally enhanced food crops: progress and perspectives. Int J Mol Sci. 2015;16:3895–914.
17. Moyo S, Swanepoel F. Multifunctionality of livestock in developing communities. In: Swanepoel F, Stroebel A, Moyo S, editors. The role of livestock in developing communities: enhancing multifunctionality. Cape Town: University of the Free State and the Technical Centre for Agricultural and Rural Cooperation; 2010. p. 1–10.
18. Randolph TF, Schelling E, Grace D, Nicholson CF, Leroy JL, Cole DC, Demment MW, Omore A, Zinsstag J, Ruel M. Role of livestock in human nutrition and health for poverty reduction in developing countries. J Anim Sci. 2007;85:2788–800.
19. Smith J, Sones K, Grace D, MacMillan S, Tarawali S, Herrero M. Beyond milk, meat, and eggs: role of livestock in food and nutrition security. Anim Front. 2013;3:6–13.
20. Drewnowski A. The nutrient rich foods index helps to identify healthy, affordable foods. Am J Clin Nutr. 2010;91:1095S–101S.
21. Azzarri C, Zezza A, Haile B, Cross E. Does livestock ownership affect animal source foods consumption and child nutritional status? Evidence from rural Uganda. J Dev Stud. 2015;51:1034–59.
22. Sanchez P, Palm C, Sachs J, Denning G, Flor R, Harawa R, Jama B, Kiflemariam T, Konecky B, Kozar R, Lelerai E, Malik A, Modi V, Mutuo P, Niang A, Okoth H, Place F, Sachs SE, Said A, Siriri D, Teklehaimanot A, Wang K, Wangila J, Zamba C. The African Millennium Villages. Proc Natl Acad Sci USA. 2007;104:16775–80.
23. Dixon JA, Gibbon DP, Gulliver A. Farming systems and poverty: improving farmers' livelihoods in a changing world. Rome: Food and Agriculture Organization of the United Nations; 2001.
24. Denning G, Kabambe P, Sanchez P, Malik A, Flor R, Harawa R, Nkhoma P, Zamba C, Banda C, Magombo C, Keating M, Wangila J, Sachs J. Input subsidies to improve smallholder maize productivity in Malawi: toward an African Green Revolution. PLoS Biol. 2009;7:e1000023.
25. Remans R, Pronyk PM, Fanzo JC, Chen J, Palm CA, Nemser B, Muniz M, Radunsky A, Abay AH, Coulibaly M, Mensah-Homiah J, Wagah M, An X, Mwaura C, Quintana E, Somers MA, Sanchez PA, Sachs SE, McArthur JW, Sachs JD. Multisector intervention to accelerate reductions in child stunting: an observational study from 9 Sub-Saharan African countries. Am J Clin Nutr. 2011;94:1632–42.
26. WHO. Physical status: the use and interpretation of anthropometry. WHO Tech Rep Ser. 1995;854:161–262.
27. de Bruyn J, Wong J, Bagnol B, Pengelly B, Alders R. Family poultry and food and nutrition security. CAB Rev Perspect Agric Vet Sci Nutr Natl Resour. 2015;10:1–9.
28. Goyder H, Mang'anya M. Livestock platform baseline study. Research into Use Malawi Programme—Malawi. 2009.
29. MAAIF. The national livestock census: a summary report of the national livestock census, 2008. Kampala: Ministry of Agriculture, Animal Industry and Fisheries, Uganda and the Uganda Bureau of Statistics; 2009.
30. PRC. The future of the global Muslim population. Washington, DC: Pew Research Centre; 2011.
31. Beveridge M, Thilsted S, Phillips M, Metian M, Troell M, Hall S. Meeting the food and nutrition needs of the poor: the role of fish and the opportunities and challenges emerging from the rise of aquaculture. J Fish Biol. 2013;83:1067–84.
32. Verduzco-Gallo I, Ecker O, Pauw K. Changes in food and nutrition security in Malawi. Washington, DC: International Food Policy Research Institute; 2014.
33. Loibooki M, Hofer H, Campbell KL, East ML. Bushmeat hunting by communities adjacent to the Serengeti National Park, Tanzania: the importance of livestock ownership and alternative sources of protein and income. Environ Conserv. 2002;29:391–8.
34. Belluco S, Losasso C, Maggioletti M, Alonzi CC, Paoletti MG, Ricci A. Edible insects in a food safety and nutritional perspective: a critical review. Compr Rev Food Sci Food Saf. 2013;12:296–313.
35. Leroy JL, Frongillo EA. Can interventions to promote animal production ameliorate undernutrition? J Nutr. 2007;137:2311–6.
36. Schaetzel T, Antal M, Guyon A. Household decision-making on homestead food production: perceptions on planting, production, and purchases in Bangladesh. Arlington: United States Agency for International Development; 2014.
37. Jin M, Iannotti LL. Livestock production, animal source food intake, and young child growth: the role of gender for ensuring nutrition impacts. Soc Sci Med. 2014;105:16–21.
38. Mosites EM, Rabinowitz PM, Thumbi SM, Montgomery JM, Palmer GH, May S, Rowhani-Rahbar A, Neuhouser ML, Walson JL. The relationship between livestock ownership and child stunting in three countries in Eastern Africa using national survey data. PLoS ONE. 2015;10:e0136686.
39. Barrett-Connor E. Nutrition epidemiology: how do we know what they ate? Am J Clin Nutr. 1991;54:182S–7S.
40. Haggblade S, Duodu KG, Kabasa JD, Minnaar A, Ojijo NK, Taylor JR. Emerging early actions to bend the curve in Sub-Saharan Africa's nutrition transition. Food Nutr Bull. 2016;37:219–41.
41. FAO. The state of food insecurity in the World. Rome: Food and Agriculture Organization of the United Nations; 2015.
42. Hirvonen K, Taffesse AS, Worku Hassen I. Seasonality and household diets in Ethiopia. Public Health Nutr. 2016;19:1723–30.
43. Beegle K, De Weerdt J, Friedman J, Gibson J. Methods of household consumption measurement through surveys: experimental results from Tanzania. J Dev Econ. 2012;98:3–18.
44. Bargain O, Donni O, Kwenda P. Intrahousehold distribution and poverty: evidence from Côte d'Ivoire. J Dev Econ. 2014;107:262–76.
45. Guèye EF. The role of family poultry in poverty alleviation, food security and the promotion of gender equality in rural Africa. Outlook Agric. 2000;29:129–36.
46. Jones AD, Ngure FM, Pelto G, Young SL. What are we assessing when we

measure food security? A compendium and review of current metrics. Adv Nutr. 2013;4:481–505.
47. Kassie M, Ndiritu SW, Stage J. What determines gender inequality in household food security in Kenya? Application of exogenous switching treatment regression. World Dev. 2014;56:153–71.
48. Kariuki J, Njuki J, Mburu S, Waithanji E. Women, livestock ownership and food security. Nairobi: International Livestock Research Institute and International Development Research Centre; 2013.
49. Valdivia C. Gender, livestock assets, resource management, and food security: lessons from the SR-CRSP. Agric Hum Values. 2001;18:27–39.
50. Wu G, Bazer FW, Cudd TA, Meininger CJ, Spencer TE. Maternal nutrition and fetal development. J Nutr. 2004;134:2169–72.
51. Mosites E, Mwangi T, Otiang E, Garland-Lewis G, Sammons M, Onyango C, Eng A, Noecker C, Manor O, Hilton S, Call D, Kariuki N, Zambriski J, Wasserheit J, Walson J, Palmer G, Montgomery J, Borenstein E, Omore R, Rabinowitz P. Characterising the taxonomic composition of children and livestock gut microbiomes and of environmental samples and the potential role for household-level microbiome sharing in western Kenya. Lancet Glob Health. 2016;4:S20.
52. Ngure FM, Humphrey JH, Mbuya MNN, Majo F, Mutasa K, Govha M, Mazarura E, Chasekwa B, Prendergast AJ, Curtis V, Boor KJ, Stoltzfus RJ. Formative research on hygiene behaviors and geophagy among infants and young children and implications of exposure to fecal bacteria. Am J Trop Med Hyg. 2013;89:709–16.
53. Checkley W, Buckley G, Gilman RH, Assis AM, Guerrant RL, Morris SS, Molbak K, Valentiner-Branth P, Lanata CF, Black RE. Multi-country analysis of the effects of diarrhoea on childhood stunting. Int J Epidemiol. 2008;37:816–30.
54. Headey D, Hirvonen K. Exploring child health risks of poultry keeping in Ethiopia: insights from the 2015 Feed the Future survey. Washington, DC: International Food Policy Research Institute and Ethiopian Development Research Institute; 2015.

6

The association of test diameter and gonad weight with some toxic trace metals level in black sea urchin (*Stomopneustes variolaris*)

H. K. S. De Zoysa[1,2*], B. K. K. K. Jinadasa[3], E. M. R. K. B. Edirisinghe[2] and G. D. T. M. Jayasinghe[3]

Abstract

Background: Quantification of toxic trace metals in biota is very important to monitor both ecosystem health and public health. *Stomopneustes variolaris* is a widespread species of sea urchin in the world that have edible gonads that are in high demand. This study attempted to examine the levels of accumulated toxic trace metals and their relationship between the test diameter and the gonad weight of the black sea urchin (*S. variolaris*) in Sri Lanka.

Results: As *S. variolaris* is highly abundant in southwest coast of Sri Lanka, three sampling sites were selected in this region (Mount-Lavinia, Beruwala and Tangalle). The concentration of toxic trace metals (Hg, Cd, Pb and As) in the gonads of *S. variolaris* was determined in randomly selected subset from the above samples. The levels of above metals in gonads did not exceed the permitted standard levels of the EU and Sri Lanka. The mean size and weight of *S. variolaris* collected from Mount-Lavinia was 5.55 ± 1.04 cm and 101.40 ± 57.76 g, respectively; for Beruwala it was 6.54 ± 0.86 cm and 147.90 ± 50.40 g, respectively, and for Tangalle it was 6.41 ± 1.05 cm and 150.50 ± 59.45 g respectively. There was a significant relationship between all the analysed trace metals and the test diameter and gonad weight of *S. variolaris* in Sri Lanka.

Conclusions: This study revealed that the trace metal levels in gonads of *S. variolaris* did not exceed the harmful level for human consumption.

Keywords: Accumulation, Echinodermata, Roe, Indian Ocean

Background

Heavy metals that are less soluble in water tend to accumulate in marine ecosystems as they do not degrade [1]. Therefore, these toxic trace metals will accumulate along the trophic levels which will affect the entire marine community. The levels of trace metal accumulation in marine organisms depend on their detoxification mechanism in metabolism. The harmfulness of these toxic trace metals is due to their role in biochemical and metabolic processes. Excessive consumption of sea food with high levels of accumulated toxic trace metals can lead to metabolic dysfunctions. Therefore, quantification of toxic trace metals in the all biota is very important to monitor both ecosystem health and public health. Currently, sea urchin gonads are a delicacy in many countries, and thus, there is a very high demand. However, nothing is known about the levels of accumulated toxic trace metals in sea urchins [1, 2].

The black sea urchin, *Stomopneustes variolaris,* is one of the edible species of sea urchin in the world. In Sri Lanka, *S. variolaris* is restricted to the western and southern coastal areas than other coastal areas [3, 4]. The black sea urchin (*S. variolaris*) is one of the warm water sea urchin species that belongs to the family Stomopneustidae. They bore into the rock, reefs and other substrates in littoral regions up to a depth of 18 m, and

*Correspondence: dezoysahks@yahoo.com
[1] Department of Bioprocess Technology, Faculty of Technology, Rajarata University of Sri Lanka, Mihintale, Sri Lanka
Full list of author information is available at the end of the article

they prefer to live in the shady areas. *S. variolaris* is an omnivore species, and also they are known to mainly feed on algae and seaweeds [5–13]. *S. variolaris* species has a widespread distribution [12, 14, 15], in tropical and subtropical regions of the Indo-Pacific from the east African coast and Madagascar to New Caledonia and Samoa to the Bonin Islands to the north [12]. It is an abundant edible species in the Indian Ocean, in the geographical regions such as Lakshadweep Islands, Andaman Islands, India and Sri Lanka [16]. The growth rate of sea urchins and natural mortality mainly depend on temperature and food availability [6]. The sea urchins accumulate nutrients in the gonad, which is a highly sought-after delicacy in the world [17–19].

The trace metals are accumulated in most body parts of echinoderms including coelomic fluid, spines, gut, gonads and Aristotle's lantern. Hence, levels of trace metals in sea urchins could be used as a bio-indicator to determine the level of environmental pollution [20]. Currently, nothing is known about the levels of trace metals in the gonads of sea urchins in Sri Lanka, which has high demand in the marine fisheries export industry [18, 19]. Therefore, this study was conducted to get an idea on the levels of toxic trace metals in *S. variolaris* in Sri Lanka.

Methods

A total of 197 individuals of *S. variolaris* were collected from three selected sites in the western and southern coasts of Sri Lanka, namely Mount-Lavinia ($n=43$), Beruwala ($n=99$) and Tangalle ($n=55$) reefs (Fig. 1) from May to November in 2014. All collected individuals were transported immediately with seawater to analytical chemistry laboratory at the Institute of Post-Harvest Technology (IPHT), National Aquatic Research Agency (NARA) within 24 h. After that, total body weight to nearest 0.01 g and the horizontal test diameter was measured to the nearest 0.02 mm using a vernier caliper. Horizontal test diameter was measured twice perpendicularly to the first measurement and the averages of the two measurements were taken. Finally, gonads were removed from each individuals and weighed after removing excess water using blotting papers.

To determine the toxic trace metal levels in gonads of *S. variolaris*, individuals with test/shell diameter between 3 and 9 cm were selected and 6, 13 and 11 individuals were selected from Mount-Lavinia, Beruwala and Tangalle, respectively [21]. The levels of toxic trace metals, Hg (mercury), Pb (lead), Cd (cadmium) and As (arsenic) were assayed in the atomic absorption spectrophotometer (AAS) (Varian 240 FS, Springvale, Australia). The standard solutions of Hg, Pb, Cd and As at 1000 mg/L (ppm) were obtained from Sigma-Aldrich (Dorset, United Kingdom) which were used to calibrate. All standards and reagents were prepared using ultra-pure water. One gram of homogenized sea urchin gonad samples were taken into a Teflon microwave digestion tube and all the samples were performed in duplicates. Acid digestions of gonad samples were performed by using the 10 mL of HNO_3 (65% nitric acid, AR) allowed to 10–15 min and capped the microwave digestion tubes. The sea urchin gonad samples were digested using the CEM/MARS XP-1500+microwave oven (CEM, Matthews, USA). Microwave accelerated system and allowed to cool room temperature and pressure was released carefully by opening the valve. Then the digested gonad samples were transferred into 50 mL of volumetric flasks and volumed up using the de-ionized water was used to determine the Hg, Pb, Cd and As of digested sea urchins' gonads samples. A graphite tube atomizer (Varian GTA-120) was used to determine the Pb and Cd, for the determinations of As palladium nitrate ($Pd(NO_3)_2$) was used as a modifier. The Hg concentration of sea urchins' gonads were determined by using the cold vapour techniques (Varian VGA 77) using the $SnCl_2$ as a reduction [3, 4, 22].

The Pearson's correlation coefficient (r) and significant correlation tests were conducted to assess the relationship between gonadal trace metals levels and test diameter and gonad weight of *S. variolaris*. Data were analysed using the Minitab 16.0 version and Microsoft Excel 2010 version at the level of significance at $\alpha=0.05$ (5%).

Results

Toxic trace metals levels in gonads

Levels of toxic trace metals in gonads of *S. variolaris* were determined separately from selected sites and the average values of accumulated concentration (ppm and ppb) and the total amount of trace metals with their average test diameter and gonad weight in selected individuals' are determined.

The average concentration (ppb) of Hg in selected individuals from Mount-Lavinia reef ($n=11$) ranged from (minimum to maximum) 0.00 ($\pm$0.00) to 14.40 ($\pm$5.33) μg/kg and total concentration in gonads varied from 0.00 to 0.27 μg (Table 1). The trace metal Cd ranged from 47.81 ($\pm$19.58) to 336.10 ($\pm$21.46) μg/kg and total concentration in gonads varied from 0.08 to 6.28 μg. Pb was not detected in the selected individuals from Mount-Lavinia. However, concentration of As ranged from 0.00 ($\pm$0.00) to 1805.87 ($\pm$104.54) μg/kg, and total accumulation by gonads varied from 0.00 to 33.75 μg.

For the Beruwala reef population, accumulated toxic trace metals ($n=13$) are shown in Table 2. The average concentration of Hg in the Beruwala reef population ranged from 34.40 ($\pm$1.62) to 104.39 ($\pm$2.18) μg/kg, and the total level in the gonads varied from 0.07 to 0.44 μg. Cd levels ranged from 150.72 ($\pm$1.81) to 1149.55

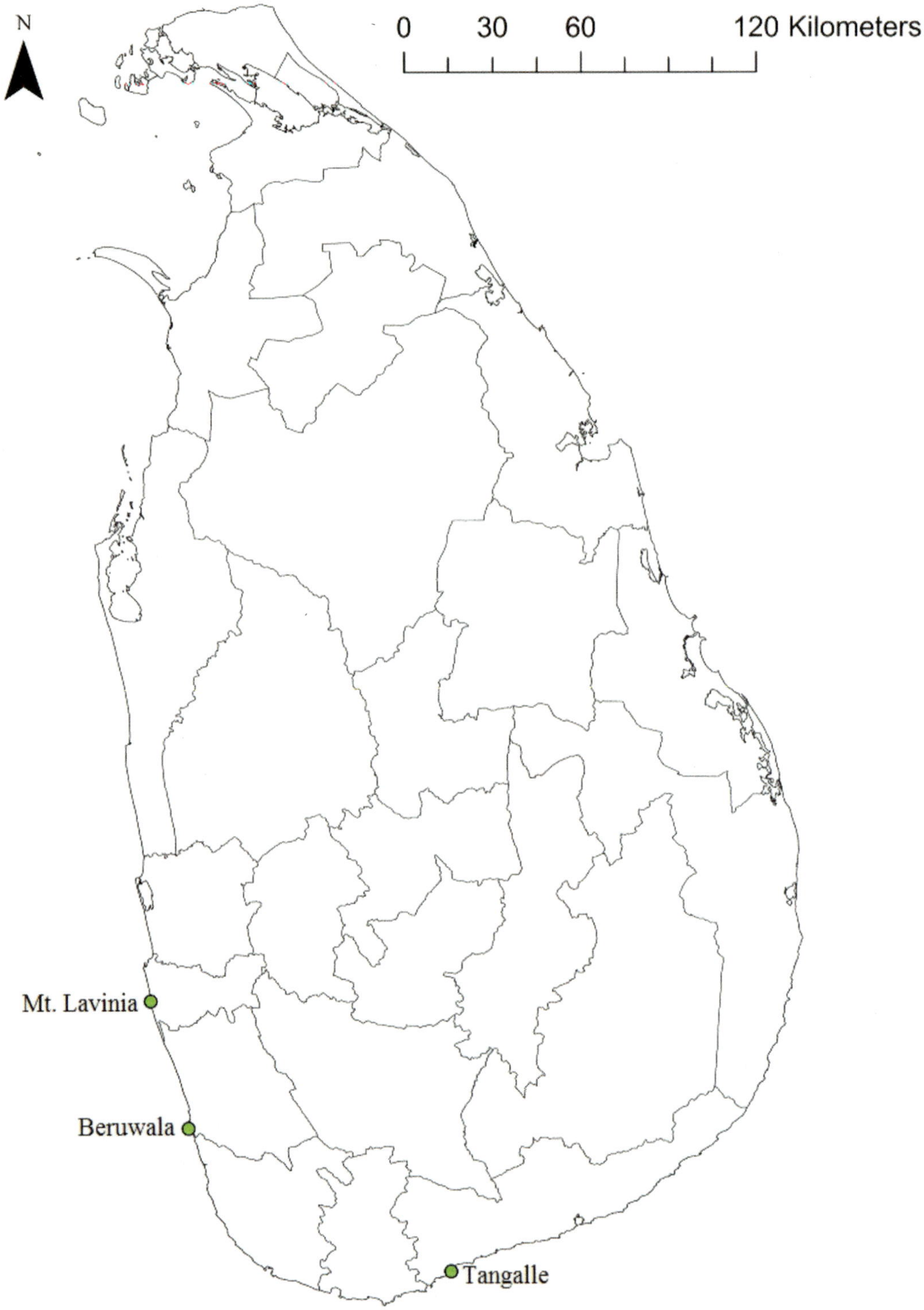

Fig. 1 Sampling sites of *S. variolaris*

(±90.07) μg/kg and total levels in gonads varied from 0.17 to 9.86 μg. Pb ranged from 0.00 (±0.00) to 206.27 (±90.07) μg/kg and total level in gonads varied from 0.00 to 0.36 μg. Also As content in the selected individuals ranged from 1142.52 (±83.51) to 4635.66 (±90.07) μg/kg and total accumulation by gonads varied from 2.99 to 22.23 μg.

Table 1 Toxic trace metals (Hg, Cd, Pb and As) concentration and total amount of particular trace metals present in *S. variolaris* at Mount-lavinia reef ($n = 6$) with standard deviation (± SD)

Test diameter (cm)	Total gonad weight (g)	Hg		Cd		Pb		As	
		Average (Hg) (µg/kg)	Total Hg (µg)	Average (Cd) (µg/kg)	Total Cd (µg)	Average (Pb) (µg/kg)	Total Pb (µg)	Average (As) (µg/kg)	Total As (µg)
3.70	1.77	0.00 (±0.00)	0.00	47.81 (±19.58)	0.08	0.00 (±0.00)	0.00	0.00 (±0.00)	0.00
4.50	1.92	0.00 (±0.00)	0.00	48.38 (±9.42)	0.09	0.00 (±0.00)	0.00	1637.16 (±396.96)	3.14
5.10	3.10	9.68 (±0.82)	0.03	156.69 (±2.63)	0.49	0.00 (±0.00)	0.00	50.27 (±156.08)	0.16
6.20	10.07	0.00 (±0.00)	0.00	158.03 (±2.73)	1.59	0.00 (±0.00)	0.00	305.31 (±54.61)	3.07
7.10	8.63	4.74 (±0.91)	0.04	217.00 (±0.78)	1.87	0.00 (±0.00)	0.00	434.28 (±324.83)	3.75
8.90	18.69	14.40 (±5.33)	0.27	336.10 (±21.46)	6.28	0.00 (±0.00)	0.00	1805.87 (±104.54)	33.75

Table 2 Toxic trace metals (Hg, Cd, Pb and As) concentration and total amount of particular trace metals present in *S. variolaris* at Beruwala reef ($n = 13$) with ± SD

Test diameter (cm)	Total gonad weight (g)	Hg		Cd		Pb		As	
		Average (Hg) (μg/kg)	Total Hg (μg)	Average (Cd) (μg/kg)	Total Cd (μg)	Average (Pb) (μg/kg)	Total Pb (μg)	Average (As) (μg/kg)	Total As (μg)
4.90	1.15	61.91 (±7.58)	0.07	150.72 (±1.81)	0.17	0.00 (±0.00)	0.00	2600.20 (±1.14)	2.99
5.30	2.13	54.38 (±23.32)	0.12	254.07 (±15.25)	0.54	0.00 (±0.00)	0.00	3043.89 (±15.87)	6.48
5.80	4.94	34.40 (±1.62)	0.17	231.60 (±30.46)	1.14	36.45 (±60.40)	0.18	1142.52 (±83.51)	5.64
5.90	2.91	102.26 (±3.65)	0.30	858.61 (±225.66)	2.50	11.94 (±49.54)	0.03	4483.33 (±272.99)	13.5
6.00	1.73	91.12 (±23.50)	0.16	270.28 (±25.63)	0.47	206.27 (±5.51)	0.36	2170.50 (±168.73)	3.75
6.00	3.59	89.19 (±1.87)	0.32	1149.55 (±90.07)	4.13	41.24 (±13.39)	0.15	4480.31 (±312.27)	16.08
6.30	3.00	85.69 (±1.61)	0.26	885.50 (±34.07)	2.66	91.36 (±33.28)	0.27	1928.18 (±187.36)	5.78
6.40	4.04	31.15 (±1.22)	0.13	308.68 (±31.87)	1.25	7.57 (±34.02)	0.03	1772.28 (±135.50)	7.16
6.50	5.59	60.56 (±13.97)	0.34	331.47 (±5.63)	1.85	61.45 (±53.93)	0.34	2040.64 (±250.12)	11.41
6.70	3.86	85.09 (±0.09)	0.33	835.22 (±39.26)	3.22	24.78 (±3.28)	0.10	4193.81 (±84.00)	16.19
7.00	8.81	44.73 (±6.13)	0.39	1119.41 (±21.88)	9.86	22.03 (±19.29)	0.19	2543.45 (±233.71)	22.23
7.30	3.48	104.39 (±2.18)	0.36	915.82 (±35.99)	3.19	11.92 (±5.12)	0.04	4635.66 (±813.92)	16.13
7.70	8.72	50.23 (±2.55)	0.44	395.64 (±3.12)	3.45	21.60 (±45.87)	0.19	1691.77 (±185.39)	14.75

Table 3 **Toxic trace metals (Hg, Cd, Pb and As) concentration and total amount of particular trace metals present in *S. variolaris* at Tangalle reef ($n = 11$) with ± SD**

Test diameter (cm)	Total gonad weight (g)	Hg		Cd		Pb		As	
		Average (Hg) (µg/kg)	Total Hg (µg)	Average (Cd) (µg/kg)	Total Cd (µg)	Average (Pb) (µg/kg)	Total Pb (µg)	Average (As) (µg/kg)	Total As (µg)
4.10	0.75	0.00 (±0.00)	0.00	83.60 (±1.61)	0.06	0.00 (±0.00)	0.00	0.00 (±0.00)	0.00
4.10	0.80	188.12 (±24.29)	0.15	289.72 (±18.01)	0.23	415.03 (±30.92)	0.33	1866.37 (±325.00)	1.49
4.80	1.12	164.31 (±7.12)	0.18	201.39 (±35.36)	0.23	772.27 (±211.65)	0.86	3050.80 (±145.32)	3.42
5.00	1.46	92.92 (±6.75)	0.14	247.80 (±33.81)	0.36	309.80 (±18.40)	0.45	2785.16 (±168.17)	4.07
5.70	3.31	51.51 (±2.21)	0.17	796.99 (±19.78)	2.64	93.70 (±11.10)	0.31	1574.48 (±97.97)	5.21
6.50	7.71	12.90 (±1.80)	0.10	997.91 (±647.59)	7.69	2.15 (±9.97)	0.02	869.85 (±44.62)	6.71
6.50	8.73	56.61 (±8.09)	0.49	559.31 (±19.37)	4.88	42.93 (±26.77)	0.37	1621.43 (±37.61)	14.16
7.20	11.80	6.27 (±4.70)	0.07	379.09 (±3.39)	4.47	0.00 (±0.00)	0.00	3037.02 (±34.40)	35.84
7.30	3.66	115.49 (±1.98)	0.42	4590.83 (±155.96)	16.80	87.68 (±16.02)	0.32	4219.97 (±104.78)	15.45
7.70	6.61	11.87 (±0.13)	0.08	500.57 (±8.29)	3.31	0.00 (±0.00)	0.00	1696.57 (±198.02)	11.21
8.30	10.72	65.66 (±1.38)	0.70	1499.39 (±25.00)	16.07	118.34 (±35.89)	1.27	4563.32 (±175.59)	48.92

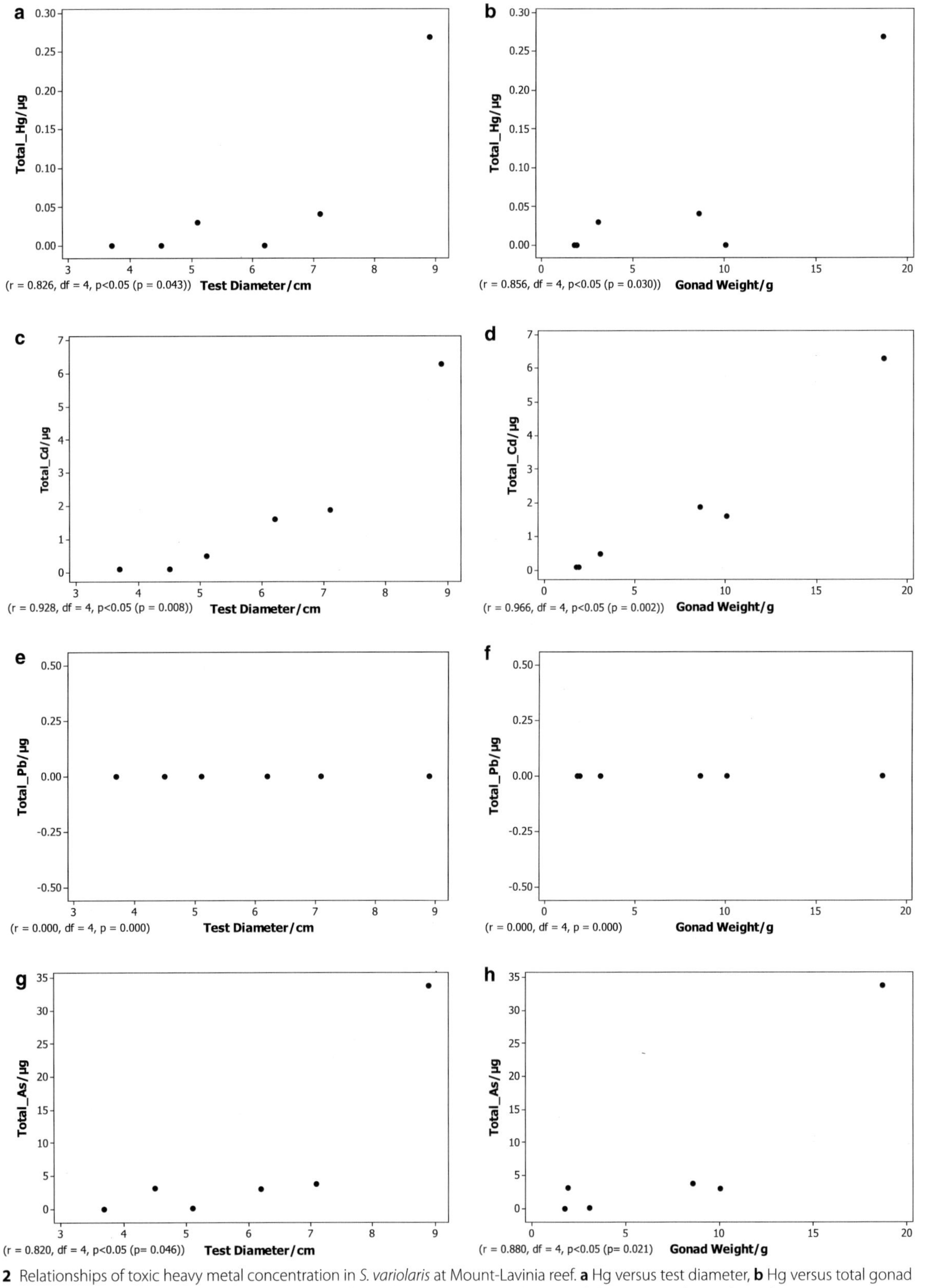

Fig. 2 Relationships of toxic heavy metal concentration in *S. variolaris* at Mount-Lavinia reef. **a** Hg versus test diameter, **b** Hg versus total gonad weight, **c** Cd versus test diameter, **d** Cd versus total gonad weight, **e** Pb versus test diameter, **f** Pb versus total gonad weight, **g** As versus test diameter, **h** As versus total gonad weight

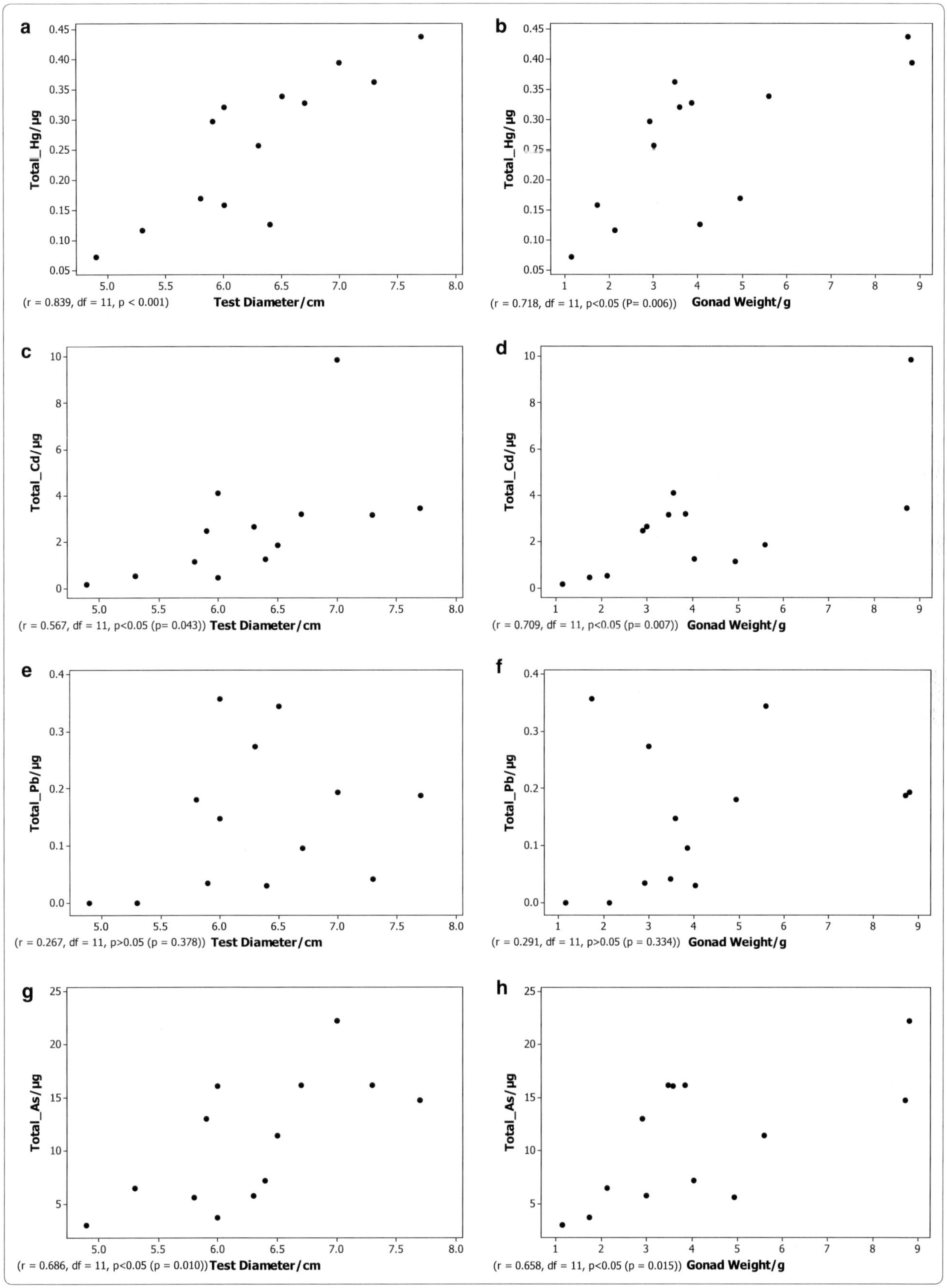
a
Total_Hg/µg
Test Diameter/cm
(r = 0.839, df = 11, p < 0.001)
b
Total_Hg/µg
Gonad Weight/g
(r = 0.718, df = 11, p<0.05 (P= 0.006))
c
Total_Cd/µg
Test Diameter/cm
(r = 0.567, df = 11, p<0.05 (p= 0.043))
d
Total_Cd/µg
Gonad Weight/g
(r = 0.709, df = 11, p<0.05 (p= 0.007))
e
Total_Pb/µg
Test Diameter/cm
(r = 0.267, df = 11, p>0.05 (p = 0.378))
f
Total_Pb/µg
Gonad Weight/g
(r = 0.291, df = 11, p>0.05 (p = 0.334))
g
Total_As/µg
Test Diameter/cm
(r = 0.686, df = 11, p<0.05 (p = 0.010))
h
Total_As/µg
Gonad Weight/g
(r = 0.658, df = 11, p<0.05 (p = 0.015))

(See figure on previous page.)
Fig. 3 Relationships of toxic heavy metal concentration in *S. variolaris* at Beruwala reef. **a** Hg versus test diameter, **b** Hg versus total gonad weight, **c** Cd versus test diameter, **d** Cd versus total gonad weight, **e** Pb versus test diameter, **f** Pb versus total gonad weight, **g** As versus test diameter, **h** As versus total gonad weight

Hg levels in *S. variolaris* individuals from Tangalle reef ($n=11$) ranged from 0.00 (±0.00) to 188.12 (±24.29) µg/kg and total level in gonads varied from 0.00 to 0.70 µg. Cd levels ranged from 83.60 (±1.61) to 4590.83 (±155.96) µg/kg and total level in gonads varied from 0.06 to 16.07 µg. Pb ranged from 0.00 (±0.00) to 772.27 (±211.65) µg/kg and total levels in gonads varied from 0.00 to 1.27 µg. In addition, As in selected individuals ranged from 0.00 (±0.00) to 4563.32 (±175.59) µg/kg and total levels in gonads varied from 0.00 to 48.92 µg (Table 3).

Relationships with test diameter and gonad weight for trace metal accumulation

The results revealed that there is a positive correlation with between gonadal Hg levels and the test diameter ($r=0.826$, $df=4$, $p=0.043$) and also a between gonad weight ($r=0.856$, $df=4$, $p=0.030$) (Fig. 2a, b) for *S. variolaris* population in Mount-Lavinia. Gonadal Cd levels also had a significant relationship with test diameter ($r=0.928$, $df=4$, $p=0.008$), but strong positive r and have significant relationship with gonad weight ($r=0.966$, $df=4$, $p=0.002$) (Fig. 2c, d). Total Pb levels showed no correlation with the test diameter and gonad weight (Fig. 2e, f). Arsenic levels showed a significant positive relationship with test diameter ($r=0.820$, $df=4$, $p=0.046$) and also with the gonad weight ($r=0.880$, $df=4$, $p=0.021$) (Fig. 2g, h).

Hg levels of the Beruwala population showed a significant positive relationship with test diameter ($r=0.839$, $df=11$, $p<0.001$) and gonad weight ($r=0.718$, $df=11$, $p=0.006$) (Fig. 3a, b). Similarly, Cd levels also showed a significant positive relationship with test diameter ($r=0.567$, $df=11$, $p=0.043$) and gonad weight ($r=0.709$, $df=11$, $p=0.007$) (Fig. 3c, d). Pb has a small positive r value and has no significant relationship with test diameter ($r=0.267$, $df=11$, $p=0.378$) and gonad weight ($r=0.291$, $df=11$, $p=0.334$) (Fig. 3e, f). Levels of the trace metal As also had a significant positive relationship with test the diameter ($r=0.686$, $df=11$, $p=0.010$) and as well as with the gonad weight ($r=0.658$, $df=11$, $p=0.015$) (Fig. 3g, h).

The analysis of the Tangalle population revealed that there is a significant positive relationship between Hg levels and the test diameter ($r=0.542$, $df=9$, $p=0.085$) and with gonad weight ($r=0.414$, $df=9$, $p=0.205$) (Fig. 4a, b). Similarly, cadmium levels showed a strong significant positive relationship with the test diameter ($r=0.764$, $df=9$, $p=0.006$) as well as with the gonad weight ($r=0.509$, $df=9$, $p=0.110$) (Fig. 4c, d). Levels of As also had a significant positive relationship with test diameter ($r=0.779$, $df=9$, $p=0.005$) and gonad weight ($r=0.825$, $df=9$, $p=0.002$) (Fig. 4e, f). Similarly, Pb levels also did not show a significant relationship with test diameter ($r=0.151$, $df=9$, $p=0.657$) and gonad weight ($r=0.048$, $df=9$, $p=0.888$) (Fig. 4g, h).

Discussion

There were low levels of trace metals in the gonads of *S. variolaris* in all the studied locations. The significant finding was that there were undetectable levels of Pb in *S. variolaris* gonads collected from the Mount-Lavinia reef. Further, at this location *S variolaris* population had accumulated As (0.00 ppm–1.81 mg/kg) at a high level (Table 4) compared to other studied metals (Hg and Cd) though this did not exceeded the European Union (EU) (Table 4) and Sri Lanka permitted level. *S. variolaris* populations in the Beruwala reef was not highly contaminated by Hg (0.03–0.10 mg/kg) and Pb (0.00–0.21 mg/kg), but highly contaminated by As (1.14–4.64 mg/kg) and Cd (0.15–1.15 mg/kg), which exceeds the permitted level of the 1.00–2.00 mg/kg (Table 4). According to the EU, there were very low levels of Hg (0.00–0.12 mg/kg) in the *S. variolaris* population of Tangalle reef. However, there were higher levels of Cd (0.08–4.60 mg/kg) and As (0.00–4.563 mg/kg) in this population though they were below the permitted levels. Accumulated Pb levels in *S. variolaris* (0.00–0.77 mg/kg) varied between permitted level (0.30–1.50 mg/kg) (Table 4). The low levels of toxic trace metals could be seen in the Mount-Lavinia

(See figure on next page.)
Fig. 4 Relationships of toxic heavy metal concentration in *S. variolaris* at Tangalle reef. **a** Hg versus test diameter, **b** Hg versus total gonad weight, **c** Cd versus test diameter, **d** Cd versus total gonad weight, **e** Pb versus test diameter, **f** Pb versus total gonad weight, **g** As versus test diameter, **h** As versus total gonad weight

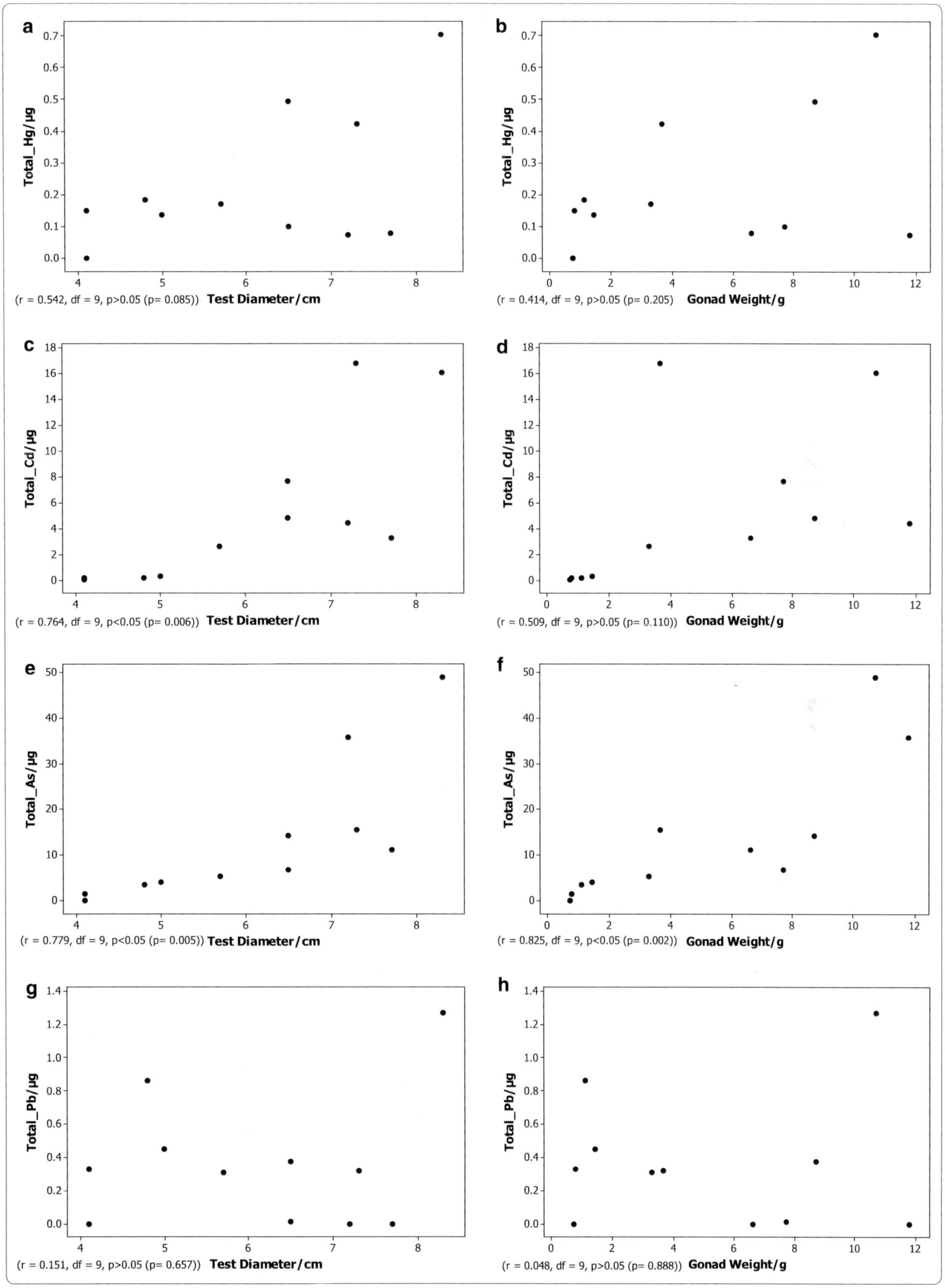
a
Total_Hg/µg
(r = 0.542, df = 9, p>0.05 (p= 0.085)) Test Diameter/cm
b
Total_Hg/µg
(r = 0.414, df = 9, p>0.05 (p= 0.205) Gonad Weight/g
c
Total_Cd/µg
(r = 0.764, df = 9, p<0.05 (p= 0.006)) Test Diameter/cm
d
Total_Cd/µg
(r = 0.509, df = 9, p>0.05 (p= 0.110)) Gonad Weight/g
e
Total_As/µg
(r = 0.779, df = 9, p<0.05 (p= 0.005)) Test Diameter/cm
f
Total_As/µg
(r = 0.825, df = 9, p<0.05 (p= 0.002)) Gonad Weight/g
g
Total_Pb/µg
(r = 0.151, df = 9, p>0.05 (p= 0.657)) Test Diameter/cm
h
Total_Pb/µg
(r = 0.048, df = 9, p>0.05 (p= 0.888)) Gonad Weight/g

Table 4 Maximum level of toxic trace metals contaminants of selected seafoods species according to EU and Sri Lanka regulations [22–24]

Contaminants	Food types	Maximum level of contaminants/(mg/kg wet weight)
Pb	Muscle meat of fish	0.30
	Crustaceans	0.50
	Bivalve, molluscs	1.50
Cd	Muscle meat of fish	0.05
	Crustaceans	0.50
	Bivalve, molluscs	1.00
Hg	Fish products and muscle meat	1.00
As	Fish, crustacea, molluscs	1.00–2.00

population when compared to the other populations. The levels of accumulated trace metals in sea urchin gonads represent the environmental condition and thus can be used as bio-indicator to determine the trace metal pollution of that particular coast. As a consequence, most sea urchin species are used as bio-indicators in the world (*Paracentrotus lividus*) [20, 23]. However, there is currently no literature available for trace metal contamination in sea urchins, specifically for Sri Lanka,.

The average levels of accumulated toxic trace metals in sea urchins were in the order from As > Cd > Pb > Hg in Beruwala population, Cd > As > Pb > Hg in Tanagalle population and As > Cd > Hg > Pb in Mount-lavinia population. According to the above order, both Beruwala and Tangalle reefs had the lowest Hg contamination than the Mount-Lavinia sample.

The accumulated levels of all of the tested toxic trace metals (Hg, Cd and As) significantly correlated with the test diameter of *S. variolaris* population in the Beruwala reef. This indicates that the accumulation of Hg, Cd and As increases with a size of the test. Nevertheless, accumulated Pb levels did not significantly correlate with average test diameter and total gonad weight. Therefore, it seems that the accumulation of Pb does not depend on the total gonad weight or the average test diameter. In the Tangalle sample, Cd and As metals showed a significant relationship between these toxic trace metal accumulations with test diameter while Hg and Pb metals did not significantly correlate with test diameter. However, total gonad weight of *S. variolaris* showed that accumulation of As significantly correlated though Hg, Cd and Pb did not. The Mount-Lavinia samples showed that there was a significant relationship between toxic trace metals accumulation with average test diameter with Hg, Cd and As metals and Pb has not significantly correlated with average test diameter. Similarly, the total gonad weight of *S. variolaris* revealed a significant relationship with Hg, Cd and As metals and Pb did not because Pb accumulation was not at detectable level. Therefore, it could not be seen any relationship between average test diameter and total gonad weight of Pb accumulation. Finally, As was the only metal which showed a clear significant relationship between both average test diameter and total gonad weight in all the selected sites.

These relationships of *S. variolaris* revealed that toxic trace metals accumulation increase with their age. This would be due to long time accumulation as a result of exposure to the polluted environment in the lifetime. In addition, accumulation of excess trace metals in gonads of sea urchins showed a relationship with their age and a similar result was also reported by Mostafa and Collins [20], because the maximum value of trace metal concentration resulted in the high average test diameter recorded individuals than the low average test diameter recorded individuals. The study conducted by Mostafa and Collins [20] revealed that accumulation of Cd and Pb showed the negative linear relationship between their wet weight and accumulated concentration in the species *P. lividus*.

The frequency distribution of shell diameter and weight of *S. variolaris* range from a minimum of 3.30 to a maximum of 8.90 cm and weight from a minimum of 30.15 g to a maximum of 346.56 g for all the population selected from Sri Lanka. But Smith and Kroh [7] recorded *S. variolaris* having maximum shell diameter of 11 cm from the Visakhapatnam Coast (India). However, 8.90 cm is the highest value recorded for the Sri Lankan *S. variolaris*.

Conclusion

The findings from this study will be helpful to the get an idea about the trace metal accumulation *S. variolaris*. Most importantly, it also indicates that the accumulated levels of trace metals did not exceed the harmful level for human consumption. Hence, due to safe levels of trace metals in *S. variolaris* gonads, there is possibility to export the gonads and gonad products from Sri Lanka, which will be a highly profitable industry in the future.

Abbreviations

IPHT: Institute of Post-Harvest Technology; NARA: National Aquatic Resources Research and Development Agency; Hg: mercury; Pb: lead; Cd: cadmium; As: arsenic; AAS: atomic absorption spectrophotometer; $Pd(NO_3)_2$: palladium nitrate; ppb: parts per billion; EU: European Union.

Authors' contributions
All authors contributed to the manuscript improvement. All authors read and approved the final manuscript.

Author details
[1] Department of Bioprocess Technology, Faculty of Technology, Rajarata University of Sri Lanka, Mihintale, Sri Lanka. [2] Department of Physical Sciences, Faculty of Applied Sciences, Rajarata University of Sri Lanka, Mihintale, Sri Lanka. [3] Institute of Post-Harvest Technology (IPHT), National Aquatic Resources Research and Development Agency (NARA), Colombo, Sri Lanka.

Acknowledgements
The authors are grateful to the National Aquatic Resources Research and Development Agency (NARA). Also, we thank to K. Ukuwela for his invaluable comments on manuscript.

Competing interests
The authors declare that they have no competing interests.

Funding
This study was supported in part by a grant from the National Aquatic Resources Research and Development Agency (NARA), Crow Island, Colombo 15, Sri Lanka.

References

1. Jakimska A, Konieczka P, Skóra K, Namieśnik J. Bioaccumulation of metals in tissues of marine animals, part I: the role and impact of heavy metals on organisms. Poli J Environ Stud. 2011;20(5):1117–25.
2. Hernández OD, Gutiérrez ÁJ, González-Weller D, Lozano G, Melón EG, Rubio C, Hardisson A. Accumulation of toxic metals (Pb and Cd) in the sea urchin *Diadema aff. antillarum* Philippi, 1845, in an oceanic island (Tenerife, Canary Islands). Environ Toxicol. 2010;25(3):227–33.
3. De Zoysa HKS, Jinadasa BKKK, Edirisinghe EMRKB. Black sea urchin (*Stomopneustes variolaris*): nutritional composition and trace metals accumulation of edible sea urchin of Sri Lanka. In: Baker R, editor. LAP LAMBERT Academic Publishing; 2016.
4. Jinadasa BKKK, De Zoysa HKS, Jayasinghe GDTM, Edirisinghe EMRKB. Determination of the biometrical parameters, biochemical composition and essential trace metals of edible sea urchin (*Stomopneustes variolaris*) in Sri Lanka. Cogent Food Agric. 2016;2(1):1–12.
5. González SJ, Cáceres WC, Ojeda FP. Feeding and nutritional ecology of the edible sea urchin *Loxechinus albus* in the northern Chilean coast. Rev Chil Hist Nat. 2008;81:575–84.
6. Reynolds JA, Wilen JE. The sea urchin fishery: harvesting, processing and the market. Mar Resour Econ. 2000;15:115–26.
7. Smith AB, Kroh A. The Echinoid Directory Natural History Museum in United Kingdom: World Wide Web electronic publication 2011 [cited 2014-12-06]. http://www.nhm.ac.uk/research-curation/projects/echinoid-directory.
8. Kato S, Schroeter SC. Biology of the red sea urchin, *Strongylocentrotus franciscanus*, and its fishery in California. Mar Fish Rev. 1985;47(3):1–20.
9. Morse BL, Hunt HL. Effect of unidirectional water currents on displacement behaviour of the green sea urchin *Strongylocentrous droebachiensis*. J Mar Biol Assoc UK. 2013;93(7):1923–8.
10. Siikavuopio SI. Green sea urchin (*Strongylocentrotus droebachiensis*, Müller) in aquaculture: the effects of environmental factors on gonad growth. University of Tromsø, Faculty of Biosciences, Fisheries and Economics, Department of Arctic and Marine Biology; 2009. https://www.researchgate.net/profile/Sten_Siikavuopio2/publication/40272651_Green_sea_urchin_Strongylocentrotus_droebachiensis_Muller_in_aquaculture_the_effects_of_environmental_factors_on_gonad_growth/links/58e256a3a6fdcc41bf991006/Green-sea-urchin-Strongylocentrotus-droebachiensis-Mueller-in-aquaculture-the-effects-of-environmental-factors-on-gonad-growth.pdf.
11. Kroh A. *Stomopneustes variolaris* (Lamarck, 1816). In: Kroh A, Mooi R, editors. Retrieved 12-06-2014, from Accessed through: Kroh, A. & Mooi, R. (2014) World Echinoidea Database at http://www.marinespecies.org/echinoidea/aphia.php?p=taxdetails&id=212440 (2014).
12. Giese AC, Krishnaswamy S, Vasu BS, Lawrence J. Reproductive and biochemical studies on a sea urchin, *Stomopneustes variolaris* from Madras Harbor. Comp Biochem Physiol. 1964;13(4):367–80.
13. James DB. Ecology of intertidal echinoderms of the Indian seas. J Mar Biol Assoc India. 1982;24(1& 2):124–9.
14. Jayakody S. Provisional checklist of sea urchins (Echinodermata: Echinoidea) of Sri Lanka. In: Weerakoon DK, Wijesundara S, editors. The Nat red list 2012 of Sri Lanka. Conservation Status of the Fauna and Flora. Colombo, Sri Lanka: Ministry of Environment; 2012. p. 370–2.
15. James DB. Indian echinoderms their resources biodiversity zoogeography and conservation. Glim Aqu Biol. 2008;7:120–32.
16. James DB. Research on Indian echinoderms—a review. J Mar Biol Assoc India. 1983;25(1 & 2):91–108.
17. James P, Siikavuopio S. A guide to the sea urchin reproductive cycle and staging sea urchin gonad samples; 2011. p. 1–20. https://www.nofima.no/filearchive/guide-to-sea-urchins_lowres.pdf.
18. Salon NA. Echinoderm fisheries of the world: a review. Rotterdam: A A Balkema; 1985. p. 109–24.
19. Scheibling RE, Mladenov PV. The decline of the sea urchin, *Tripneustes ventricosus*, fishery of barbados: a survey of fishermen and consumers. Mar Fish Rev. 1987;49(3):62–9.
20. Mostafa HM, Collins KJ. Heavy metal concentrations in sea urchin tissues from Egypt, Ireland and United Kingdom. Chem Ecol. 1995;10(1–2):181–90.
21. Dincer T, Cakli S. Chemical composition and biometrical measurements of the Turkish Sea Urchin (*Paracentrotus lividus*, Lamarck, 1816). Crit Rev Food Sci Nutr. 2007;47(1):21–6.
22. European Commission. Commission Regulation (EC) No 1881/2006 of 19 December 2006 setting maximum levels for certain contaminants in foodstuffs. Off J Eur Union. 2006;L364:5–24.
23. Sumner J. Hazards affecting Australian seafood. In: Part 2 supporting information. Adelaide, Australia: SafeFish and the Australian Seafood Cooperative Research Centre; 2011. p. 46.
24. Jinadasa BKKK, Jörundsdóttir HÓ, Gunnlaugsdóttir H. Preparation of a standard operation procedure for validation of laboratory methods for trace metal analysis in seafood for national aquatic resources research and development agency (NARA), Sri Lanka. 2010.

7

Conservation-based breeding program design for genetic improvement in Gumz sheep in the western lowlands of Ethiopia

Y. Dagnew[1,2*], M. Urge[2], Y. Tadesse[2] and S. Gizaw[3]

Abstract

Background: Genetic improvement in livestock in developing countries like Ethiopia remains challenging; selection-based breeding schemes have the potential to provide resource-poor producers with access to improved animals that could ensure increased productivity of animals and hence contribute to reduced poverty. Therefore, the aim of this study was to compare alternative breeding program designs to improve productivity of Gumz sheep for sustainable utilization and conservation of genetic resource. In this paper, two village-based and two central nucleus-based sheep breeding schemes were modeled and evaluated in terms of their biological and economic efficiencies. The schemes were scheme-1: a village-based breeding scheme with existing lambing, scheme-2: a village-based scheme with improved lambing, scheme-3: central nucleus-based scheme with 5% nucleus size and scheme-4: central nucleus-based scheme with 10% nucleus size.

Results: The annual genetic gains per year in 6-month weight (kg) were differed across schemes and ranged from 0.154 to 0.171 in village-based scheme, and 0.334 to 0.336 in central-based schemes. The annual genetic gain per year in number of lambs born per ewe bred ranged from 0.0017 to 0.0036% in both village- and central nucleus-based scheme. The genetic gain in the proportion of lambs weaned per ewe was comparable across central nucleus-based scheme but little differed in village-based schemes and ranged from 0.0015 to 0.0016%.

Conclusion: Central nucleus-based schemes resulted in the highest genetic gain and economic efficiencies in all breeding objectives compared to village-based schemes. However, this scheme requires establishing a very large central nucleus flocks. Village-based scheme with estrous synchronization was the best option. The expected genetic gains and profits were higher as it promotes participation of farmers and achieving concentrated lambing which in turn increase selection intensity and genetic progress. This suggested the possibility for sustainable Gumz sheep improvement and conservation through village-based schemes and estrous synchronization.

Keywords: Conservation, Nucleus breeding, Sheep, Smallholder farmer, Village breeding

Background

Sheep production is an important component of livestock sector to address food insecurity and reduce poverty among smallholder farmers in the developing countries like Ethiopia. Because of their unique adaptation to marginal environments with low level of input, indigenous sheep contributes much for sustaining rural livelihoods by producing a wide range of products. They provide cash income, food (meat and milk), skin and manure for their keepers. They also serve as a means of risk mitigation during crop failures and monitory saving in addition to their sociocultural and ceremonial purposes [1, 2]. The demand for sheep mutton has increased due to increased human population and urbanization. Importance of sheep in the national economy of Ethiopia was also boosted by both domestic and international markets. Ethiopia possess many adapted indigenous sheep

*Correspondence: yohanesdagnew@yahoo.com
[2] School of Animal and Range Sciences, College of Agriculture and Environmental Science, Haramaya University, PO Box 138, Dire Dawa, Ethiopia
Full list of author information is available at the end of the article

breeds, which are important components of smallholder livestock production systems in Ethiopia [3, 4]. The sheep population of the country is diverse consisting of about 14 sheep types in six major groups and nine breeds [5].

Northwestern lowland of Amhara region is among the potential areas in livestock production and consequently in the supply of live animals to the Sudan market [6]. This offers an opportunity for farmers to increase their income by timely supplying and targeting appropriate market. This shows that sheep production could be one potential source for employment opportunity to alleviate poverty in the country. Therefore, the rising demand for live sheep or sheep product in and outside the country will necessitate producing market-oriented sheep breed through improvement in the current inefficient sheep production and marketing systems in the country.

The Gumz sheep breed, on which this study focused, is one of the indigenous sheep breed distributed in the western and northwestern lowlands of Ethiopia [7, 8]. Gumz sheep is resistant to disease with better survival rate, while having a substantial economic, social and cultural importance [7]. Gumz sheep are also superior in most of the reproductive traits as compared to the Rutana sheep breed [9]. However, it has lower growth performance and body weights compared to the exotic Rutana sheep. Rutana sheep was introduced from the Sudan due to its fast growth rate and big body size. These characteristics make the Rutana sheep and their crossbreds to be economically more important and have been influencing the farmers' opinions toward the Gumz sheep.

Even though, Gumz sheep breed is highly prolific and adaptable to the hot environmental condition, farmers are more interested in Rutana sheep and their crossbreds. However, this is seriously threatening the existence of indigenous Gumz sheep breed. Reports indicated that the population size of Gumz breed is declining [5]. Hence, the current drive for rapid livestock development through crossbreeding and the threat status of indigenous sheep [10] requires research to design and implement suitable breeding strategies so as to improve productivity and conserve indigenous breeds.

Lack of effective, sustainable breeding programs for local breeds is one reason that such breeds lose their competitive advantage, especially where production systems or external conditions are subject to change [11]. Conservation of Gumz sheep genetic resources could be imperative as these have been ensuring most to sheep genetic diversity in Ethiopia and reduce poverty among smallholder farmers. This requires optimal breeding program to conserve and improve productivity of Gumz sheep. Given the current harsh production circumstances and the potential of Gumz sheep to meet future changes in production conditions and production goals, the useful functional and adaptive traits provided by indigenous Gumz sheep could be secured through development of conservation-based breeding programs. The objective of this study was to compare alternative breeding program designs to improve productivity of Gumz sheep for sustainable utilization and conservation of genetic resource.

Methods

Study areas

The study was conducted in Metema and Quara districts of the Amhara Regional State of Ethiopia. Metema and Quara districts were purposively selected based on distribution of Gumz, Rutana and their crossbred's sheep breeds. Metema district is located at 900 km northwest of Addis Ababa and at 180 km west of Gondar. It has an altitude range of 550–1680 masl, mean range temperature of 22–43 °C and mean range annual rainfall of 850–1100 mm [7, 12] . Quara district is located at 1123 km northwest of Addis Ababa and 324 km southwest of Gondar. The district is situated at an altitude ranging from 528 to 654 masl. The mean annual rainfall of the area ranges from about 600 mm to around 1200 mm while the average daily temperature is 31°C with ranges from 19.2 to 41.1 °C minimum and maximum temperature, respectively [13] (Fig. 1).

Description of production and breeding system of Gumz sheep

The existed production system, production objective, breeding management, infrastructure and other relevant information were studied using participatory approaches [14]. A brief account of production system and breeding management obtained from the previous survey results is summarized below. A detailed description of production system is given in article published earlier [14].

Northwestern lowland of Amhara region is among the potential areas in livestock production and consequently in the supply of live animals for both domestic and export market (Sudan market) [6, 12]. Cattle, goat, sheep and donkey were the livestock species reared by smallholder farmers. The study area is characterized as a subsistence-oriented and low-input extensive production system. The smallholder farmers possess small size of livestock number mainly relays on family labor.

The most important sheep breeds reared in the study areas were the indigenous Gumz and the introduced Rutana sheep breed. In addition, there was a crossbred sheep between Gumz and Rutana sheep. The smallholder farmers primarily keep the indigenous Gumz sheep followed by the crossbreds. As reported earlier [14], Gumz sheep are kept under low-input and low-output production system. Gumz sheep have multipurpose functions and supporting the livelihoods of smallholder farmers

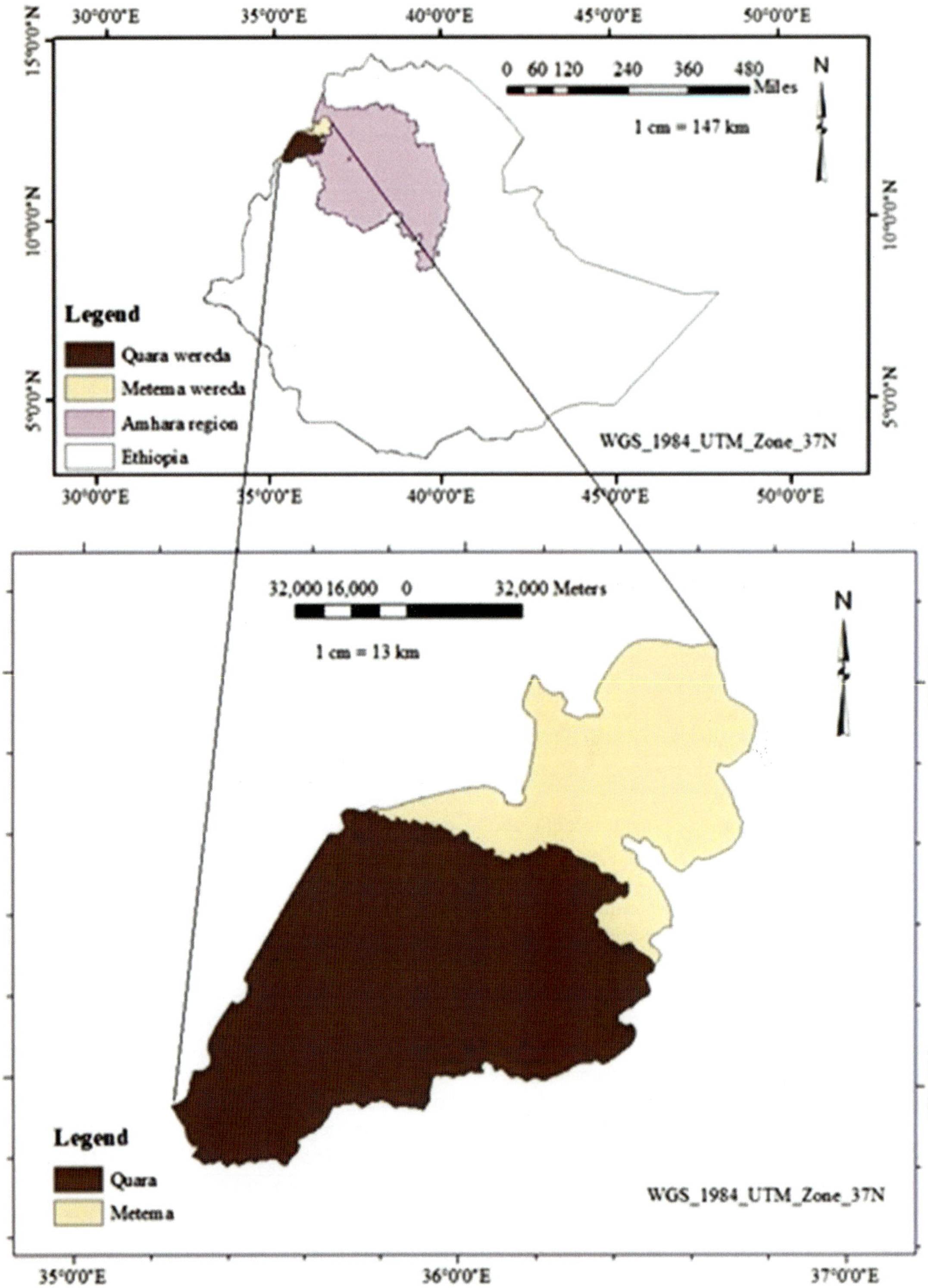

Fig. 1 Map of the study districts

by being financial and food source [14]. The typical flock sizes were small, male animals are either sold or slaughtered and their number remains small in the flock. The major type of herding practiced in the study areas was free grazing. Smallholder farmers herd sheep either together with goat or separately depending on the availability of hired labor. As reported by the farmers, diseases, labor shortage, stock theft, feed shortage and predator were the most important limiting factors for the sheep production in the area. Since farmers live with low level of education, there is no performance or pedigree recording. The involvement of other stakeholders in genetic improvement in local indigenous sheep genotype was very minimal.

Breeding goal and selection criteria

The Gumz sheep farmers breeding goal required for the design were generated from previous studies [7, 14]. The breeding-objective traits identified included 6-month weight (SWT), pre-weaning lamb survival (PWS) and litter size (LTS). Bio-economic models were constructed on Microsoft Excel spreadsheet to derive economic values of traits, considering both tangible and intangible benefits of rearing Menz sheep [15]. The Menz sheep-relative economic weight in previous studies was adopted for the current simulation study. The relative economic weights for SWT, PWS and LTS are presented in Table 2.

Selection of villages and alternative designs of breeding schemes

Multistage stratified sampling was employed to select kebeles purposively based on sheep breed distribution, sheep population and accessibility. Three kebeles were selected from each district representing smallholder production systems. Accordingly, a total of six kebeles (three from each district) representing smallholder systems were selected.

The villages/kebeles with the highest concentration of pure Gumz sheep in each district were selected to introduce the breeding program. The selection was based on the previous flock inventory results taken from each kebele/villages, secondary data from the district agriculture and rural development office, focus group discussions held with key informants, village leaders, researchers and livestock experts in the area. The criteria employed were high Gumz sheep possession, presence of communal grazing land in the villages, accessibility and willingness of the community in the village to participate in the breeding strategy.

A total of 14,057 sheep and ewes of 7591 were monitored from data available from the kebeles/villages. The breeding program was designed to serve this population. Six villages from Metema district (Shinfa, Kokit, Tumet, Achera, Agam-wuha and Kumer-afitete) and eight villages from Quara district (Bambawe, Bermile, Mehaded, Diza-Gumz, Kozera, Seliferedi, Daza and Kuza) were taken. Those kebeles/villages were selected purposely to trace the exact breeding tract of Gumz sheep.

Two main schemes were proposed for evaluating optimal breeding program: These were village-based schemes (scheme 1–2) and central nucleus-based schemes (scheme 3–4) which are being implemented by the Amhara Agricultural Research Institute (ARARI). Both schemes are being increasingly advocated for traditional production systems [16]. Each of the village-based and central nucleus-based schemes was evaluated under two scenarios: The schemes identified were village-based schemes (schemes 1–2) and central nucleus-based schemes (scheme 3–4; Table 1). The village-based scheme is supposed to be established in each cooperative breeding village, whereas the central nucleus-based scheme with one central nucleus flock is designed to address the whole Gumz sheep population in the selected villages. The four alternative schemes were simulated using ZPLAN computer program [17] developed for this purpose deterministically. The four alternative schemes are described below:

Scheme-1: village-based scheme with existing lambing distribution over a year. The distribution of lambing was recorded from a monitoring study (Fig. 2). Lambs born during the peak lambing season between July and September were considered to be available for selection every year. The whole village sheep population is involved in selection. The villagers select breeding rams from across all the flocks in the village taken as one big breeding flock and use the selected rams communally. The traditional year-round distribution of lambing could result in low selection intensity.

Scheme-2: Village-based scheme with improved lambing distribution: Through improved breeding management and estrous synchronization, it would be possible to concentrate lambing within a short period of time. The scheme was modeled with about 85% of the lambing to take place within a short period of time.

Scheme-3: A central nucleus-based scheme with 5% nucleus size: Scheme-3 is modeled with a nucleus size of 5% of the total population of ewes. A possible open-nucleus breeding scheme for the indigenous Gumz sheep consisting of a total of 360 breeding ewes and 15 rams. There will be 10 nucleus flocks and each nucleus flock will consist of 36 breeding ewes. The top 5% of the rams will be selected as the future sires for the nucleus flocks. The second top sires will be distributed to flocks of smallholder farmers participating in the breeding strategy project.

Table 1 Population, biological and economic parameters for four alternative breeding schemes for Gumz sheep

	Village-based schemes		Central nucleus-based schemes	
	Scheme-1	Scheme-2	Scheme-3	Scheme-4
Proportion of populations in				
Production unit	1.0	1.0	0.95	0.90
Central nuclei			0.05	0.10
Proportion of ewes lambing	0.48	0.80	–	–
Lifetime use (years) of:				
Rams in central nucleus	–	–	0.67	0.67
Rams in villages	2.5	2.5	2.5	2.5
Ewes in central nucleus			5	5
Ewes in villages	5.0	5.0	5	5
Mating ratio (F/M)—villages	35	35	35	35
Mating ratio—central nucleus	–	–	35	35
Lambing interval (years)	0.60	0.60	0.60	0.60
Conception rate—central nucleus	–	–	0.85	0.85
Conception rate—villages	0.85	0.85	0.85	0.85
Age at first lambing (years)	1.14	1.14	1.14	1.14
Twinning rate	1.32	1.32	1.32	1.32
Survival of rams—villages	0.85	0.85	0.85	0.85
Survival of rams—central nucleus	–	–	0.85	0.85
Survival of ewes—villages	0.85	0.85	0.85	0.85
Survival of ewes—central nucleus	–	–	0.85	0.85
Lamb weaning rate—village	0.85	0.85	0.85	0.85
Weaning rate—central nucleus	–	–	0.90	0.90
Suitability for breeding[a]	0.90	0.90	0.90	0.90
Fixed costs per ewe (Birr)	12.0	12.0	158.9	158.9
Variable costs per ewe (Birr)	12.05	12.05	11.37	11.37

[a] Proportion of proven selection candidates physically suitable for breeding

Scheme-4: A central nucleus-based scheme with 10% of nucleus size: Same as scheme-3, but the scheme was consisting of a total of 720 breeding ewes and 30 rams. Ewes will be divided into 10 flocks of 72 ewes/flocks by stratified randomization technique using body weight. About 10% of best rams each year should be selected at nucleus flocks for own use. The second best rams will be distributed to flocks of smallholder farmers participating in the breeding strategy project.

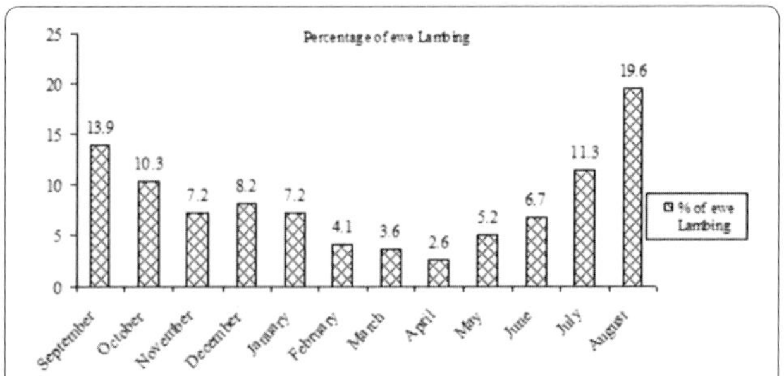

Fig. 2 Lambing distribution of Gumz ewes throughout the year

Table 2 Phenotypic standard deviations (σ_p), economic values, heritabilities along diagonal, genetic (above diagonal) and phenotypic (below diagonal) correlations used in simulated selection in Menz sheep *Source*: [18]

Traits	Economic values (birr)	Variance	SWT	LTS	PWS
SWT	34.5	4.64	0.29	0.09	0.10
LTS	60.2	0.012	0.09	0.13	−0.02
PWS	190.6	0.013	0.10	0.00	0.07

SWT six-month weight, *LTS* litter size, *PWS* pre-weaning survival

Input parameters for the breeding program

Essential input parameters for running ZPLAN are presented in Table 1. Some of the parameters required for the designs were generated from the production system, own flock ranking and on-farm monitoring studies [7, 9, 14].

Genetic and phenotypic parameters

The phenotypic standard deviations, the economic weight and the traits' heritabilities, genetic and phenotypic correlations are presented in Table 2. Since the genetic parameters are lacking for the Gumz sheep breed, the parameters estimated for Menz sheep are used [18]. Relative economic weight for each trait was based on the program for which breeding plan was designed and computed by standardizing the values with the additive genetic standard deviation (σ_A) following the guidelines of FAO [19].

Evaluation of alternative breeding programs

Alternative breeding schemes were evaluated using the computer program ZPLAN [17]. Based on the population, biological and economic parameters, ZPLAN

Table 3 Genetic gain per year in the breeding objective and its component traits achieved from selection using four alternative Gumz sheep breeding schemes

	Village-based schemes		Central nucleus-based scheme	
	Scheme-1	Scheme-2	Scheme-3	Scheme-4
Breeding objective[a]	5.71	6.31	12.29	12.37
SWT (kg)	0.154	0.171	0.334	0.336
LTS	0.0017	0.0018	0.0036	0.0036
PWS	0.0015	0.0016	0.0029	0.0029

[a] The annual monetary genetic gains for the breeding objective were calculated as a sum of the products of the respective genetic gains and economic weights of the component traits (*SWT*: six-month weight, *MWT* mature weight, *LTS* litter size and *PWS* pre-weaning survival)

calculates the annual genetic gain for the breeding objective, genetic gain for single traits and return of investment adjusted for costs (profit) using the gene flow method and selection index procedures.

Results

The annual monetary genetic gains of breeding-objectives traits are shown in Table 3 for the different breeding schemes. The result showed that the central nucleus-based schemes resulted in fastest genetic progress than those from village-based schemes. Monetary genetic gains of 12.29 and 12.37 for central nucleus-based schemes were predicted while the corresponding values for village-based schemes were 5.71 and 6.31 (Table 3). Monetary genetic gains for the central nucleus-based scheme in the breeding objective, which is the total gain from the three traits expressed in monetary units, were higher than the village-based schemes. This was because selection assumed accurate breeding values of the candidate rams, unlike in the village-based schemes where candidates were selected based on phenotypes. The highest genetic gain of 0.336 kg per year for 6 month's weight was predicted from a larger central nucleus-based scheme (scheme-4) but the profit was lower than the smaller central nucleus-based scheme (scheme-3) since the discounted costs of maintaining the larger nucleus flock were higher (15.89) than the smaller nucleus flock (7.94). Among the village-based schemes, the highest value of 0.171 kg annual genetic gain of the 6-months weigh was simulated from scheme-2 whereas the lowest value of 0.154 kg was simulated from scheme-1. The results for the two village-based schemes showed that both the monetary genetic gains and profits were highest if most of the ewes (about 85%) were synchronized to lamb in restricted months rather than throughout the year.

The lowest annual genetic gains of twining rate with the range of 0.0017–0.0036% were predicted from all schemes. Relatively lower annual genetic gains of 0.0017 and 0.0018% were predicted from scheme-1 and 2, respectively. The annual genetic gain of lamb survival at different schemes ranged from 0.0015 to 0.0029%. Relatively higher annual genetic gain of 0.0029% survival rate was predicted from central nucleus-based schemes.

The annual returns, costs and profit per ewe of the population for different alternative schemes are presented in Table 4. The highest annual return of 64.97 Birr/ewe/year for scheme-4 was predicted from central nucleus-based scheme while the lowest 41.65 was obtained in scheme-1 from village-based schemes (Table 4). The annual costs were calculated per ewe in the whole population. These costs were higher for village-based schemes than central nucleus-based schemes. This is because the cost for the central nucleus-based scheme was distributed throughout the whole population of about 7200 ewes. The profit/ewe/year was in the range of 18.51 Birr in scheme-1 to 49.08 Birr in scheme-4. Among the village-based schemes, scheme-2 was the profitable design but it was less profitable compared to the central nucleus-based schemes. Scheme-3 was more profitable (50.57 Birr/ewe) than the other central nucleus-based scheme (scheme-3).

Discussion

All scenarios simulated as alternatives had advantages in terms of genetic gain and profit. Results show that a slow genetic progress was obtained in village-based scheme with existing lambing distribution over a year (scheme-1) compared to village-based scheme with improved lambing distribution (scheme-2). This was because only few selection candidates would be available in each round of selection due to dispersed lambing resulting in low selection intensity and thus slow genetic progress. A study in the Ethiopian highlands [20] also reported that year-round distributed lambing could have implication in

Table 4 Returns, costs and profits per ewe per year (Birr) obtained from selection in Gumz sheep using four alternative breeding schemes

	Village-based schemes		Central nucleus-based scheme	
	Scheme-1	Scheme-2	Scheme-3	Scheme-4
Return/ewe/year	41.65	46.03	58.52	64.97
Cost/ewe/year	23.10	23.14	7.94	15.89
Fixed costs per ewe	11.27	11.27	7.38	14.77
Variable costs	11.83	11.87	0.56	1.12
Profit/ewe/year	18.51	22.88	50.57	49.08

genetic improvement and flock productivity. This is due to the dispersed lambing and small number of candidates which lead to lower selection intensity and hence slow genetic progress.

Village-based scheme with estrous synchronization could be another alternative for achieving concentrated lambing with a short period of time. Estrous synchronization would help to increase the number of lambs and help in adjusting lambing time to a better season [20]. Currently, introduction of estrous synchronization and artificial insemination (AI) in village flocks has been suggested for achieving planned and concentrated lambing [21, 22]. In the present study, faster genetic progress and profit from village-based scheme was predicted through improved breeding management and estrous synchronization. Similarly, a previous study on mature Merino ewes [23] reported 54.32% lambing rate over natural synchronized lambing after cervical artificial insemination following a progestagen (MAP)-PMSG (375IU) treatment. In these scenarios, higher selection intensity is exploited to result in higher predicted annual genetic gain and profit. Another study [20] also reported that estrous synchronization had 17.1–25.2% more lambing over natural synchronized lambing. The advantage of estrous synchronization over selection under natural estrous synchronization was also given 160.5–190.1% more profit [20]. Estrous responses to hormonal synchronization ranging from 55 to 93.2% have been reported for Ethiopian sheep breeds [24, 25]. A review [26] of estrous responses in Nigerian goat breeds showed a range from 20 to 100% success.

The genetic gain in 6-month weights in village-based scheme with improved lambing distribution was highest compared to village-based schemes with existing lambing distribution over a year (Table 3). By improving breeding management and introducing estrous synchronization into the current breeding practices, genetic gains and profits from scheme-2 were relatively higher than those from scheme-1. This implies that village-based scheme with improved lambing (scheme-2) is a better alternative breeding scenario to scheme-1 in the village-based breeding programs. The predicted genetic gain in 6-month weight ranged from 0.154 to 0.171 kg. These values were found in the ranges of genetic gain of 0.120 to 0.179 kg in 6-month weight of Menz sheep [21]. The genetic gain of 6-month weight predicted in this study is lower than the range of the predicted annual genetic gain of 6-month weight of western lowland and Abergelle goats [27].

Very low genetic gains of twining rate were predicted from all alternatives ranging from 0.0017 to 0.0036%. The differences of genetic gain of litter size between different alternatives were very small. The genetic gains of lamb survival at different scenarios ranged from 0.0015 to 0.0029%. The present study revealed that genetic progress in litter size and lamb survival was low and comparable among village-based schemes and central nucleus-based schemes. Compared to the current study, lowest genetic gain results for lamb survival with the range of 0.009–0.01% were predicted for Ethiopian Afar sheep breed [28]. The low genetic gain in reproductive traits is attributed to their low heritability and genetic correlations with growth traits [28–30] in the breeding objective.

The central nucleus-based scheme with 10% nucleus size (scheme-4) had most efficient genetic gain compared to central nucleus-based scheme with 5% nucleus size (scheme-3), but it was less profitable. The highest genetic gain in 6-month weigh was obtained from scheme-4, as scheme-4 benefited from higher selection intensity compared to scheme-3. Similarly, previous study [31] reported that applying higher selection intensity would result in even higher genetic gain on the one hand and higher discounted returns on the other hand. The predicted genetic gain in 6-month weight (0.336 kg) of central nucleus-based scheme was comparable to those of [15, 32] who reported 0.29 and 0.34 kg genetic progress, respectively, in 6-month weight of nucleus flocks of Menz sheep. Comparison of central nucleus-based schemes and village-based schemes showed that a highest genetic gain was predicted from central nucleus-based schemes. The study by [30] also noted that central nucleus-based scheme with selection on breeding values had an advantage over a village-based scheme where selection was on candidates' phenotypes. The central nucleus-based scheme was also highest in profit compared to the village-based schemes. However, some central nucleus-based breeding programs have failed under smallholder production systems [33, 34]. The failure of central nucleus-based breeding programs could be due to technical limitations, absence of infrastructures to support the breeding programs [1].

Conclusion

The central nucleus-based scheme resulted in the highest genetic gain and economic efficiencies compared to village-based schemes. However, this scheme requires establishing a large central nucleus flocks. Running such a large nucleus flock may not be sustainable due to the long-term financial support and commitment required. Village-based scheme is the best alternative breeding scenario that needs to be optimized for genetic improvement in Gumz sheep. This study result suggested scheme-2 (village-based scheme with estrous synchronization) is better than scheme-1. The possibility of introducing hormonal estrous synchronization in village flocks is promising, as the results indicate effectiveness of the technology.

Abbrevaitions

AI: artificial insemination; ARARI: Amhara Agricultural Research Institutes; LTS: litter size; PMSG: pregnant mare serum gonadotropin; PWS: pre-weaning lamb survival; SWT: six-month weight.

Authors' contributions

YD and SG collected, analyzed and interpreted the data. SG, MU and YT helped to draft the manuscript. All authors read and approved the final manuscript.

Author details

[1] Department of Animal and Range Sciences, College of Agriculture, Bule Hora University, PO Box 144, Bule Hora, Ethiopia. [2] School of Animal and Range Sciences, College of Agriculture and Environmental Science, Haramaya University, PO Box 138, Dire Dawa, Ethiopia. [3] Herd Health, Livestock Expert, International Livestock Research Institute, PO Box 5689, Addis Ababa, Ethiopia.

Acknowledgements

I would like to thank Ministry of Education (MOE) for the scholarship and financial support. I am very grateful to all assistances, knowledge and experiences we have received from the farmers in Metema and Quara districts.

Competing interests

The authors declare that they have no competing interests.

Funding

This research was conducted with the financial support of Ethiopian Ministry of Education (MOE).

References

1. Kosgey IS, Baker HMJ, Udod JAM. Successes and failures of small ruminant breeding programmes in the tropics: a review. Small Ruminat Res. 2006;61:13–28.
2. Getachew T, Haile A, Tibbo M, Sharma AK, Sölkner J. Herd management and breeding practices of sheep owners in a mixed crop–livestock and a pastoral system of Ethiopia". Afr J Agric Res. 2010;5(8):685–91.
3. Haile A, Wurzinger M, Mueller J, Mirkena T, Duguma G, Mwai MA, Sölkner J, Rischkowsky B. Guidelines for setting up community-based sheep breeding programs in Ethiopia. 2011.
4. Edea Z, Haile A, Tibbo M, Sharma AK, Sölknerand J, Wurzinger M. Sheep production systems and breeding practices of smallholders in western and south-western Ethiopia: implications for designing community-based breeding strategies. Livest Res Rural Dev. 2012;24:7.
5. Gizaw S. Sheep resources of Ethiopia: genetic diversity and breeding strategy. Ph.D. thesis, Wageningen University, Netherlands; 2008.
6. Mulugeta E, Gebremedhin B, Hoekstra D. Analysis of the Ethio-Sudan Cross-Border Cattle Trade. IPMS (improving productivity and market success) of Ethiopian farmers project working paper 2. ILRI (International Livestock Research Institute), Nairobi, Kenya; 2007.
7. Abegaz S, Hegde BP, Taye M. Growth and physical body characteristics of Gumuz sheep under traditional management systems in Amhara Regional State, Ethiopia". Livest Res Rural Dev. 2011;23:1–5.
8. Lemma S. Phenotypic characterization of indigenous sheep breeds in the Amhara national regional state of Ethiopia. M.Sc. thesis, Haramaya University, Ethiopia; 2009.
9. Dagnew Y, Urge M, Tadesse Y, Gizaw S. Growth performance of Gumz, Rutana and Gumz-Rutana crossbred sheep under on-farm conditions in north western lowlands of Amhara, Ethiopia. East Afr J Vet Anim Sci. 2017;2(1):1–13.
10. Gizaw S, Getachew T, Edea Z, Mirkena T, Duguma G, Tibbo M, Rischkowsky, B., Mwai, O., Dessie T, Wurzinger, M., Solkner, J and Haile A. Characterization of indigenous breeding strategies of the sheep farming communities of Ethiopia. A basis for designing community-based breeding programs. International Center for Agricultural Research in the Dry Areas (ICARDA); 2013. 40 pp.
11. Hiemstra SJ, Drucker AG, Tvedt M, Louwaars N, Oldenbroek JK, Awgichew K, Abegaz S, Bhat PN. What's on the menu? Options for strengthening the policy and regulatory framework for the exchange, use and conservation of animal genetic resources. Anim Genet Resour. 2007;4(3):65–74.
12. Desalew T, Tegegne A, Nigatu L. Rangeland Condition and feed resources in Metema District, North Gondar Zone, Amhara Region, Ethiopia. IPMS (improving productivity and market success) of Ethiopian farmers project working paper 25. Nairobi, Kenya, ILRI; 2010.
13. Habtamu T, Bekele A. Habitat association of insectivores and rodents of Alatish National Park, northwestern Ethiopia. Trop. Ecol. 2008;49:1–11.
14. Dagnew Y, Urge M, Tadesse Y, Gizaw S. Sheep production and breeding systems in north western lowlands of Amhara region, Ethiopia: implication for conservation and improvement of Gumz sheep breed. Open J Anim Sci. 2017;7(2):179–97.
15. Gizaw S, van Arendonk JAM, Valle-Zárate A, Haile A, Rischkowsky B, Dessie T, Okeyo AM. Breeding programs for smallholder sheep farming systems: I. Evaluation of alternative designs of breeding schemes. J Anim Breed Genet. 2014;131(5):341–9.
16. Wuletawu Z, Ayalewu W, Solkiner J. Breeding scheme based on analysis of community breeding objectives for cattle in north-western Ethiopia. Ethiop J Anim Prod. 2006;6(2):53–66.
17. Willam A, Nitter G, Bartenchlager H, Karras K, Niebel E. ZPLAN mannual for a PC-program to optimize livestock selection schemes. Stuttgart: Institute of Animal Production in the Tropics and Subtropics. Universität Hohenheim; 2008.
18. Gizaw S, Joshi BK. Estimates of genetic parameters for growth traits in Menz and Awassi x Menz crossbred sheep in Ethiopia. Indian J. Anim. Sci. 2004;74(8):864–7.
19. FAO. Breeding strategies for sustainable management of animal genetic resources. FAO animal production and health guidelines. No. 3. Rome; 2010.
20. Gizaw S, Tesfay Y, Mekasha Y, Mekuriaw Z, Gugsa T, Ebro A, Gebremedhin B, Hoekstra D. Hormonal oestrus synchronization in four sheep breeds in Ethiopia: impacts on genetic improvement and flock productivity. LIVES working paper 25. Nairobi, Kenya: International Livestock Research Institute (ILRI); 2016.
21. Galukande EH, Mulindwa H, Wurzinger M, Roschinsky R, Mwai O. Crossbreeding cattle for milk production in the tropics: achievements, challenges and opportunities. J Anim Genet Resour. 2013;52:111–25.
22. Getachew T, Haile A, Wurzinger M, Rischkowsky B, Gizaw S, Abebe A. Review of sheep crossbreeding based on exotic sire and among indigenous breeds in the tropics: an Ethiopian perspective. Afr J Agric Res. 2016;11(11):901–11.
23. Simonett L, Ramos G, Gardon JC. Effect of estrus synchronization and artificial insemination on reproductive performance of Merino sheep. Braz J Vet Res Anim Sci. 2002;39(3):143–6.
24. Mekuriaw Z. Effect of time of pregnant mare serum gonadotrophin administration on oestrus synchronization efficiency and fertility in black-head ogaden ewes, Ethiop. J Anim Prod. 2009;9:135–45.
25. Mekuriaw Z, Assefa H, Tegegne A, Muluneh D. Estrus response and fertility

of Menz and crossbred ewes to single prostaglandin injection protocol. Trop Anim Health Prod. 2016;8(1):53–7.

26. Omontese BO, Rekwot PI, Ate IU, Ayo JO, Kawu MU, Rwuaan JS, Nwannenna AI, Mustapha RA, Bello AA. An update on oestrus synchronisation of goats in Nigeria. Asian Pac J Reprod. 2016;5(2):96–101.
27. Abegaz S, Sölkner J, Gizaw S, Dessie T, Haile A, Mirkena T, Getachew T, Wurzinger M. Optimizing alternative schemes of community-based breeding programs for two Ethiopian goat breeds. Acta Agrar Kaposváriensis. 2014;18(1):47–55.
28. Mirkena T, Duguma G, Willam A, Wurzinger M, Haile A, Rischkowsky B, Okeyo MA, Tibbo M, Sölkner J. Community-based alternative breeding plans for indigenous sheep breeds in four agro-ecological zones of Ethiopia. J Anim Breed Genet. 2012;129:244–53.
29. Safari E, Fogarty NM, Gilmour AR. A review of genetic parameter estimates for wool, growth, meat and reproduction traits in sheep. Livest Prod Sci. 2005;92:271–89.
30. Gizaw S, Rischkowsky B, Valle-Zarate A, Haile A, Van Arendonk JAM, Okeyo AM, Dessie T. Breeding programs for smallholder sheep farming systems: I. Evaluation of alternative designs of breeding schemes. J Anim Breed Genet. 2014;131:341–9.
31. Jembere T, Kefelegn K, Rischkowsky B, Haile A, Okeyo MA, Dessie T. Recording birth weight has no significance in village based genetic improvement programs of small ruminants. Livest. Res. Rural Dev. 2016;28:8.
32. Gizaw S, Lemma S, Komen H, van Arendonk JAM. Breeding programmes for smallholder sheep farming systems: II. Optimization of cooperative village breeding schemes. J Anim Breed Genet. 2007;14:1–8.
33. Kosgey IS, Van Arendonk JAM, Baker RL. Economic values for traits in breeding objectives for sheep in the tropics: impact of tangible and intangible benefits. Livest Prod Sci. 2004;88(1–2):143–60.
34. Singh SK, Singh MK, Rai B. Evaluation of different goat production systems in India and breeding strategy for improvement. In: Proceeding of eighth national conference on animal genetics and breeding; national livestock breeding policy, 8–10 March 2005, Central Institute for Research on Goats, Makhdoom, Farah-281 122, Mathrua (UP) India.

8

Determinants of crop–livestock diversification in the mixed farming systems: evidence from central highlands of Ethiopia

Wuletaw Mekuria[1*] and Kindu Mekonnen[2]

Abstract

Background: This paper devotes to determinants of crop–livestock diversification in the mixed farming systems in the *Gudo Beret* watershed, central highlands of Ethiopia. In the highland agro-ecologies, the mixed farming-dominated agriculture is persistently confronted with food insecurity, feed scarcity and land degradation. During heavy rains or drought periods, farmers become vulnerable to food and feed shortfalls. To fill such gaps, the government of Ethiopia has made efforts to enhance the productive capacity of soils and thereby increase productivity on diversified farms in different agro-systems. Anthropogenic and natural calamities adversely affect the livelihoods of smallholders in general and crop–livestock systems in particular. The aim of this study is to examine the determinants of crop–livestock diversification and evaluate the link between farm activities and household food supplies.

Methods: Both qualitative and quantitative data were collected from 211 randomly selected household-heads in *Gudo Beret* watershed, central highlands of Ethiopia. Descriptive statistics, inferential tests, and econometric models were employed for analysis.

Results: Tobit model results revealed that livestock holding is positive and significantly influence the extent of crop–livestock diversification at 1% significant level. Extension contact and irrigated land are also positive and significantly influence the extent of diversification at 5% level, while land rent-out, improved seed and soil fertility status are negative and significantly influence the crop–livestock diversification at 1, 5, and 5% significant levels. Ordinary least square regression results show that barley, wheat- and faba bean-based farming have positive correlation and significant effects on the household food energy, while vegetables and highland fruits production have negative correlation with the household food supply and significant at 1% level.

Conclusions: Crop–livestock diversification is the best potential livelihood strategy for farm households. Barley, wheat and faba bean are the most dominant crops that have immense contribution for human nutrition and crop residue mainly used to feed livestock. Therefore, crop–livestock producers, agricultural practitioners, the Ethiopian government and partner organizations need to give due attention for agricultural extension service, investment for supplementary irrigation and integrated mixed crop–livestock systems.

Keywords: Crop–livestock diversity, Margalef index, Mixed farming, Tobit model, Watershed

*Correspondence: wuletaw.mekuria@aau.edu.et
[1] Centre for Rural Development, College of Development Studies, Addis Ababa University, P. o. b 1176, Addis Ababa, Ethiopia
Full list of author information is available at the end of the article

Background

Mixed farming is a process by which farmers produce crops and rear livestock simultaneously to ensure sustainable agriculture [1, 2]. The production of one or more crops and livestock on available resources is crop–livestock diversification [3]. Diversified farms can ensure food security, conserve biodiversity, improve dietary preference, increase household income, reduce vulnerability to shocks and create job opportunity [4–6]. Mixed production systems also enhance land productivity and improve water use efficiency [4, 7]. Farm diversification is a means to minimize risks and insurance against crop failures [4, 5, 8, 9]. Various farm activities provide a wide range of responses to uncertain conditions and increase household income and resilience [4, 10]. Households who practice crop–livestock systems have improved 50% of productivity and farm income in the highlands of Ethiopia compared to smallholders that only raise crops [4].

Many research findings maintain the theory of agricultural intensification [1, 11–14]. However, intensified farming and continuous cultivation with limited soil amendments and conservation practices resulted in soil erosion and nutrient depletion [15]. Although the experience of crop–livestock integration is an evidence for agricultural intensification, specialized cereal cultivation could not generate desired benefits for the livelihoods of households in Ethiopia [16]. Other researchers have tried to mediate the debates between farm diversification and intensification [2, 10, 17, 18]. They have suggested that agricultural intensification is an appropriate strategy for large-scale, location specific and capital-intensive enterprises. It is also possible to intensify farm activities without reduction of crop and livestock components. For instance, use of high yielding varieties, fertilizers, pesticides along with soil and water conservation practices are considerable shifts from extensive traditional farming to intensive agriculture in the highlands of Ethiopia.

Subsistence consumption, source of household income, animal feed, fuel energy, export earnings, national returns and improvement of soil fertility are the major contributions of crops. Crop production is heavily dependent on climate variables, seeds, water, soil nutrients, biodiversity and technical knowledge of farmers [19]. Livestock also provide food, fuel, manure, draught power, offspring, sociocultural benefits, and transport services [20]. Ethiopia has a large livestock resource being the first in Africa and the tenth in the world [21]. In Sub-Saharan Africa, the country has the largest livestock density after Uganda in terms of cattle, sheep and goat [22]. Livestock contribute about 13–16% of export earnings [23], and 7.9% of the total gross domestic product to the national economy [24]. Maintaining the proper stocking rates on pasturelands can optimize livestock production and maximize animal health and its performance [25].

Despite mixed farming is one of the main livelihood strategies of farmers, Ethiopia has been challenged with anthropogenic and natural factors. Boserup [26] and Godfray et al. [19] confirmed that the intensity of food supply has positive correlation with population growth at different rates and increasing trends. However, in Ethiopia, the trend of production growth is immensely uneven compared to the growth of human population due to erratic nature of rainfall and susceptible to weather shocks. Favorable weather conditions of a year result in bumper harvest, whereas severe drought is an evident for crop failure that adversely affects agricultural production [16, 24]. Population growth aggravates the competitive demands between crop production and livestock husbandry [15]. High stocking rate, low farm productivity and land degradation are the key problems in the high lands of Ethiopia [27–29].

Land degradation, food insecurity and feed scarcity are still adversely affecting the livelihoods of farmers and landscape situations of the study area [10, 30–32]. Farmlands could not produce adequate food for the rising population and exhibited large rates of malnutrition [5]. Food insecurity and lose of natural vegetation will intensify underfeeding unless increasing pressures and competing demands over the land are wisely managed [33]. Despite significant efforts have been made on mixed farming in managing production risks, studies on farm diversification in Ethiopia is minimal and largely focused on crops and incomes. To the best of our knowledge, there has been no study conducted on crop–livestock diversification in central highlands of Ethiopia. Thus, this research could provide information on determinants of diversification and its contribution for household food security. Therefore, the objectives of this study were to identify factors that influence the probability and extent of farmers' decision on crop–livestock diversity and to examine the link between farm diversification and food supply in response to various adverse effects in the *Gudo Beret* watershed, central highlands of Ethiopia.

The next sections of the paper describe methods including area description, sampling procedures, source of data, methods of data collection, and specifications of the theoretical and empirical models. Demographic characteristics of the interviewed household-heads, regression estimates and the link between crop–livestock diversification and household food security are reported in the results and discussion parts. The paper also concludes key findings with policy recommendations.

Methods

Description of the study area

The study was conducted in *Gudo Beret* watershed of the *Basona* district in central highlands of Ethiopia (Fig. 1). The watershed is geographically located in latitudes between 9°76′ and 9°81′ North, and longitudes 39°65′ and 39°73′ East at a distance of 160 km Northeast of Addis Ababa and 32 km Northeast of *Debre Berhan* town on the way to *Dessie*. More than 90% of the topography has medium and steep slopes. The altitude of the study area ranges between 2828 and 3700 m above sea level elevated from west to east. The dominant land use/cover types were cropland and woodlot followed by grazing land and bush lands. *Regosols*, *Cambisols*, and *Lithosols* were the three dominant soil classes that exist in the watershed [31].

The watershed is moist and cold climate zone characterized by 55.7% of *Dega*, and 44.3% of high *Dega* agro-ecologies. The mean daytime temperature ranges between 2.4 °C during winter and 19.2 °C during summer seasons. The watershed has bimodal rainfall pattern with annual rainfall between 1278 and 2060 mm. The temperature has increased by 0.13 °C between 1995 and 2014, while the average annual rainfall has declined by 4.9 mm in those years. Household in the watershed practice predominantly mixed crop–livestock activities. Barley, faba bean, field pea, wheat, vegetables, potatoes, lentil, linseed, and oats were the main crops grown, while sheep, cattle, and donkey were major livestock types reared in the study area.

Sampling procedures

Two-stage sampling techniques were employed. In the first stage, the study watershed was selected purposively owing to the high livestock density and production potential for the mixed crop–livestock systems. In the second stage, 211 household-heads were selected by systematic random sampling technique in probability to proportional size. The sample size was determined and respondents were selected within the sampling frame of the study population. Interview schedule was prepared and household interview was conducted from May to July 2016.

Sources of data and methods of data collection

The primary cross-sectional data were collected from selected households through structured interview that were administered by trained enumerators. The collected data include demographic characteristics, crop species, livestock breeds, household income, biophysical features,

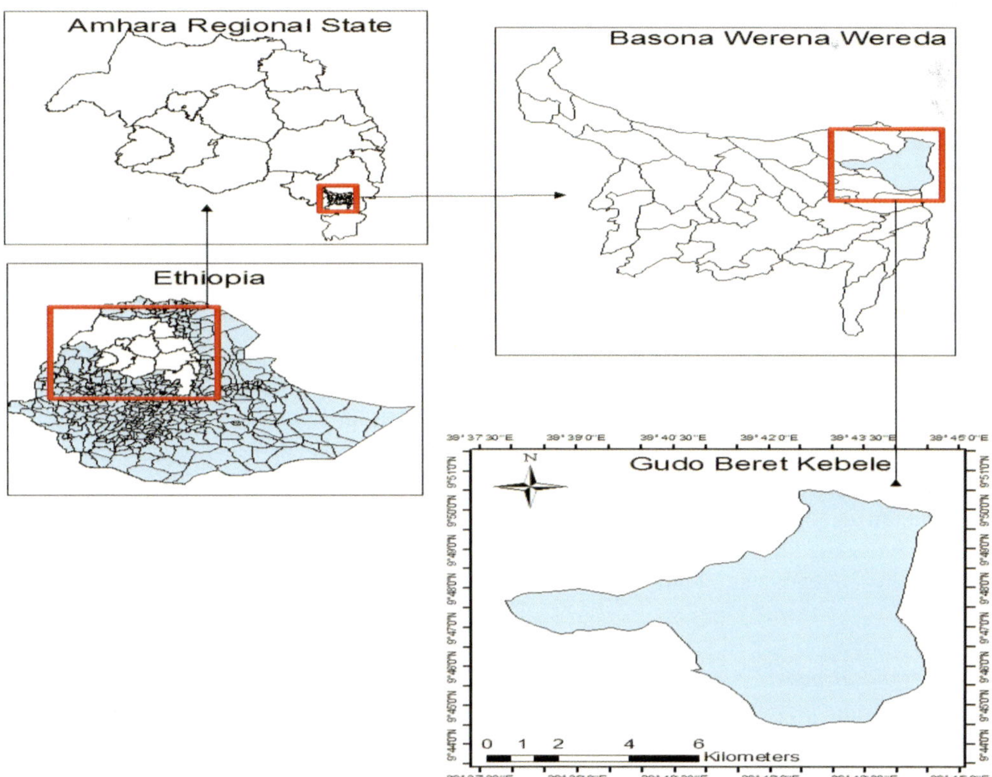

Fig. 1 Location maps of the study watershed

infrastructures, institutions, and socioeconomics. Secondary data such as agro-climates and other supportive information were collected from office of agriculture and related empirical literatures.

Methods of data analysis

Theoretical model

The utility or profit maximizing behavior was employed to analyze decisions of households [34, 35]. The assumption is that household-heads choose practices when only they want to maximize utility using diversified farms greater than monoculture. Following Greene [36], the latent utility of a *i*th farmer for the *j*th diversification process is denoted by U_{ji}, where $j=1$, otherwise 0 if the *i*th farmer is diversify or not.

$$D_j = \beta_j X_i + \varepsilon_j \quad \text{and} \quad D_k = \beta_k X_i + \varepsilon_k \tag{1}$$

D_j and D_k are perceived attributes of diversified farms *j* and *k*; X_i is the vector of independent variables that influence the perceived interest of the farms, β_k and β_j are parameters to be estimated and ε_j and ε_k are error terms assumed to be independently and identically distributed. If a farmer decides to use option *j*, it follows that the perceived utility from option *j* is greater than the utility from other options (say *k*) illustrated as:

$$D_{ij}(\beta_j x_i + \varepsilon_j) > D_{ik}(\beta_k x_i + \varepsilon_k), \quad k \neq j \tag{2}$$

The probability that a farmer will use option *j* among the set of diversified farm activities could be defined as:

$$\begin{aligned} &P(Y = {}^1/_x) = P(D_{ij} > D_{ik}) \\ &P(\beta_j x_i + \varepsilon_j - \beta_k x_i - \varepsilon_k > 0/x) \\ &P(\beta_j x_i - \beta_k x_i + \varepsilon_j - \varepsilon_k > 0/x) \\ &P(x^* x + \varepsilon^* > 0/x) = F(\beta^* x_i) \end{aligned} \tag{3}$$

where *P* is the probability function, D_{ij}, D_{ik} and X_i are as defined above, $\varepsilon^* = (\varepsilon_j - \varepsilon_k)$ is a random disturbance term, $\beta^* = (\beta_i - \beta_j)$ is a vector of unknown parameters that can be interpreted as a net influence of a vector of independent variables influencing the decision to use farm diversities, and $F^*(\beta^x X_i)$ is a commutative distribution function of ε^* evaluated at $(\beta^x X_i)$. Based on this theoretical assumption, an empirical model was employed to analyze factors influencing farmers' decision on probability and extent of diversification as an important livelihood strategy.

Specification of the empirical model

Descriptive statistics (frequency, mean, standard deviation), inferential tests (*Chi-square test* and *t* test), and econometric models were used to analyze the collected data. Descriptive methods were employed to describe households, livestock types, crop species, and diversification indices. Subsequent to descriptive analysis, the Tobit econometric model was employed to identify determinants of crop–livestock diversification. In addition, ordinary least square was employed to identify farm activities that influence dietary food energy. Excel, Statistical Package for Social Sciences (SPSS) version 20 and Stata Version 12 were used for data analysis.

Measuring crop–livestock diversification In order to compute a particular crop and livestock of interest, we adapted index for crop–livestock diversification. Margalef index was employed to measure crop–livestock diversification owing to some advantages compared to other indices. The index has a good discriminating ability and better goodness of fit [37]. It can capture different units of animal and plant species [38–41]. The formula was adapted in Magurran [42].

$$D_{\text{Mg}} = \frac{S-1}{\ln(N)} \tag{4}$$

where D_{Mg} is Margalef index; *S* is the total number of each crop species planted and livestock types reared by a household; *N* is number of individuals, i.e., cropland and total count of animals held by a household; ln is the natural logarithm.

If the value of Margalef index is zero, a household tends to specialize in very few numbers of crop species per limited land or number of animals per total population. In other words, the value zero indicates complete specialization and a value greater than zero denotes some measures of diversification [39]. Ten types of crops (barley, wheat, faba bean, field pea, lentil, linseed, vegetables, potato, oats, and highland fruits) and eleven livestock types (cow, ox, bull, heifer, calf, sheep, goat, horse, donkey, mule, and poultry) were commonly practiced in the study area. The Margalef index is left censored for the value of *N* equal to one because any number divided by the natural logarithm, ln(1), is undefined (Eq. 4). In order to avoid such problems, crops grown were computed in terms of meter square and number of animals in number.

Determinants of crop–livestock diversification The generated information in modeling supports household decisions and serves for early warning systems in agriculture. Logit and Probit econometric models explain only probabilities of independent variables on non-continuous dependent variables [14]. Ordinary least square regression also only ascertain relationships between many independent and single continuous dependent variables, but one could not be sure about the underlying casual mechanisms [43]. The linear regression (ordinary least square—OLS) model would apply if all households diversified their crop–livestock activities.

In the study area, not all households participated in diversified farm activities. Hence, ordinary least square was assumed to create a sample selectivity bias because the model can exclude non-diversifiers from analysis [39]. Therefore, Tobit model is an appropriate model for censored data that gives joint decisions of probability and intensity as the same time [44–46], regardless of the number of censored observations.

The concern of this study was analysis of diversification among diversifiers and non-diversifiers. If the dependent variables were two or more diversification patterns, the most appropriate model could be a multivariate seemingly unrelated regression model. In the study area, the majority of households were diversifiers, while the least part of households were non-diversifiers due to several reasons. Examining factors among diversifiers and non-diversifiers help to suggest solutions for policymaking. The limited dependent variable regression model (Tobit) was specified to estimate the joint estimates of factors affecting crop–livestock diversification and a set of independent variables. The formula is adapted in Greene [36].

$$\begin{aligned} Y_i^* &= \beta x_i + u_i \\ y_i &= y_i* \quad \text{if} \quad y_i^* > 0 \\ &= 0 \quad \text{if} \quad y_i^* \leq 0 \end{aligned} \tag{5}$$

Y_i is observable and y^* is latent dependent variable. A latent variable can be observable whenever it is positive. Once the latent variable is negative, the observation becomes censored, and one can simply observe, $y=0$. In this study, the data are left-censored. The subscript i run from 1 to n which was used to index the observations of a sample. The total number of observations denoted by n. X_i is the vector of independent variables, β is a vector of unknown coefficient, and u_{i} is an independent distributed error term or unobservable variable that affect y^*assumed to be normal with a mean zero and constant variance. However, y_i^* is observed if $y_i^*>0$ and is not observed if $y_i^* \leq 0$. Then, the observed y_i is defined as:

$$\begin{aligned} Y_i = y_i^* &= \beta x_i + u_i; \quad \text{if} \quad y_i^* > 0 \quad i = 1, 2, 3 \\ &0 \quad \text{if} \quad y_i^* \leq 0 \end{aligned} \tag{6}$$

where Y_i is observed dependent variable, in this case the value of diversity; X_i is independent variables affecting the dependent variable and its intensity; β is coefficients, and u_i are residuals that are independently and normally distributed with mean zero and a common variance. The model parameters were estimated by maximizing the Tobit likelihood function.

$$L = \prod_{y>0} f \frac{1}{\sigma} \left(\frac{y_i - \beta x_i}{\sigma} \right) \prod_{y \leq 0} F \left(-\frac{\beta x}{\sigma} \right) \tag{7}$$

where f and F are the distribution and density function of y_i^*, respectively; $\prod_{y \leq 0}$ is product over those i for which $yi* \leq 0$ and $\prod_{y>0}$ is the product over those i for which $yi* > 0$.

In the model, the independent variables were computed in terms of probability and intensity marginal effects. The change in explanatory variables (X_i) affects the conditional mean of Y_i in the positive part of the distribution and the probability that the observation fallen in that part of the distribution. The marginal effect on the latent dependent variable is expressed as:

$$y^* \frac{\partial E(Y_i)}{\partial X_i} = F(z)\beta_i \quad \text{where,} \ \frac{\beta_i X_i}{\sigma} \ \text{is denoted by } z \tag{8}$$

The other marginal effect is an independent variable on the actual value of the dependent variable. In this case, the probability of an observation is different from zero. Thus, the reported Tobit coefficients show a one-unit change in an independent variable (X_i) alters the latent (*censored*) dependent variable.

Marginal effect on the expected value for y (*censored and uncensored*); the change in the probability of a dependent variable, i.e., changes in the dependent variable as independent variable X_i change was:

$$\frac{\partial E(y)}{\partial xk} = \Phi \left(\frac{X_i \beta}{\delta} \right) \beta k \tag{9}$$

Marginal effects on the expected value of y for *uncensored* observation; the change in intensity of a dependent variable with respect to a change in an independent variable among diversifiers explained as follows. This indicated how a one-unit change in an independent variable xk affected uncensored observation.

$$\frac{\partial E[y/y > 0]}{\partial xk} = \beta k \left[1 - \lambda(\alpha)(\frac{X_i \beta}{\delta} + \lambda(\alpha) \right] \tag{10}$$

Description of variables used for analysis Crop–livestock diversification is the dependent variable computed using Margalef index. The ranges of weighted indices for crop–livestock diversity were from 0 to 1.0. A list of independent variables were identified and selected based on the previous empirical literatures and available database, which were expected to influence the choices of diversification decisions [3, 14, 39, 47, 48].

Variance inflation factor (VIF) was employed to test multicollinearity for continuous and contingency coefficient (CC) for discrete independent variables. Collinearity occurs when two or more variables are highly correlated and have the same effect on the dependent variable [49]. Highly correlated variables have high standard errors, low t-ratios, and wide confidence intervals in regression [44]. The minimum cutoff points for CC and VIF are 0.75 and 10, respectively [50]. There was no multicollinearity problem among discrete independent variables. However, there was a problem of collinearity between distance to the nearest market place and distance to the *Kebele* office from the residence of households. Of the two collinear variables, distance to the market place was excluded from analysis because its VIF value was higher than distance to the *Kebele* office. Demographic, economic, institutional, biophysical, and social factors are presented in Table 1.

Household food balance Prior to analyze the link between household food supply and farm activities, availability of food was measured. Food balance model was employed to compute the net available food in calories. The analytical method was adapted from FAO-WFP [51]. The data were collected through household survey except post-harvest losses. Robin [52] has estimated post-harvest loss of grain crops in Ethiopia from 1994 to 2010, in which the crop yield was estimated with 10% loss. The quantity of estimated food supply was converted from kg to equivalent calories for each household. Household food balance model is expressed as:

$$N_{ij} = (O_{ij} + P_{ij} + R_{ij} + A_{ij}) - (L_{ij} + S_{ij} + M_{ij} + T_{ij}) \quad (11)$$

where N =net available food; O= own produced; P=food purchased; R =food obtained through remittance or gift; A= food received through food aid; L = food loss to household; S =crops reserved for seed at home; M= amount of food marketed or sold; and T=food transfer to others. Estimations (Eq. 11) were made for total food of household i in a given year j. The

Table 1 Hypothesized variables, descriptions, and measurements in relation to crop–livestock diversity *Source*: Asante et al. [3]; Manyong et al. [14]; Rehima et al. [39]; Iiyama et al. and Obayelu et al. [47, 48]

Acronyms of variables	Variable class	Variables, units and measurement	Hypothesized Sign
Dependent variable (Y_i)			
Crop–livestock diversity	Continuous	Crop–livestock diversification index	
Independent variable (X_i)			
AGE	Continuous	Age of a household head in years	(−/+)
SEX	Discrete	Household head, 1 = male, 0 = female	(Male +)
EDUC	Discrete	Education in year of schooling	(+)
HHLAB	Continuous	Household labor in adults	(+)
LAND	Continuous	Size of land in ha	(−)
LIVES	Continuous	Livestock holding in TLU	(+)
NONFINC	Continuous	Annual non-farm income, ETB yr^{-1}	(±)
FINC	Continuous	Annual farm income, ETB yr^{-1}	(±)
LROUT	Continuous	Land rent-out in ha	(−)
OFERT	Continuous	Amount of organic fertilizer used in ton	(+)
IRRIG	Continuous	Cultivated irrigation land in ha	(+)
IFERT	Continuous	Inorganic fertilizer applied in kg	(+)
IMPSEED	Continuous	Improved seed in kg	(+)
CREDIT	Discrete	Access to credit, Yes = 1, No = 0	(+)
EXTCONT	Continuous	Extension contact in days per month	(+)
LTSECUR	Discrete	Land tenure security Yes = 1, 0 otherwise	(Secure +)
SOIL	Discrete	Fertility of soil (high, medium, or low)	+
DMKT	Continuous	Distance to the local market in hr	(−)
DROAD	Continuous	Distance to the asphalt road in hr	(−)
DKEBELE	Continuous	Distance to the *Kebele* office in hr	(−)
SLOPE	Discrete	Slope class (steep, medium, flat)	(−/+)
SOCIAL	Discrete	Social status, Yes = 1, No = 0	(+)

The official exchange rate of money during data collection as 1 US dollar = 21.5 Ethiopian birr available food items include grains, potato, vegetables, fruits, and animal products (milk, meat, egg, and honey).

To find out the link between diversified farm activities and household food supplies, ordinarily least squares regression was employed. The model was specified after farm activities measured in standardized units using Z-score and household food energy quantified in calories.

$$Y_i = \beta_0 + \beta_1 X_{i1} + \cdots \beta_n X_{in} + \varepsilon \quad (12)$$

where Y_i = household food energy in calories; X_i = farm activities for ith farmers; β_0 = intercept; $\beta_1 \cdots \beta_n$ = are coefficients; and ε = the error term.

Results and discussion

Descriptive statistics

Nearly 29% of households were female-headed. About 21% of the household-heads were illiterate, whereas 43% of the household-heads were at least able to read and write. The average size of household members was 4–5 and 3.9 adult equivalents per household. The average sizes of farmlands and livestock holdings were 1.2 ha and 4.0 TLU per household. Almost every household produced annual crops, while 94% of households reared livestock in the study watershed. Households in the study area keep cattle, equine, sheep and goat, or some combinations of these as well as cultivate crops. Crop production, animal husbandry, forestry, non-farm and off-farm activities are the main livelihood strategies of smallholders. The proportions of crops and livestock holdings varied among farmers according to growth potentials, socioeconomic attributes and biophysical factors. The mean crop–livestock production diversity of the study area (8.76) was lower than (10.2) the national average [41].

Three-quarters of the cattle population were indigenous breeds, while 25% were improved breeds. Oxen were the highest cattle population, while sheep were the highest number in livestock population. Sheep and chicken were the predominant livestock types in number followed by ox, donkey and cow. Cattle were the leading livestock population (61%) followed by 20% of sheep and goat, 18% of equines and 1% of poultry in terms of TLU (Fig. 2).

In this study, diversifiers refer to households who produced more than one crop or livestock activities, while non-diversifiers represent households who did not grow crops and/or rear animals, or they produced only a single enterprise in the crop–livestock production systems. The χ^2 test showed that there was no statistical difference between diversifiers and non-diversifiers on discrete independent variables. The t test results revealed that the mean differences for some continuous independent variables were statistically significant between diversifiers and non-diversifiers at different probability levels. The most continuous independent variables have had larger mean differences for diversifiers than non-diversifiers except non-farm income and extension contact (Table 2). Age of household head, farm income and livestock were statistically significant at 1% probability levels, while improved seed was significant at 5% level.

The distribution of diversification index

Our result indicated high degree of diversification. The average crop–livestock diversification index was 0.57 with standard deviation of 0.22. The average crop–livestock diversification index of this study was almost comparable with the finding of Asante et al. [3] found 0.59 in Ghana. Figure 3 depicted normal distribution in

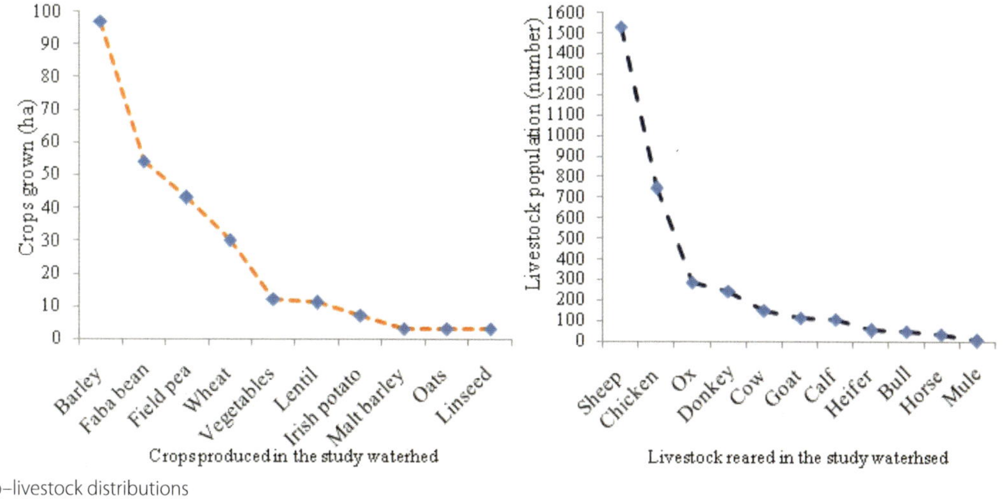

Fig. 2 Crop–livestock distributions

Table 2 Summarized descriptive statistics for independent variables by sample farmers

Variable acronyms	Total mean ($N=211$)	Diversifiers mean ($N=207$)	Non−diversifiers mean ($N=4$)	t/χ^2 test	p value
AGE	43.8	43.9	37.8	−5.100***	0.000
SEX†	68.7	67.8	0.9	0.665	0.591
EDUC†	79.2	77.3	1.9	1.074	0.582
HHLAB	2.8	2.8	1.9	−2.499*	0.078
LAND	1.3	1.4	0.7	−2.727*	0.068
LIVES	4.0	4.1	0.6	−12.370***	0.000
NONFINC	944.8	916.0	2437.5	1.059	0.366
FINC	3892.6	3955.8	625.0	−6.490***	0.000
LROUT	0.1	0.1	0.0	−1.103	0.323
OFERT	87.4	88.5	31.3	−2.850*	0.055
IRRIG	0.0	0.0	0.0	−0.465	0.465
IFERT	61.3	61.7	37.5	−0.996	0.388
IMPSEED	15.6	15.9	3.8	−2.781**	0.035
CREDIT†	40.3	39.8	0.5	0.396	0.650
EXTEN	1.3	1.3	1.5	0.717	0.717
LTSECUR†	47.4	46.5	0.9	0.011	0.449
SOIL†	86.3	84.9	1.4	0.436	0.627
DROAD	18.4	18.6	9.8	−0.286	0.286
DKEBELE	28.0	28.3	14.3	−0.093*	0.093
SLOPE†	5.7	5.7	0.0	0.246	0.620
SOCIAL†	25.2	24.2	0.9	1.342	0.263

***, **, *Significant at 1, 5, and 10%; † is designated for discrete independent variables

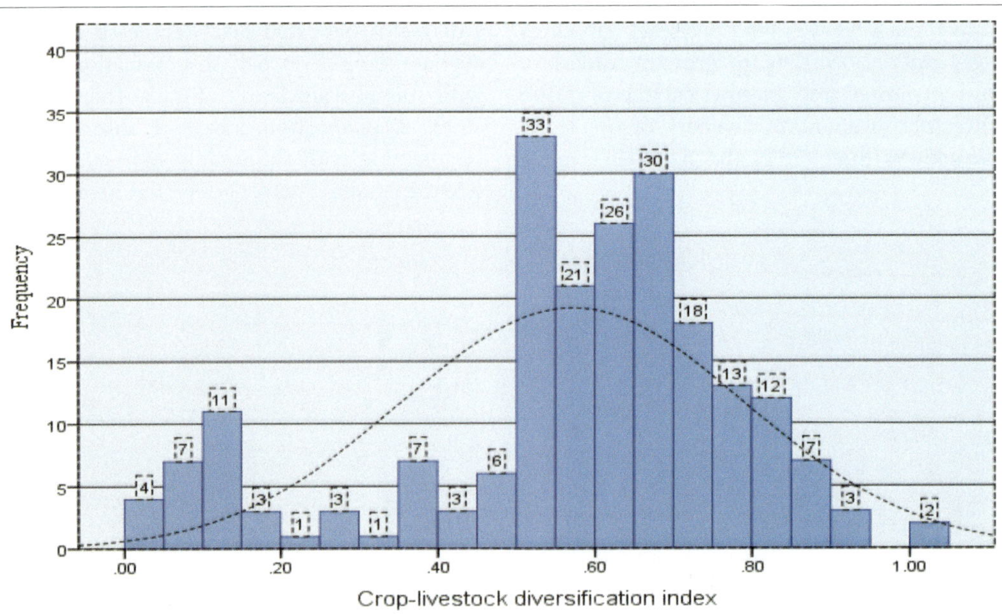

Fig. 3 Distribution of crop–livestock diversification index

crop–livestock diversification though it was moderately skewed to the left. More than three-fourth (79%) of the households having indices 0.5 and above, suggest a high level of crop–livestock diversification among farmers. The estimated index was significant among diversifiers (t value $=-5.471$; $p<0.01$).

Nearly 98% of households have diversified crop–livestock activities and 58% of them have practiced higher than the mean index of diversifiers. Limited number of households (2%) did not diversify farm activities, because of landlessness, labor dependency or specialized in monoculture. It implies that the level of diversification was quite diverse among household. Variations for proportions of crop–livestock activities might happen due to differences in farm resources, priorities of the farm activities, and ability to manage their available resources. These conditions forced them not to look for ways to engage them in diversified activities. As a result, they have rented out their land and involved in non-farm activities with or without producing a particular crop or a single animal. The diversity of livestock (0.86 ± 0.39) was tremendously higher than the diversity of crops (0.28 ± 0.12) with reasonable symmetrical distribution. About 6.2% of households did not own livestock and other 4.7% of households specialized in a single animal. Similarly, 0.9% of the households did not produce crops and other 3.8% of the producers specialized in growing only a single crop.

Households were classified into three equal parts based on crop–livestock diversification index. The average indices of the lower, middle and high categories were 0.17, 0.54, and 0.80. About 14, 54, and 32% of the households have diversified their activities in the lower, middle, and high category of indices. The ages of 61% households were between 31 and 50 years. The highest age of diversity index was between 31 and 40 years. Majority of (60%) the female-headed households were in the lower category. High proportion of households (77%) held farmlands less than 1.3 ha within the lower category. Almost half of the farmers accessed farmlands on average 0.8 ha. Three-quarters (77%) of households possessed ≥ 1.0 ha of land, while 18% of farmers held between 0.5 and 1.0 ha of land. Five percent of households held < 0.5 ha of land. Household-heads with educational level of grade 9 and above were 5% (Table 3).

Determinants of farm diversification

The Tobit model output suggests that land holding size, livestock, non-farm income, land rented-out, irrigated land, extension contact, soil fertility, road distance, *Kebele* distance, and improved seed aer the major factors significantly affected crop–livestock diversification. After Tobit regression, marginal effects were computed. The parameter estimates of latent coefficients and marginal effects of the latent variable for continuous decision in Tobit regression were almost similar. Sichoongwe et al. [43], Dube and Guveya [53] reported that coefficients of the probability and intensity of Tobit results are the same. Latent coefficients showed directions and level of significance for each independent variable, while marginal effects indicated not only directions and level of significance, but also offered information regarding probabilities and extent of influence on the dependent variable. The probability coefficients revealed the likelihood of diversity in the discrete choice of crop–livestock diversification. The regression results in Table 4 indicated that the probability coefficients for diversification attributed by the independent variables were insignificant. Because, the likelihood of households to diversify their crop–livestock activities was 99.9%, that showed insignificant difference from the descriptive results (98%).

It implies the contribution of independent variables to the choice of decisions on crop–livestock diversification was less important. One of the main reasons could be the limited number of households who did not diversify their farm activities. The intensity coefficients for some independent variables were significant among diversifiers. Altogether, 81% of the households could diversify their farm activities more than the constant value (0.41) owing to independent variables. Thus, intensity coefficients infer how a one-unit change in an independent variable alters the dependent variable of crop–livestock diversity.

Farm size

More land size increased the capacity of farmers to diversify farm activities. As -anticipated, landholding size is found to have positive relationship and significant effect ($p < 0.1$) on the extent of diversification though the probability of diversity is insignificant. An additional unit of land can increase the probability of diversification by 0.1%. It means that as land size increases by one ha, crop–livestock diversification increases by 5.1%. An increase in land size could increase the index category of crop–livestock diversification. For example, a farmer having an index of 0.33 in lower category can increase to 0.35 of the middle category if the land size increases by one ha. This means, more landholdings enabled farmers to allocate their farm activities in multiple productions than small farms to minimize production risks. This result is consistent with the findings of many studies [3, 39, 43, 54]. However, the possibility of getting more land is unlikely unless a household acquires through rent-in, inheritance, or sharecropping.

Non-farm income

This variable influenced the continuous decision of diversification negatively and significantly ($p < 0.1$). Getting more non-farm income was unlikely to diversify crop–livestock activities. As non-farm income increases by one birr, the crop–livestock diversity decreases by less than 1.0%. Thus, the extent of decrease in diversity due to non-farm income was negligible. Higher incomes from non-farm activities could affect the interest of farmers and led to divert their resources away from farming. Previous

Table 3 Basic characteristics of crop–livestock diversifiers by Margalef index

Socioeconomic characteristics	Index			
	0.0–0.33 ($N=30$)	0.34–0.67 ($N=114$)	0.68–1.00 ($N=63$)	Total ($N=211$)
Age of household head (years)				
<30	23.3	13.2	15.9	15.5
31–40	33.3	29.8	34.9	31.9
41–50	30.0	33.3	20.6	29.0
51–60	6.7	14.0	17.5	14.0
Above 61	6.7	9.7	11.1	9.6
Total	100.0	100.0	100.0	100.0
Sex of household head				
Male	40.0	71.0	77.8	68.6
Female	60.0	29.0	22.2	31.4
Total	100.0	100.0	100.0	100.0
Size of landholding (ha)				
0.0–1.30	76.7	45.6	44.4	49.8
1.31–2.60	23.3	51.8	52.4	47.8
2.60–4.00	10.0	2.6	3.2	2.4
Total	100.0	100.0	100.0	100.0
Household labor (AE)				
0.0–2.30	63.3	47.4	46.0	49.3
2.31–4.60	23.3	41.2	42.9	39.1
4.61–6.90	13.4	11.4	11.1	11.6
Total	100.0	100.0	100.0	100.0
Livestock holding (TLU)				
0.0–4.20	100.0	40.4	50.8	52.2
4.21–8.40	0.0	51.8	42.9	41.6
8.41–12.6	0.0	7.8	6.3	6.2
Total	100.0	100.0	100.0	100.0
Education (years of schooling)				
Not literate	30.0	20.2	19.1	21.3
Grade 1–4	33.3	63.2	57.1	57.0
Grade 5–8	23.3	13.2	19.1	16.4
Grade 9–12	10.0	2.6	4.7	4.3
Diploma and above	3.4	0.8	0.0	1.0
Total	100.0	100.0	100.0	100.0

Size of landholding, household labor, and livestock holding are classified into three (low, middle, and high) categories

studies have found that non-farm/off-farm income per capita has non-harmonious effects. It could have either negative or positive influence on diversification [3] depending on the amount of return and extent of farmers' engagement. Our finding was consistent with studies conducted in Indonesia, India, Zimbabwe, Ethiopia, Kenya, Ghana and Nigeria on farm and livelihood diversifications [3, 41, 51, 55]. Asante et al. [3] have found that off-farm income has positive and significant influence on the probability of crop–livestock diversity, whereas their finding revealed that the extent of diversification caused by non-farm income is negative.

Livestock

This variable had positive and significant influence on probability ($p<0.1$) and intensity ($p<0.01$) of crop–livestock diversity in the farming systems. As livestock size increases by one TLU, the level of diversification increases by 3%. The positive coefficient of this result is similar with other findings [14, 51].

Table 4 Tobit regression results for determinants on crop–livestock diversity

Variables	Marginal effects			
	Prob. coef.	P-level	Inten. coef.	P-level
AGE	0.0000	0.366	0.0012	0.319
EDUC†	−0.0005	0.390	−0.0295	0.410
SEX†	0.0010	0.883	0.0048	0.881
LAND	0.0010	0.173	0.0510*	0.076
HHLAB	−0.0002	0.420	−0.0099	0.384
FINC	−0.0000	0.310	−0.0000	0.250
NONFINC	−0.0000	0.179	−0.0000*	0.083
LIVES	0.0006*	0.062	0.0302***	0.000
OFERT	−0.0000	0.396	−0.0002	0.356
LROUT	−0.0034*	0.093	−0.1715***	0.007
CREDIT†	0.0004	0.523	0.0005	0.513
IFERT	0.0000	0.484	0.0001	0.458
EXTEN	0.0006	0.130	0.0309**	0.032
IRRIG	0.0083	0.133	0.4267**	0.035
SOCIAL†	0.0009	0.170	0.0544	0.127
SOIL†	−0.0011*	0.082	−0.0880**	0.027
DROAD	−0.0000	0.176	−0.0021*	0.079
DKEBELE	0.0000	0.157	0.0019*	0.059
SLOPE†	−0.0012	0.595	−0.0439	0.460
SEED	−0.0000	0.143	−0.0010**	0.045
LTSECUR†	0.0005	0.454	0.0254	0.423
-cons	0.4148***			

4 observations are left censored

Log likelihood = 49.14; LR $\chi^2 = 83.36$ (p value = 0.000)

Pseudo $R^2 = -5.59$

Significant level ***($p < 0.01$), **($p < 0.05$), and *($p < 0.1$) and † denotes discrete variables

Land rent-out

According to descriptive statistics, a small number of farmers (7%) were dependent on land rent-out for their subsistence of food demands. Land rent-out was negative and significant on crop–livestock diversity. As a household increases land rent-out by one ha, the probability of diversity declines by 0.3% ($p < 0.1$). The intensity of crop–livestock diversity also decreases by 17.2% ($p < 0.01$). A household that rented out his/her land was not likely to diversify farm activities, as he/she had to use the land to produce more crops and harvest forage or keep animals on it. According to Daniel et al. [55], land transfer adversely affects long-term investment in agriculture and results in low returns. The same authors have substantiated that transfer rights and threats of expropriation influence negatively long-term investments in agriculture that perpetuate low growth and poverty. Otsuka [56] also reported that land use rights and strengthening of land tenure security in Ethiopia enhance the incentives of farm households to invest more in the acquisition of new knowledge and land improvements.

Extension service

Agricultural extension service is found to have significant ($p < 0.05$) and positive effect on the extent of crop–livestock diversification. As extension contact increases by 1 day in monthly bases, the level of crop–livestock diversification increases by 3%. Households who gained knowledge and acquired skill from development agents might optimistically influence the intensity of crop–livestock diversification. Evidences from other empirical studies revealed that extension service has mixed effects on crop–livestock diversification. Rehima et al. [39] and Asante et al. [3] pointed out that extension service has significant contributions in increasing different farm enterprise options. Our finding contradicted with Wondimagegn [50] who found that the frequency of contact of development agents has adverse effects on farm diversities.

Irrigation land

An increase in irrigation land is found to have positive and significant influence on the extent of crop–livestock diversity ($p < 0.05$). Small-scale irrigation used to grow vegetables, animal feed, and some grain crops such as lentil. Irrigation enables to increase crop–livestock diversity as it supplement water at the time of rain shortage. Other studies also reported similar findings [21, 51]. Moisture retained through irrigation can increase both the probability and intensity of crop–livestock diversification [14]. Access to irrigable lands may not necessarily suggest area expansion. It could be the use of stored moisture and increasing the frequency of cropping per year on existing lands.

Soil fertility

The model result showed that soil fertility is a biophysical factor found to have negative relationship between the probability ($p < 0.1$) and intensity ($p < 0.05$) of diversification. As poor soils change to fertile status, a decision on the probability and intensity of diversification declines by 0.1% and 8.8%, respectively. According to the household survey and field observation, farmers were more likely to practice diversified farms on degraded plots than fertile lands. Households who own fertile lands could not motivated to diversify more crops and livestock activities rather they want to produce crops that are more profitable. Some researchers [39, 51] reported similar findings that fertile soils reduce the extent of crop diversification because fertile plots tend to be specialized on limited varieties. Households in *Gudo Beret* have grown diverse crops and eucalyptus trees to diversify household incomes. The relationship between organic fertilizer and diversification was also negative though its level of probability was insignificant. It implies that soils that are more fertile grow crops with low level of diversity. Households tend to cultivate only crops as long as the crops provide high returns for subsistence needs regardless of marketing opportunities and nutritional benefits.

Road distance

The extent of influence of road distance on crop–livestock diversity was negative and significant ($p < 0.1$). Short distance between household's residence and the main tarmac road had favorable effect on the crop–livestock diversity. Households dwell far away from the main road were less likely to diversify crop–livestock activities. The result revealed that the extent of diversity declines by 0.2% as a farmer travels 1-min of time. Various farm activities such as irrigation, seedling nurseries, and other infrastructures are located near to the main asphalt road. Ashfaq et al. [55] have found similar results in India where road distance affects farm diversification negatively and significantly. Road distance is negative on crop diversity though the extent of influence is insignificant in eastern and southern parts of Ethiopia [39, 50]. Households reside in near to the main road reduce transport and marketing costs. Farmers located farther away from the main road might produce only to meet their subsistence and nutritional needs [3, 43]. Various findings reported unlike relationship between road distance and crop diversification in accordance with the nature of crop production [43].

Kebele distance

The time of travel from household's residence to the nearest *Kebele* office affected the intensity of crop–livestock diversification positively and significantly ($p < 0.1$). An increase in time in 1-min travel between household's residence and the *Kebele* office, the extent of diversification increases by 0.2%. The effect of distance between Kebele office and household's residence is positive in relation to some crops and negative on other crops grown in Nigeria [46]. In our study, most of households settled near to the centre of the *Kebele,* employed in non-farm/off-farm activities. Some households who reside adjacent to *Gudo Beret* town have limited farmlands and involved in nonfarm activities. The nature of settlement patterns close to the town is highly populated and less likely to obtain adequate grazing fields and croplands. The other probable reason for positive relationship between distance to the office and farm diversification is social and administrative responsibilities. One-quarters of household-heads have had different social responsibilities being involved in administrative and social obligations. Those households who played social and political affairs and settled near to the office might have limited participations on diversified farm activities.

Improved seed

The improved crop seed cultivars had negative effect on the discrete and continuous decisions of crop–livestock diversity, suggesting that it is less important for subsistence producers. The likelihood of a household to diversify crop–livestock activities attribute to improved seed was not significant. Even though improved seed is significant ($p < 0.05$) on the crop–livestock diversity, the extent of influence was not considerable. As improved seed variety increases by one kg, the extent of crop–livestock diversity decreases by 0.1%. The probable reason for negative relationships could be households tend to use improved seeds may give less attention to livestock as a result crops compete with livestock for land resources. In addition, high cost and delay to access for improved seed imply adverse effects on diversification.

The results of relationship between the dependent variable and some independent variables were contrasted with prior hypotheses. Age, sex, credit, fertilizer, social status, and land tenure security were positive with crop–livestock diversity though they are insignificant. Positive relationship of age with crop–livestock diversity implies extent of diversification is higher among older household-heads. The positive result for fertilizer is in agreement with many studies [39, 43]. The coefficients for education, household labor, farm income, and slope of farmlands were negative with crop–livestock diversity. It implies the household labor did not affect the level of crop–livestock diversity [3]. In this study, education is negative but insignificant on diversification. Several reports also noted that education has negative relationships with farm diversifications [39, 43, 46, 48, 50, 51]. The negative relationship for education indicates literate farmers are less likely to diversify farm activities where they are liable to engage in off-farm/non-farm livelihood strategies.

Crop–livestock diversification and household food security nexus

F-statistics for linear regression was significant at 1%, implying that the crop–livestock activities included in the model were jointly significant in influencing the dependent variable -food energy. The value of R^2 is 0.304. This means that 30.4% of the variation in food energy was explained by farm activities included in the model. Substantial differences were existed among households for food energy, as they have pursued diverse crop–livestock activities. Results showed that barley, wheat, and faba bean positively influence food energy, while vegetables and highland fruits had a negative influence on food energy at 1% (Table 5). Households with more croplands for barley, wheat and faba bean were more likely to have better food energy in terms of calories. On the other hand, a negative influence of vegetable production and highland fruits on food energy showed that households with higher fruits and vegetables are less food secure.

Conclusions and policy recommendations

Diversified farms remain a source of income, risk reduction strategy, and means of livelihoods for smallholders. Crop production and animal husbandry are the two subsystems in the broad mixed crop–livestock system. This study aimed to examine determinants of crop–livestock diversification and the effect of crop–livestock activities on household food energy in central highlands of Ethiopia. The study used cross-sectional data to find out farmers' decision on crop–livestock diversity. Food balance sheet was employed to measure food energy. The study included demographic, economic, institutional, biophysical, topographic and social factors to investigate crop–livestock diversification decisions.

Table 5 OLS results on household food energy in calories

Farm activities	Coef.	*t* value	*p* value
Cow	− 184.748	− 0.621	0.535
Ox	233.853	1.379	0.169
Bull	246.277	0.563	0.574
Heifer	− 93.087	− 0.173	0.863
Calf	916.771	0.642	0.521
Horse	457.177	1.119	0.265
Mule	1488.291	1.537	0.126
Donkey	− 563.249	− 1.414	0.159
Sheep	− 297.408	− 1.154	0.250
Goat	− 1454.018*	− 1.679	0.095
Chicken	1745.954	0.609	0.543
Bee colonies	21.672	0.399	0.690
Barley	1618.349***	2.874	0.005
Wheat	3522.539***	4.245	0.000
Faba bean	2359.257***	3.754	0.000
Field pea	− 1128.296*	− 1.868	0.063
Lentil	− 16.441	− 0.022	0.983
Linseed	− 1118.350	− 0.392	0.695
Potato	1221.641	1.191	0.235
Vegetables	− 5858.333***	− 2.793	0.006
Highland fruits	− 162.991***	− 3.746	0.000
-cons	1905.135***	5.363	0.000
R^2	0.304		
Adjusted R^2	0.227		
F	3.933		
N	211		

***, **, * are significant at 1, 5 and 10%, respectively

The results of descriptive statistics revealed that the majority of farmers produced on average 8.76 types of crop–livestock activities per household. Margalef index of diversity showed a wide variation between zero to one with a mean value of 0.57 attributed to household resources, farmers' preference, ability to manage farm activities, and other natural factors. Crop–livestock diversifiers have better socioeconomic and demographic characteristics compared to non-diversifiers. The Tobit results showed that farmers decided to diversify their farm activities with little influence of discrete choices from independent variables, whereas the extent of diversification was influenced mainly by economic factors. Farmers who accessed irrigation land, more livestock and extension contact were found to have higher level of diversity compared to have-nots. This means that farmers access to more lands from rain-fed and irrigation farming, more livestock and frequent extension support were liable to have higher level of crop–livestock diversities. Conversely, the relationship between land rent-out and crop–livestock diversity was negative and significant, implying that land rent-out reduced both the likelihood and extent of crop–livestock diversity. In addition, results of ordinary least square regression showed that barley, wheat, and faba bean were significant crops that can influence household food energy positively.

The results imply that more farmland, irrigation land, larger livestock, and effective extension service were decisive factors for crop–livestock diversification. Grain crops were the most significant farm activities for household food energy. In conclusion, rain-fed farming supplemented with irrigation and effective agricultural extension services are crucial factors for mixed production. Crop–livestock integration contributes for livelihoods, household food security, and biodiversity conservation. Therefore, among diversified farm activities, grain crops and small ruminants have immense contributions for household food energy. The results of this study suggest that agricultural policies need to assure crop–livestock diversification, which is a means of livelihoods. We also recommend further research on crop–livestock diversification to find out the best patterns of diversity so that households can ensure food security and cash incomes.

Authors' contributions

WM generated the idea, designed the study, prepared interview schedule, made preliminary survey, and took part in data collection, data entry, data analysis and write up. KM participated in the study design, data analysis, support statistical techniques, and drafted the paper. Both authors read and approved the final manuscript.

Author details

[1] Centre for Rural Development, College of Development Studies, Addis Ababa University, P. o. b 1176, Addis Ababa, Ethiopia. [2] Crop-Livestock Systems, International Livestock Research Institute, P.o.b 5689, Addis Ababa, Ethiopia.

Acknowledgements
The authors acknowledge respondent farmers and data collectors. Addis Ababa University, University of Gondar, and Africa RISING project (ILRI) were also acknowledged for their financial support in accomplishing this paper. The content is solely the responsibility of author/s and does not necessarily represent funding organizations.

Competing interests
The authors declare that they have no competing interests.

Funding
This research was supported by funding from Addis Ababa University, University of Gondar, and Africa RISING project (ILRI).

References

1. Iiyama M, Kaitibie S, Kariuki P, Morimoto Y. The status of crop–livestock systems and evolution toward integration. Ann Arid Zone. 2007;46(3&4):1–23.
2. Todaro MP, Smith SC. Population growth and economic development: causes, consequences, and controversies. Eleventh ed., Economic developmentBoston: Pearson Addison Wesley; 2012.
3. Asante BO, Villano RA, Patrick IW, Battese GE. Determinants of farm diversification in integrated crop–livestock farming systems in Ghana. Renew Agric Food Syst. 2017. https://doi.org/10.1017/s1742170516000545.
4. Liniger H, Mekdaschi SR, Hauert C, Gurtner M. sustainable land management in practice: guidelines and best practices for Sub-Saharan Africa: field application. Rome: FAO; 2011.
5. Herrero M, Thornton PK, Notenbaert A, Msangi S, Wood S, Kruska R, et al. Drivers of change in crop–livestock systems and their potential impacts on agro-ecosystems services and human wellbeing to 2030. A study commissioned by the CGIAR system wide livestock Programme, Addis Ababa; 2012.
6. Moraine M, Duru M, Nicholas P, Leterme P, Therond O. Farming system design for innovative crop–livestock integration in Europe. Animal. 2014;8(8):1204–17. https://doi.org/10.1017/S1751731114001189.
7. Berndes G, Bird N, Cowie A. Bio-energy: land use change and climate change mitigation., IEA bioenergyDallas: ExCo; 2010. p. 03.
8. Alexandratos N, Bruinsma J. World agriculture towards 2030/2050: the 2012 revision. ESA Working Paper No. 12-03. FAO: Rome; 2012.
9. Darnhofer I, Gibbon D, Dedieu B. Farming systems research: an approach to inquiry. In: Gibbon D, editor. Farming systems into the 21st century: the new dynamic. Dordrecht: Springer; 2012. p. 3–31.
10. Kuria A, Lamond G, Pagella T, Aster G, Hadgu K, Sinclair F. Local knowledge of farmers on opportunities and constraints to sustainable intensification of crop–livestock-trees: mixed systems in Lemo Woreda, Southern Nations Nationalities and People Region (SNNPR), Ethiopian highlands. A field study report; 2014. https://cgspace.cgiar.org/bitstream/handle/10568/41669/Lemo_ARmay.pdf?sequence=1&isAllowed=y. Accessed 8 Mar 2018
11. Boserup E. The conditions of agricultural growth: the economics of agrarian change under population pressure. Piscataway: Transaction Publishers; 1965.
12. McIntire J, Bourzat D, Pingali P. Crop–livestock interaction in Sub-Saharan Africa., Regional and sectoral studiesWashington: The World Bank; 1992.
13. Shideed KH, Mohammed EME, editors. Adoption and impact assessment of improved technologies in crop and livestock production systems in the WANA region. The development of integrated crop/livestock production in low rainfall areas of Mashreq and Maghreb Regions (Mashreq/Maghreb Project). Aleppo: ICARDA; 2005.
14. Manyong VM, Okikeb I, Williamsc T. Effective dimensionality and factors affecting crop–livestock integration in West African savannas: a combination of principal component analysis and Tobit approaches. Agric Econ. 2006;35:145–55.
15. IFAD. Integrated crop–livestock farming systems. Livestock thematic papers. Tools for project design. 2010.
16. Alemayehu S, Dorosh P, Sinafikeh A. Crop production in Ethiopia: regional patterns and trends. International Food Policy Research Institute, Ethiopia Strategy Support Program II (ESSP II), Working Paper No. 16. 2011.
17. Lambin EF, Geist HJ, Lepers E. Dynamics of land-use and land-cover change in tropical regions. Annu Rev Environ Resour. 2003;28:205–41. https://doi.org/10.1146/annurev.energy.28.050302.105459.
18. Daniel D. Sustainable land management technologies and approaches in Ethiopia. MoARD; 2010. Addis Ababa, Ethiopia.
19. Godfray HCJ, Beddington RJ, Crute IR, Haddad L, Lawrence D, Muir JF, Pretty J, Robinson S, Thomas SM, Toulmin C. Food security: the challenge of feeding 9 billion people. Science. 2010;327(812):812–8. https://doi.org/10.1126/science.1185383.
20. Belay D, Azage T, Hegde BP. Smallholder livestock production system in Dandi District, Oromia Regional State, Central Ethiopia. Glob Vet. 2012;8(5):472–9.
21. World Bank. Ethiopia's productive safety net program (PSNP) integrating disaster and climate risk management. Case study. 2013.
22. Helina T, Schmidt E. A spatial assessment of livestock production and market access. Summary of ESSP Working Paper 44, "spatial analysis of livestock production patterns in Ethiopia". 2013.
23. Bewket W, Radeny M, Mungai C. Agricultural adaptation and institutional responses to climate change vulnerability in Ethiopia. CCAFS Working Paper No 106. Research Program on Climate Change, Agriculture and Food Security. Copenhagen. 2015.
24. NPC. Growth and transformation plan II (2015/16-2019/20). Vol. I. The main text. Addis Ababa; 2016.
25. Thorne MS, Stevenson MH. Stocking rate: the most important tool in the toolbox. Cooperative extension service, College of tropical agriculture and human resources, University of Hawai'i at Manoa. Pasture and range management, PRM-4, 2007.
26. Boserup E. Population and technological change: a study of long-term trends. Chicago: University of Chicago; 1981.
27. Eleni Y, Wagner W, Exner-kittridge M, Dagnachew L, Blöschl G. Identifying land use/cover dynamics in the Koga Catchment, Ethiopia, from multi-scale data, and implications for environmental change. ISPRS Int J Geo-Inf. 2013;2:302–23. https://doi.org/10.3390/ijgi2020302.
28. IFAD. Smallholders, food security and the environment. 2013.
29. Nigussie H, Atsushi T, Nyssen J, Poesen J, Tsubo M, Dereje D, et al. Soil erosion and conservation in Ethiopia: a review. Prog Phys Geogr. 2015;39(6):750–74. https://doi.org/10.1177/0309133315598725.
30. Tigist B. Assessment of surface water resource and irrigation practices in Gudo Beret Kebele, Amhara Region, Ethiopia. Thesis Presented to Addis Ababa University, Ethiopia. 2016.
31. Mekonnen K, Jogo W, Bezabih M, Mulema A, Thorne P. Determinants of survival and growth of tree lucerne (*Chamaecytisus palmensis*) in the crop–livestock farming systems of the Ethiopian highlands. Agrofor Syst. 2017. https://doi.org/10.1177/0309133315598725.
32. Lulseged T, Zenebe A, James E, Tesfaye Y, Kifle W, Kindu M, et al. Mapping soil erosion hotspots and assessing the potential impacts of land management practices in the highlands of Ethiopia. Geomorphology. 2017;292:153–63. https://doi.org/10.1016/j.geomorph.2017.04.038.
33. UNEP. Managing increasing pressures on land. Based on GEO-5 land chapter. 2012.
34. Feder G, Just RE, Zilberman D. Adoption of agricultural innovation in developing countries. A survey. Working Papers No 542. The World Bank, Washington; 1982.
35. Rahm MM, Huffman WE. The adoption of reduced tillage: the role of human capital and other variables. Am J Agric Econ. 1984. p. 405–413
36. Greene W. Econometric analysis. 5th ed. Upper Saddle River: Prentice-hall, Inc.; 2002.
37. Das S, Lyla PS, Khan SA. Marine microbial biodiversity: present status and advanced statistical paradigms. In: McNeely JA, McCarthy TM, Smith A, Olsvig-Whittaker L, Wikramanayake ED. editors. Application of statistical tools: a case study. published by the society for conservation biology asia section and resources himalaya, Kathmandu, Nepal, 2006. p. 455.
38. Margalef R. Information theory in ecology. Int J Gen Syst. 1958; 3:36–71. www.vliz.be/imisdocs/publications/139371on. Accessed 30 Oct 2016.
39. Rehima M, Belay K, Dawit A, Rashid S. Factors affecting farmers' crop diversification: evidence from SNNPR, Ethiopia. Int J Agric Sci. 2013;3(6):558–65.
40. FAO. Livelihood diversification and vulnerability to poverty in Rural Malawi. ASA working paper no 15-02. Rome; 2015.

41. Kibrom T, Krishina V, Qaim M. Production diversity and dietary diversity in smallholder farm households. Proc Natl Acad Sci. 2015;112(34):10657–62. https://doi.org/10.1073/pnas.1510982112.
42. Magurran AE. Measuring biological diversity. Hoboken: Blackwell; 2004.
43. Sichoongwe K, Mapemba L, Ng'ong'ola D, Tembo G. The determinants and extent of crop diversification among smallholder farmers. A case study of southern province, Zambia. Washington: Int Food Policy Res Inst.; 2014. p. 12.
44. Maddala GS. Introduction to econometrics. Second edition, Macmillan publishing company, New York. Maxwell Macmillan Canada, Toronto; Maxwell Macmillan International New York, Oxford, Singapore, Sydney. 1988.
45. Davidson R, Mackinnon JG. Bootstrap testing in nonlinear models. Int Econ Rev. 1999;40(2):487–508.
46. Rahman S, Chima CD. Determinants of food crop diversity and profitability in southeastern Nigeria: a multivariate tobit approach. Agriculture. 2016;6(14):1–14.
47. Iiyama M, Maitima J, Kariuki P. Crop–livestock diversification patterns in relation to income and manure use: a case study from a rift valley community, Kenya. Afr J Agric Res. 2007;2(3):058–66.
48. Obayelu OA, Akintunde OO, Obayelu AE. Determinants of on-farm cassava biodiversity in Ogun State, Nigeria. Int J Biodivers Sci Ecosyst Serv Manag. 2015;11(4):298–308. https://doi.org/10.1080/21513732.2015.1071282.
49. Gujarati DN. Basic econometrics. 4th ed. New York: The McGraw-Hili Companies Inc; 2003.
50. Wondimagegn M, Bekabil F, Jemal H. Pattern trends and determinants of crop diversification: empirical evidence from smallholders in Eastern Ethiopia. J Econ Sustain Dev. 2011;2(8):78–90.
51. FAO-WFP. Joint guidelines for crop and food security assessment missions, Rome; 2009.
52. Robin B. Grains post-harvest loss assessment in Ethiopia: final report. Working Paper No 2377; 1998. http://gala.gre.ac.uk/10758. Accessed on 10 Nov 2017.
53. Dube L, Guveya E. Factors influencing crop diversification: a case study of Manicaland and Masvingo Provinces in Zimbabwe. Int J Reg Dev. 2016;3(2):1–25. https://doi.org/10.5296/ijrd.v3i2.9194.
54. Ashfaq M, Hassen S, Ceer NR, Baig MZ, Asma J. Factors affecting farm diversification in rice–wheat. Pak J Agric Sci. 2008;45(3):91–4.
55. Daniel A, Ali DS, Gautam M. Property rights in a very poor country: tenure insecurity and investment in Ethiopia. World Bank. Policy research. Working Paper No 4363. 2007.
56. Otsukan K. Strategy for agricultural growth in Ethiopia. Briefing note prepared for the DFID funded study "understanding the constraints to continued rapid growth in Ethiopia: the role of agriculture". 2008.

9

Assessment on the management of draft horses in selected areas of Awi Zone, Ethiopia

Bimrew Asmare[1*] and Zemenu Yayeh[2]

Abstract

Background: Ethiopia has many equines, which are important for the livelihood of many smallholder farmers in the country. However, there has been no study on the management of draft horses under farmers condition in this area of the country; hence, this study was conducted with the objective to determine the status of horse management and feeding practices in the area. The study areas were selected purposively based on the availability of horses. The study was conducted by administering a semi-structured questionnaire to horse owners and with complimentary information from key informants and secondary data sources.

Results: The majority of household heads were in the age range of 41–50 years in Banja Shekudad district while 51–60 years in Ankesha Guagusa district. Regarding education status, 38 and 34% of household heads in Banja Shekudad district and Ankesha Guagusa district, respectively, were illiterate. The finding of this study elucidates that horses were used in both districts for many kinds of activities including land cultivation. The purpose of horse keeping in Banja Shekudad districts is ploughing only (32%) followed by packing of goods (21%) while packing of goods (31%) followed by riding only (22%) in Ankesha Guagusa district. Overall respondents (100%) noted that horses breed early in rainy season in both districts, which might be associated with availability of feed resources during the rainy season. The majority of horse owners have no training in relation to horses husbandry, while few (24.12% in Banja Shekudad and 33.78% in Ankesha Guagusa districts) have got some horse-specific training mainly related to feeding and health care of horses. Horses were kept mainly for ploughing in both districts. Horses were provided supplementary feeds by 49 and 74.5% in Banja Shekudad and Ankesha Guagusa districts, respectively. Watering of horses was commonly practiced twice a day during dry season while once during wet season. In both districts, the majority of horse owners had no formal training on horse husbandry. Lack of knowledge on horse management and feeding were the major problems observed.

Conclusion: In the study area, draft horses are very important in the livelihood of smallholder farmers. However, the overall husbandry of draft horses and feeds and feeding in particular of horses was found to be poor. The reason for such poor husbandry was mainly due to shortage of resources and lack of awareness about management practices. Hence, it is concluded that adequate credit service, training and finding alternative feeds and feeding practices should be addressed to utilize full potential of working horses.

Keywords: Feeds, Grazing, Horse, Husbandry, Supplementation, Working equines

Background

Equines are extensively spread throughout the world for sports competition, recreation, draft power and food. In developed countries, many are kept as pets, as companions, or for work such as land ploughing and packing, in occupational therapy programme [1]. There are about 90 million equines in the developing world, with highest population found in Central Asia and North and East Africa [2]. Of the total equine population, over 95% of donkeys and mules and 60% of horses are found in developing countries [3], where they are kept mainly for work. According to [4], there are about 2.03 million horses,

*Correspondence: limasm2009@gmail.com
[1] Department of Animal Production and Technology, College of Agriculture and Environmental Sciences, Bahir Dar University, P.O. Box 79, Bahir Dar, Ethiopia
Full list of author information is available at the end of the article

7.3 million donkeys, 0.3 million mules and 1.16 million camels found in Ethiopia. Mengistu [5] and Ameni [6] noted that in Ethiopia, the rugged terrain characteristics, absence of well-developed modern transport networks and the prevailing low economic status of the community necessitate the use of equines for transportation.

Horses have a prominent position in the agricultural and transport systems of Ethiopia, mainly as draft power, pack and riding animals. In urban and peri-urban set-ups where electromotive power is absent or does not satisfy the transport needs, carts with the aid of horse power are common forms of transportation [7]. Although horses are considered as powerful animals with multiple functions, the management applied in them is at its infant state in Ethiopia. Similarly, the overall management and the status of horses' feeding practice were not described yet in the study area. Thus, this initial survey based was conducted to assess the management of horses, husbandry practice with particular emphasis on feeds and feeding in the study area.

Fig. 1 Study districts: Banja Shekudad (A) and Ankesha Guagusa

Methods

Study area

Banja Shekudad district is one of the districts in Awi administrative zone of Amhara Regional State. The district is situated at latitude of 11°10′ north and longitude of 36°15′ east in a distance of 122 km away from the regional town Bahir Dar to the south and 447 km from Addis Ababa to the north with an altitude range of 1900–3300 above sea levels. The annual rainfall of the district ranges from 2200 to 2400 mm daily temperature of 15–26 °C the overall area coverage of 47,915.809 thousand hectares. In terms of land-use pattern, about 65% of the area is cultivated land and planted settlements and wastelands account for about 10, 8 and 17%, respectively [8].

Ankesha Guagusa district is one of the districts in Awi zone, Amhara Regional State. The district is situated at altitude of the coordinates of 10°31′46″ to 10°41′32″ north and 36°36′18″ to 36°59′33″ east and 139 km away from the regional town (Bahir Dar) and 421 km from Addis Ababa. The annual rainfall of the area ranges from 1000 to 2000 mm with the daily temperature of 15–20 °C. The altitude of the district ranges from 1500 to 2800 m above sea level with total area of 95,503 ha (Fig. 1).

Methods of data collection and analysis

Sample selection and study design

Households possessing at least one horse in the study districts were selected randomly as the study population. A cross-sectional and retrospective type of study was conducted using questionnaire survey, focus group discussion and observation to collect data on characteristics and practices of smallholder horse husbandry systems. The sample size was calculated based on the formula recommended by Poete and Daplyn [9]:

$$N = (ZC/X)^2$$

where N is the estimated sample size, Z is the confidence level, C is coefficient of variation in the population and X is the accuracy level. With 95% of confidence level, 50% coefficient of variation and 5% of accuracy level, a total of 198 households were required for the study.

Thus, from available population of horse owners in the study area households were selected randomly selected for this study.

Data collection

The study was conducted from December 2015 to July 2015 in selected districts (Banja and Ankesha Guagusa districts) of north-western Ethiopia. Data were collected using questionnaire interview, group discussion with community who own horses and personal observation. The data collected include landholding and land-use pattern, livestock herd size and composition, purposes of horse husbandry, housing condition of horse, horse feeds and feeding systems, breeding practices, reproductive and productive performance of horse and horse production constraints. Data collected through observation of the farm environment were assessing housing conditions, feeding and feed storage practices, and health status of horses and availability of health care services in the area.

Data analysis

Information collected through questionnaire was extracted and summarized for analysis using SPSS statistical software computer programs (SPSS version 20). Descriptive statistics like percentage and mean were used to describe the farming system management practices of horse production levels and constraints in the study area.

Results

Household characteristics of respondents

The household characteristics of respondent household's heads are indicated in Table 1. The majority of respondents in both districts were male headed. Regarding age structure, in Banja Shekudad district the majority (33.12%) of respondents were at mid-age 41–50 years, while in Ankasha most (41.9%) respondent were at old age.

Educational characteristics of respondents

The educational status of horse owners in the study (Table 2) noted that the majority (38% for Banja Shekudad and 34% for Ankesha Guagusa) of respondents were illiterate. This high illiterate class has got negative impact on livestock production and to accept new technologies regarding management and animal utilization improvement [10]. Therefore, improving education level of farmers has paramount importance in overall agricultural production.

Table 1 Social characteristics of respondents by district (*N* = 198). *Source*: Survey 2015

Parameters	Banja Shekudad (*N* = 99)	Ankesha Guagusa (*N* = 99)
Sex of HHs		
Male	87.5%	97%
Female	12.5%	3%
Total	100%	100%
Age (years)		
18–30	23.11%	8.5%
31–40	17.8%	10%
41–50	33.12%	21.5%
51–60	19.3%	41.9%
> 60 years	6.67	12%
Average family size (*N*)	8.0	6.0
Total	100%	100%

Table 2 Educational characteristic respondents by district (*N* = 198). *Source*: Survey 2015

Level of educational	Banja Shekudad (*N* = 99) (%)	Ankesha Guagusa (*N* = 99) (%)
Illiterate	38	34
Read and write	22	19
Grades 1–8 complete	20	16
Grades 9–10	12	18
Grades 11–12	5	4
Certificate and above	3	9
Total	100	100

Land and livestock holding

The characteristics of farming system in both districts were mixed crop–livestock with more or less similar features. The average landholding of respondent in Banja Shekudad districts was 0.84 ha/household, while 1.2 ha/household was observed in Ankesha Guagusa district. Types of livestock owned by respondents in the two districts were cattle, sheep and goat, mules, horses and donkeys chicken and honey bee colony. The average livestock herd size and composition in the study area are indicated in Table 3. The average livestock holding per house hold in Banja Shekudad and Ankesha Guagusa districts were 7.62 and 8.15 TLUs, respectively. Out of the total TLU of the districts, horses occupy 1.96 and 1.68 TLUs for Banja Shekudad and Ankesha Guagusa districts, respectively. This elucidates that horses are important in the farm household in both districts.

Purpose of horse husbandry in the study districts

The finding of this study elucidates that horses are used in both districts for many kinds of activities including land cultivation (Table 4). The initial sources of horses in all respondents (100%) in both districts are purchase from market. The purpose of horse keeping in Banja Shekudad districts is ploughing only (32%) followed by

Table 3 Average livestock holding (TLU) of respondents by district (*N* = 198). *Source*: Survey 2015

Livestock species	Banja Shekudad (*N* = 99)	Ankesha Guagusa (*N* = 99)
Cows	1.04	1.77
Oxen	1.23	1.1
Sheep + goat	1.4	1.4
Horses	1.9	1.68
Mules	1.3	1.2
Donkeys	0.75	1
Total	7.62	8.15

TLU tropical livestock unit

packing of goods (21%) while packing of goods (31%) followed by riding only (22%) in Ankesha Guagusa district.

Feeds and feeding of horses

Livestock feeding was both partial grazing and home feeding in both districts (Table 5). In Ankesha Guagusa district, 44.4% of respondents use half a day grazing and home feeding using cut fodder and crop residues, while 55.6% of respondents follow whole-day grazing and night-time supplementation of their grazing stock. In Banja Shekudad district, 63.4% of respondents follow half-a-day grazing followed by stall feeding of cut fodder and crop restudies depending on availability with relatively few respondents 36.6% follow whole-day grazing. The stall or homestead feeding practiced by some respondents is an interesting feature of livestock feeding with enormous advantage to promote fodder development and to reduce free grazing. This livestock feed practices need to be promoted to overall livestock farmers in the highland areas where grazing areas are decreasing. In both districts, all horse owners provide supplementary diets in both dry and wet seasons for all types of horses (riding, packing and ploughing). The type of supplementary feeds for all horses includes oat and barely grain. Regarding the amount of supplement provided, majority (59% in Banja Shekudad district and 71% in Ankesha Guagusa) of respondents supplied higher amount for working horses than non-working horses. The type of supplementary feeds used for grazing animals according to their importance was wheat bran (38.3%), oat grain and wheat bran (62%) and other by-products (7.7%).

Table 4 Purpose of horses for respondents by district (*N* = 198). ***Source*: Survey 2015**

Purpose	Banja Shekudad (*N* = 99) (%)	Ankesha Guagusa (*N* = 99) (%)
Ploughing only	32	19.50
Riding only	17	22
Packing of goods	21	31
ploughing and riding	17	10
Packing and riding	5	4
Ploughing and packing	8	13.10
Total	100	100

N number

Water and watering frequency of horses

The watering frequency of horses in dry and wet seasons in the study districts is presented in Table 6. The result noted that the majority of respondents in both districts provide water twice a day in dry season and once in the wet season. The availability and accessibility of water resources have a significant impact on the production and productivity of livestock in general and horses in particular. However, water availability was not reported as problems in the study districts, which might be associated with highland areas in both districts. The sole

Table 5 Feed resources of horses (% of respondents) in the study districts by season (*N* = 198). ***Source*: Survey 2015**

Dry season			Wet season		
Type of feed	Districts		Type of feed	Districts	
	Banja Shekudad (*N* = 99) (%)	Ankesha Guagusa (*N* = 99) (%)		Banja Shekudad (*N* = 99) (%)	Ankesha Guagusa (*N* = 99) (%)
TS, WB and NSC	48.31	55.11	Cut forage only	19.9	39.23
TS, WB and BH	22.79	24.9	Grazing only	37.5	27.27
BS and BH	28.9	20	Grazing and cut forage	42.6	33.5
Total	100	100	Total	100	100

TS teff straw, *WB* wheat bran, *NSC* noug seed cake, *BS* barley straw, *BH* bean halum

Table 6 Watering frequency of horses (% of respondents) in the study districts by season (*N* = 198). ***Source*: Survey 2015**

Frequency of watering	Dry season		Wet season	
	Banja Shekudad (*N* = 99) (%)	Ankesha Guagusa (*N* = 99) (%)	Banja Shekudad (*N* = 99) (%)	Ankesha Guagusa (*N* = 99) (%)
Not provide	2	4	18	24
Once a day	25.8	35.1	70	60
Twice	72.2	60.9	12	16
Total	100	100	100	100

(100%) source of water for horse drinking in respondents in the study districts was river water.

Housing, breeding and health care of horses

In this study, it was observed that all respondents (100%) in both districts provide shelter at home during night to protect from predators or other factors although the majority of respondents do not have separate houses for horses in both districts (Table 7). Breeding management of horses like other animals has very much economic benefit. However, all respondents in both districts indicated that breeding of horses is random by the horses themselves without any human intervention. Overall respondents (100%) noted that horses breed early in rainy season in both districts, which might be associated with availability of feed resources during the rainy season.

The major health problems of horses observed were cough and digestion colic. Although some of respondents (35% in Banja Shekudad district and 31.8% in Ankesha Guagusa) were used modern veterinary service and drugs to treat their horses against prevailing diseases, it is evident that most of respondents (47% in Banja Shekudad and 54.2% in Ankesha Guagusa districts) used homemade treatment for their horses.

Training on horse husbandry

The majority of horse owners have no training in relation to horses husbandry, while few (24.12% in Banja Shekudad and 33.78% in Ankesha Guagusa districts) have got some horse-specific training mainly related to feeding and health care of horses.

Constraints of horse husbandry

The major constraint noted by all respondents in both districts is shortage of land for horse grazing and feed production. Another is lack of access to credit to buy horse and inputs required for horse management, and incidence of disease is among mentioned problems during group discussion conducted in two sessions with respondents.

Table 7 Housing of horses (% of respondents) by district (*N* = 198). *Source*: Survey 2015

Type of house	Banja Shekudad (*N* = 99) (%)	Ankesha Guagusa (*N* = 99) (%)
No separate house	45.5	40.7
Separate house	40	35.1
Both	14.5	24.2
Total	100	100

Discussion

Household characteristics

The sex characteristics of respondents showed that the respondents in both districts were male-headed households that might be due to the fact that males are closely related to livestock handling and management. The male-headed household characteristics of respondents in both districts were in agreement with Asmare et al. [11] and Wolelie et al. [12] for different parts of Ethiopia. The age structure of household heads in both districts was similar in both sexes. The results of both districts were similar to the reports of Solomon et al. [13]; Asmare et al. [11] who reported the average age of respondents was 43.2 ± 1.0 years for Metekel Zone, north-western Ethiopia. The average family size of respondents in both districts in the present study was comparable with Asmare et al. [11] but lower than districts (Mandura and Pawe) of Metekel zone of Benishangul-Gumuz region ranging 6.04–6.94 [14]; the reports by Tesfaye [15] with overall mean family size of 5.7 persons in Metemma district in Northwest Ethiopia and national average (5.20).

Educational characteristics of respondents

The educational statuses of majority of respondents in both districts were found illiterate, which might be associated with poor education access in the area. The educational characteristics of horse owner respondents were almost to the reports of earlier reports in Ethiopia [11, 12, 16] for different districts of Ethiopia. This might be due to various factors such as access to education, awareness about the importance of education and other related factors. However, it is important to note that from technology adoption, this higher population of the illiterate class in both districts will have a disadvantage on the acceptance of new technologies like trainings, improved agricultural technologies and adopting them for better live improvement. In other words, adoption of improved technologies should also consider the literacy condition of farmers. Studies by Appleton and Balihuta [17] indicated that education is the main issue in agricultural development. (In particular, primary and secondary schooling had higher impact on agricultural development compared to any other level of education.)

Land and livestock holding characteristics

Mixed crop–livestock was the characteristic farming system in study districts with more or less similar features. The average landholding of Banja Shekudad districts was 0.84 ha per household, while that of Ankesha Guagusa district was 1.12 ha per household. The landholding per household in both districts was small as compared to the national average of the country. This might be hindrance to respondents in practicing improved fodder production for horses and other livestock in the area. The size of landholding per household of both districts is lower than Asmare et al. [11] for Burie-Zuria district in north-western Ethiopia, which was 1.4 ha per household. Moreover, the landholding status of both districts was lower than earlier reports and lower Yayeh et al. [18] in which the landholding of 0.98 ha for Debre Marqos district, Amare [19] with a report of 1.14 ha in Fogera district both are in the same region. Overall, the landholding in both districts is lower than the national average landholding size of 1.6 ha reported by FAO [20], which may in turn affect improved forages.

Horses are the dominant livestock species in Ankesha Guagusa, while cattle are dominant in Banja districts. This observation concurs with the report of Gizachew [21] that generally, equines are the most widely distributed animals in higher and medium highland zones. The average horse holding of respondents in both districts is higher than different reports [22, 23] for central highlands of Ethiopia.

Purpose of horse husbandry

The purpose of horses for the households in the study areas elucidates that horses are used for many activities in both districts mainly for ploughing and as pack animals. The result noted the land cultivation or ploughing using horse power has given much emphasis in the current study area, which is in agreement with earlier reports [12, 22, 24, 25]. The result was also in line the studies [26, 27] who reported equines can be used for different roles in the farm family daily chores such as transportation of goods by pack and cart due to their sturdy nature and manageable behaviour.

Feed and feeding of horses

The activity of feeding of horses in both districts was almost similar in that grazing on natural pasture and supplementation using home-grown feeds. This home/homestead feeding is an interesting feature of livestock feeding, which in turn has enormous advantage to promote fodder development and using cut-and-carry system, thereby reducing free grazing. Regarding the amount of supplement given to horses, the type and amount of feed fed requirement varies according to the workload of the horses. The finding agrees with different results [25, 28, 29]. Similar reports also made by other researchers [30, 31]. Feed shortage during the dry season and disease were the major constraints for horse production and agree with the reports of other authors for different livestock species. However, in the study area the overall horse feeding and supplementation practice were not quantified, which needs further research in the field.

Water and watering frequency of horses

The majority of respondents in both districts provide water twice a day in dry season and once in the wet season. The provision of adequate quantities of clean drinking water is a major prerequisite for satisfactory milk production, growth and animal health [32], but the minimum amount required is affected by various factors and therefore seldom known exactly. The availability and accessibility of water resources have a significant impact on the production and productivity of livestock in general and horses in particular. Source of water for horses and frequency of watering is in agreement with Ayza et al. [16]) and Beriso et al. [33] for other species of livestock in different areas of Ethiopia. Letting horses to drink water at least once a day was common feature of all respondents (100%) in both districts, which is in agreement with earlier reports [24].

Housing, breeding and health care of horses

In the study districts, the majority of respondents do not have separate houses for horses in both districts. That means horses were kept within the family house of respondents like the earlier results for horses [21, 26] and other species of livestock [16]. This might be due to lack of awareness and shortage of resources to construct separate horse houses in the study areas. The season of breeding of horses in both districts is early wet season in all respondents (100%) in both districts. Although breeding management of horses like other animals has very much economic benefit, breeding of horses in all respondents is by the horses themselves without any human intervention. Like other livestock species [34, 35] in most production systems of Ethiopia, no planning of breeding and human-assisted breeding of horses in most cases. According to respondents, horses breed early rainy season in both districts, which might be associated with availability of feed resources associated with the rainy season. The major problems of horses observed are cough and digestion colic in both districts. From overall respondent in both districts, only few get modern veterinary service, while majority use traditional medication for their horses. This might be due to lack of awareness and shortage of capital to afford veterinary services. The finding is in agreement with other equines species such as donkey health management reports in Ethiopia [36, 37] in which few equine species get annually veterinary care and owners also provide home-made treatment ailments to their animals.

Training on horse husbandry

Even though training access is crucial in improving animal husbandry including horses, majority of horse owners have got no training related specifically to horses, while few have some horse-specific training. It is noted that training on livestock husbandry is essential to livestock owners to increase production and productivity of their stock. This finding agrees with earlier reports of Asmare et al. [11] for cultivated forage production and management in selected areas of Ethiopia. It is noted by Ampaire and Rothschild [38] that training on livestock husbandry is essential to livestock owners to increase production and productivity of their stock. Overall, the importance of training of farm households in developing countries (including Ethiopia) also indicates that training has useful effect to improve farm productivity [38–41].

Constraints of horse husbandry

Feed shortage during the dry season and disease were the major constraints for horse production in both districts. Shortage of feed during dry season might be due to expansion of crop land and decreased feed or fodder production, which can be used for feeding of horses. In Ethiopia, expansion of crop farming in grazing areas was a major problem for horse grazing as reported for other species of livestock [42]. The problem of disease and illness in horses might be linked to lack of awareness on horse healthcare management and improper nutrition of horses. The overall result of this finding agrees with the reports of other authors for different livestock species [11, 13, 16]. The constraints of reported in this study call more research and intervention in horse management in the study districts to exploit the potential of horses in the livelihood of poor farmers in the areas.

Conclusion

The present finding shows that horses are important animals in the study areas, serving for numerous purposes. Horse management system is found mainly on traditional mixed crop–livestock system in which horses are allowed to graze freely throughout the day in non-working days and off-duty grazing during working days. Although lack of know-how for improved management of horses and feed shortage were the major problems of working horses, supplementation of all working horses with available feed resources was found as a common and useful practice in the study districts. Hence, it is recommended that adequate credit service, training and finding alternative feeds and feeding practices should be sought to use the horses' potential in the socio-economic value for their owners. As this study was an initial assessment on horse husbandry status of the study areas, further comprehensive studies including controlled feeding management should be conducted in future.

Abbreviations

AGDOA: Ankesha Guagusa District Office of Agriculture; BDOA: Banja District Office of Agriculture; BH: bean halum; BS: barley straw; CSA: Central Statistical Agency; FAO: Food and Agriculture Organization; ha: hectare; km: kilometre; N: number; NSC: noug seed cake; SPSS: Statistical Package for Social Sciences; TS: teff straw; WB: wheat bran.

Authors' contributions

BA contributed in research proposal writing, data collection, data analysis, data interpretation and article writing. ZY contributed in research data collection, data analysis, data interpretation and article writing. Both authors read and approved the final manuscript.

Authors' information

Bimrew is an assistant professor of Animal Feeds and Nutrition, currently teaching and conducting research in the area of animal feeds and feeding, animal production and productivity improvement in Ethiopia. Zemenu is a lecturer of Animal Production currently teaching and conducting research in animal production and productivity improvement in Ethiopia.

Author details

[1] Department of Animal Production and Technology, College of Agriculture and Environmental Sciences, Bahir Dar University, P.O. Box 79, Bahir Dar, Ethiopia. [2] Department of Animal Sciences, College of Agriculture and Natural Resources, Debre Markos University, P.O. Box 269, Debre Markos, Ethiopia.

Acknowledgements

The authors would like to acknowledge farmers who participated in interview and group discussion, the office of Agriculture of respective districts for secondary data sources. We are also grateful to Dinberu Workineh and Wasihun Alemenew who helped us during data collection at field level.

Competing interests

The authors declare that they have no competing interests.

Funding

Not applicable.

References

1. Alujia AS, Lopez F. Donkeys in Mexico. In: Fielding D, Pearson RA, editors. Donkeys, mules and horses in tropical agricultural development. Edinburgh: CTVM; 1991. p. 1–7.
2. FAO Statistical Database Website. Food and Agriculture Organization, Rome, Italy, 2003 (FAOSTATS: http://faostat.fao.org/default.aspx).
3. Fielding D. The number and distribution of equines in the world. In: Proceedings of the colloquium on donkeys, mules and horses in tropical agricultural development, Edinburgh, 3 to 6 September, 1991. p. 62–66.

4. CSA (Central Statistical Authority). Agricultural sample survey agricultural sample survey livestock and livestock characteristics (private peasant holdings), Ethiopia. 2015. p. 103.
5. Mengistu A. The genetic resources perspective of equines in Ethiopia and their contribution to the rural livelihoods. In: Proceedings of the 11th annual conference of the Ethiopian Society of Animal Production (ESAP), Addis Ababa, Ethiopia, 2003. p. 81–5.
6. Ameni G. Preliminary trial on the reproducibility of epizootic lymphangitis through experimental infection of two horses. Short communication. J Vet. 2006;172:553–5.
7. Bazezew M, Chanie M, Tesfaye T, Kassa A, Mekonnen B, Wagaw N. Lameness and associated risk factors in cart mules in northwestern Ethiopia. Glob Vet. 2014;12:869–77.
8. Banja Shekudad District office of Agriculture (BDOA), 2015. Annual Report (unpublished).
9. Poete CD, Daplyn PF. Data for agrarian development. Cambridge: Cambridge University; 1993.
10. Bekele B. The effect of illiteracy on Rural Development in Jeldu District, Oromia Regional State, Ethiopia. In: A thesis submitted to Indra Gandhi Open University for partial fulfillment of Master Degree of Arts in Rural Development. Addis Ababa, Ethiopia; 2012
11. Asmare B, Demeke S, Tolemariam T, Tegegne F, Wamatu J, Rischkowsky B. Determinants of the utilization of desho grass (*Pennisetum pedicellatum*) by farmers in Ethiopia. Trop Grassl. 2016;4:112–21.
12. Wolelie B. Smallholder cattle production systems and husbandry management in West Shewa zone of Oromia Regional State, central Ethiopia. World Sci News. 2016;53:178–88.
13. Solomon Z, Binyam K, Bilatu A, Ferede A, Gadisa M. Smallholder cattle production systems in Metekel zone, northwest Ethiopia. Res J Agric Environ Manag. 2014;3:151–7.
14. Mekuriaw Y, Urge M, Animut G. Role of indigenous bamboo species (*Yushania alpina* and *Oxytenanthera abyssinica*) as ruminant feed in northwestern Ethiopia. Liverst Res Rural Dev. 2011;23:185.
15. Tesfaye M. Characterization of cattle milk and meat production, processing and marketing system in Metema district, Ethiopia. M.Sc. Thesis, Hawassa University, Ethiopia, 2007.
16. Ayza A, Yilma Z, Nurfeta A. Characterization of milk production systems in and around Boditti, south Ethiopia. Liverst Res Rural Dev. 2013;25:183.
17. Appleton S, Balihuta A. Education and agricultural productivity: evidence from Uganda. J Int Dev. 1996;8:415–44.
18. Yayeh Z, Hailmariam M, Belhu K, Asmare B. Characterization of dairy cattle production systems in Debre Markos District, Amhara Regional State, Ethiopia. Pac J Agric Sci Res. 2014;2:42–51.
19. Amare S. Livestock production systems and available feed resources in different agro-ecologies of north Gondar zone, Ethiopia. An M.Sc. Thesis Presented to the School of Graduate Studies of Haramaya University, 2006. 80 p.
20. FAO (Food and Agriculture Organization). Country report. FAO, Addis Ababa, Ethiopia, 2008.
21. Gizachew L. Revegetation of degraded range land with *Cenchrus ciliaris* and *Chlories gayanain* middle Awash of Afar Region. An M.Sc Thesis presented to the School of Graduate Studies of Alemaya University, 2002. p. 95.
22. Hassen A, Ebro A, Kurtu M, Treydte AC. Livestock feed resources utilization and management as influenced by altitude in the Central Highlands of Ethiopia. Liverst Res Rural Dev. 2010;22:12.
23. Abera M, Tolera A, Assefa G. Feed resource assessment and utilization in Baresa Watershed, Ethiopia. Int J Sci Res. 2014;3:66–72.
24. Mekuria S, Mulachew M, Abebe R. Management practices and welfare problems encountered on working equids in Hawassa town, Southern Ethiopia. J Vet Med Anim Health. 2013;5:243–50.
25. Makki EK, Mohammed Musa EO. Management and field performance of horses and donkeys used for ploughing on smallholder farms. An example from North Kordofan, Sudan. Liverst Res Rural Dev. 2013;25:80.
26. Dinka H, Shelima B, Abalti A, Geleta T, Mume T, Chala R. Socio-economic importance and management of carthorses in the mid rift valley of Ethiopia. In: Pearson RA, Muir CJ, Farrow M, editors. The future for working equines. The fifth international colloquium on working equines. In: Proceeding of an international colloquium held at the Addis Ababa University, Ethiopia, 30th October to 2nd November 2007. The Donkey Sanctuary, Sidmouth, Devon, 2007. p. 181–188.
27. Mekuria S, Abebe R. Observation on major welfare problems of equine in Meskan district, southern Ethiopia. Liverst Res Rural Dev. 2010;22:48.
28. Harris PA. Feeding and management advice for tying up. In: Harris PA, Gomarsall G, Davidson HPB, Green R, editors. Proceedings BEVA specialist meeting on nutrition and behaviour. Newmarket: Equine Veterinary Journal; 1999. p 100–104.
29. Anderson M, Dennis R. Improving animal based transport options, approaches, issues and impact. In: Starkey P, Mwenya E, Stares J, editors. Improving animal traction technology. Proceeding of the first workshop of the animal traction network for Eastern and Southern Africa (ATNESA) held 18–23 January 1992, Lusaka, Zambia; 1994. p. 378–395.
30. Chanie M, Fentahun T, Mitiku T, Berhan M. Strategies for improvement of draft animal power supply for cultivation in Ethiopia: a review. Eur J Biol Sci. 2012;4:96–104.
31. Guyo S, Legesse S, Tonamo A. A review on welfare and management practices of working equines. Glob J Anim Sci Liverst Prod Anim Breed. 2015;3:203–9.
32. Little W, Shaw SR. A note on individuality of drinking water by dairy cows. Anim Prod. 1978;26:225.
33. Beriso K, Tamir B, Feyera T. Characterization of smallholder cattle milk production system in Aleta Chukko District, southern Ethiopia. J Adv Dairy Res. 2015;3:132.
34. Bayou E, Haile A, Gizaw S, Mekasha Y. Characterizing husbandry practices and breeding objectives of Sheko cattle owners for designing conservation and improvement strategies in Ethiopia. Liverst Res Rural Dev. 2014;26:235.
35. Zewdie B, Welday K. Reproductive performance and breeding strategies for genetic improvement of goat in Ethiopia: a review. Greener J Agric Sci. 2015;5:023–33.
36. Mohammed A. Management and breeding aspects of donkeys around Awassa, Ethiopia. In: Fielding D, Pearson RA, editors. Donkeys, mules and horses in tropical agricultural development. Edinburgh: CTVM; 1991. p. 185–8.
37. Niraj K, Fisseha K, Shishay N, Hagos Y. Welfare assessment of working Donkeys in Mekelle City, Ethiopia. Glob Vet. 2014;12:314–9.
38. Ampaire A, Rothschild MF. Effects of training and facilitation of farmers in Uganda on livestock development. Liverst Res Rural Dev. 2010;2010(22):130.
39. Ali M, Flinn JC. Profit efficiency among Basmati rice producers in Pakistan Punjab. Am J Agric Econ. 1989;71:303–10.
40. Wang J, Cramer GL, Wailes EJ. Production efficiency in Chinese agriculture: evidence from Rural Household Survey data. Agric Econ. 1996;15:17–28.
41. Seyoum ET, Battese GE, Fleming EM. Technical efficiency and productivity of maize producers in eastern Ethiopia: a study of farmers within and outside the Sasakawa Global 2000 project. Agric Econ. 1998;19:341–8.
42. Fetsum S, Solomon Z, Mikias D, Yigremachew S, Demeke M, Gizachew W, Yeshi C. Gender based participatory rural appraisal of farming systems in Wombera, Bullen and Guba Woredas. In: Gender differentials for planning agricultural research. Proceedings of the workshop on gender analysis in agricultural research, Addis Ababa, Ethiopia, 2009. p. 161–87.

10

Toward climate-smart agriculture in West Africa: a review of climate change impacts, adaptation strategies and policy developments for the livestock, fishery and crop production sectors

Robert Zougmoré[1], Samuel Partey[1*], Mathieu Ouédraogo[1], Bamidele Omitoyin[2], Timothy Thomas[3], Augustine Ayantunde[4], Polly Ericksen[5], Mohammed Said[5] and Abdulai Jalloh[6]

Abstract

Many projections of the impact of climate change on the crop, livestock and fishery production sectors of African agriculture are reported in the literature. However, they may be arguably too general to understand the magnitude of impact and to inform adaptation strategies and policy development efforts that are tailored to promoting climate-smart agriculture in the West African region alone. This paper was synthesized from several scholarly literature and aimed at providing up-to-date information on climate change impacts, adaptation strategies, policies and institutional mechanisms that each agriculture subsector had put in place in dealing with climate change and its related issues in West Africa. For each subsector (crop, fishery and livestock), the current status, climate change impacts, mitigation and adaption strategies have been analyzed. In addition, we reviewed recent policy initiatives in the region that foster the development and adoption of climate-smart agricultural options to improve resilience of farming systems and livelihoods of smallholder farmers to climate change risks. From community to national and regional levels, various strategies and policies are also being taken to guide actions and investment for climate-smart agriculture in West Africa.

Keywords: Rural livelihood, Resilience, Food security, Climate change, Finance, Africa

Background

West Africa remains vulnerable to episodic climate shocks (primarily drought). Food crises continue to hit the region (particularly in the dry areas), with resultant loss of lives and livelihoods, and a cycle of disaster relief that compete with long-term developments [1]. As a region that produces at least 30% of the food requirements of the African continent, increasing resilience of agricultural systems to climate change is a major developmental agenda in the quest to end hunger and reduce poverty while also mitigating greenhouse gas emissions [2]. The concept of climate-smart agriculture (CSA) dwells on this development agenda and aims at fostering the development and implementation of agriculture innovations that (1) sustainably increases agricultural productivity to support equitable increases in incomes, food security and development; (2) adapts and builds resilience to climate change from the farm to national levels; and (3) develops opportunities to reduce greenhouse gas emissions from agriculture compared with past trends [3]. Despite the potential benefits CSA could offer West Africa, it is confronted with many challenges. Notable among these is a clear understanding of

*Correspondence: s.partey@cgiar.org
[1] CGIAR Research Program on Climate Change, Agriculture and Food Security (CCAFS), International Crops Research Institute for the Semi-Arid Tropics (ICRISAT), BP 320, Bamako, Mali
Full list of author information is available at the end of the article

CSA concept among farmers, policy makers and potential investors; appropriate packaging of technologies that underpin CSA practices for wider adoption by farmers; how we can manage trade-offs from the farmers' and policymakers' perspectives; marginality of agro-ecological regions in West Africa that requires research on the biophysical adaptation of CSA practices across the region; institutional arrangements to upscale CSA from the farm scale to the landscape and country levels; limited technological capabilities and human resources competence; fitting CSA into the existing policy frameworks; and the development and implementation of effective risk-sharing schemes [4].

In view of the above challenges, a major question arises: How will West Africa adapt its agriculture to the impacts of climate change and variability today and in the future? Among other factors, providing answers to this question requires starting with an understanding of various agricultural subsectors and their respective current adaptation strategies; policy developments and institutional settings may foster the adoption of sustainable agricultural systems that effectively mainstream climate change in the region. This paper therefore analyzes current and future climate change impacts, adaptation initiatives and policy developments for the livestock, fishery and crop production sectors of West Africa, with a perspective of promoting climate-smart agriculture. Together with other scholarly materials, some contents of this paper were synthesized from a working paper (WP N°118-http://hdl.handle.net/10568/67103) that served as a background document during the high-level forum of climate-smart agriculture stakeholders in West Africa, organized by the Economic Community of West African States (ECOWAS) in June 2015, Bamako, Mali.

Livestock production sector

Current status of livestock production in West Africa

The livestock sector remains a major contributor to rural livelihoods and the national economies of many West African nations. At least 100 million poor people including women in West Africa rely on livestock as part of their livelihood strategy [5, 6]. In the Sahelian zone, the sector accounts for about 35% of gross domestic product (GDP) and supplies about 30% of the revenue in the agriculture sector [7]. In the arid and semiarid agro-ecological zones of West Africa, livestock husbandry provides the main source of employment for the majority of the people and is by far the most important source of revenue. For both crop farmers and pastoralists, livestock serve as a productive asset to generate income, and form a key element in food security strategies in many countries of West Africa. In addition, livestock provide draft power, skins, transport and manure, and fulfill various sociocultural functions such as payment of dowry, establishment and reinforcement of relationships and source of prestige within the pastoral society [8, 9]. Livestock production in West Africa is largely associated with exploitation of natural rangelands (i.e., pastoral and agro-pastoral systems) [10]. However, livestock are also raised in mixed smallholder systems in which crop residues are increasingly becoming important as animal feed with the expansion of crop fields into marginal (grazing) lands [11].

FAO [12] revealed an increased growth rate in milk production in West Africa (from 2000 to 2010), primarily due to the fast increase in milk production in Cape Verde (8.4 percent), Mali (12.4%) and Sierra Leone (6.9%). However, the region still lags behind Northern and Southern Africa in terms of meat production (Fig. 1). Although a major contributor to the national economies of most West African countries, the sector is faced with several challenges. Technical challenges facing livestock production in the region include unfavorable policies, low and variable forage availability, poor feed quality, access to water, low productivity of indigenous breeds and degradation of rangelands. Low available forage due to low biomass production from rangelands is particularly a major problem in arid and semiarid zones in the region. Drought is another major constraint to livestock production in West Africa, particularly in the Sahelian countries [13]. Drought affects livestock production through reduced herbage production and water scarcity which often leads to high herd mortality. Besides the aforementioned, West African livestock producers are constrained by poor infrastructures for transportation, processing and marketing and weakly enforced institutional mechanisms. For example, pastoral and agro-pastoral producers in the north, who supply 60% of cattle meat, 40% of small ruminant meat and 70% of milk [7], rely on moving their cattle in search of grazing area and water every year during the Sahelian dry season to gain access to the more humid farming areas of the south. Protection of the key transhumant corridors from north to south is key for their animals to survive [7] (Fig. 2), but increasing conflicts due to more livestock competing for resources is a concern [14]. In terms of economic competitiveness, West Africa still imports animal products. Commodity chains need better infrastructure support in linking producers and traders across value chains and harmonizing the different actors of the livestock industry to foster regional trade [7]. Livestock production systems are in a transition to both a more sedentarized but possibly more intensified mode that needs proper policy support, in order to sustain the economic contributions but to adapt to changing climate conditions.

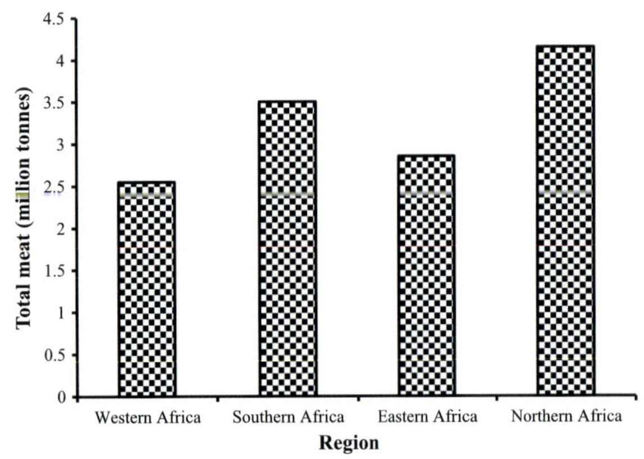

Fig. 1 Distribution of meat in Africa in 2010. *Source*: adapted from FAO [12])

Observed climate change impacts and projections for the livestock sector

The fifth assessment report of the IPCC reports that the vulnerability of livestock keeping communities may be amplified as climate change interacts with other stressors of the livestock sector such as rangeland degradation; increased variability in access to water; fragmentation of grazing areas; sedentarization; changes in land tenure from communal toward private ownership; in-migration of non-pastoralists into grazing areas; lack of opportunities to diversify livelihoods; conflict and political crisis; weak social safety nets; and insecure access to land and markets [15]. It is reported that the changing frequency of extreme climate conditions such as droughts and floods has had greater impacts on livestock and the associated livelihoods than average trends from climate change (that is, average change in precipitation and temperature) [16, 17]. For example, repeated occurrence of droughts in the Sahel has led to adoption of agro-pastoralism (combination of crop farming and livestock rearing within the same farm) among the pastoralists who were once solely depending on livestock for their livelihood [7]. Similarly, crop farmers have diversified in the past two decades into rearing livestock due to repeated crop failure associated with droughts [18].

Climate change is expected to affect livestock at both the species and breed levels, although this is a major research gap. Specific impacts of climate change on livestock may include changes in availability and quality of forage resources, access to water, species and breeds of livestock that can be kept, livestock mobility and animal

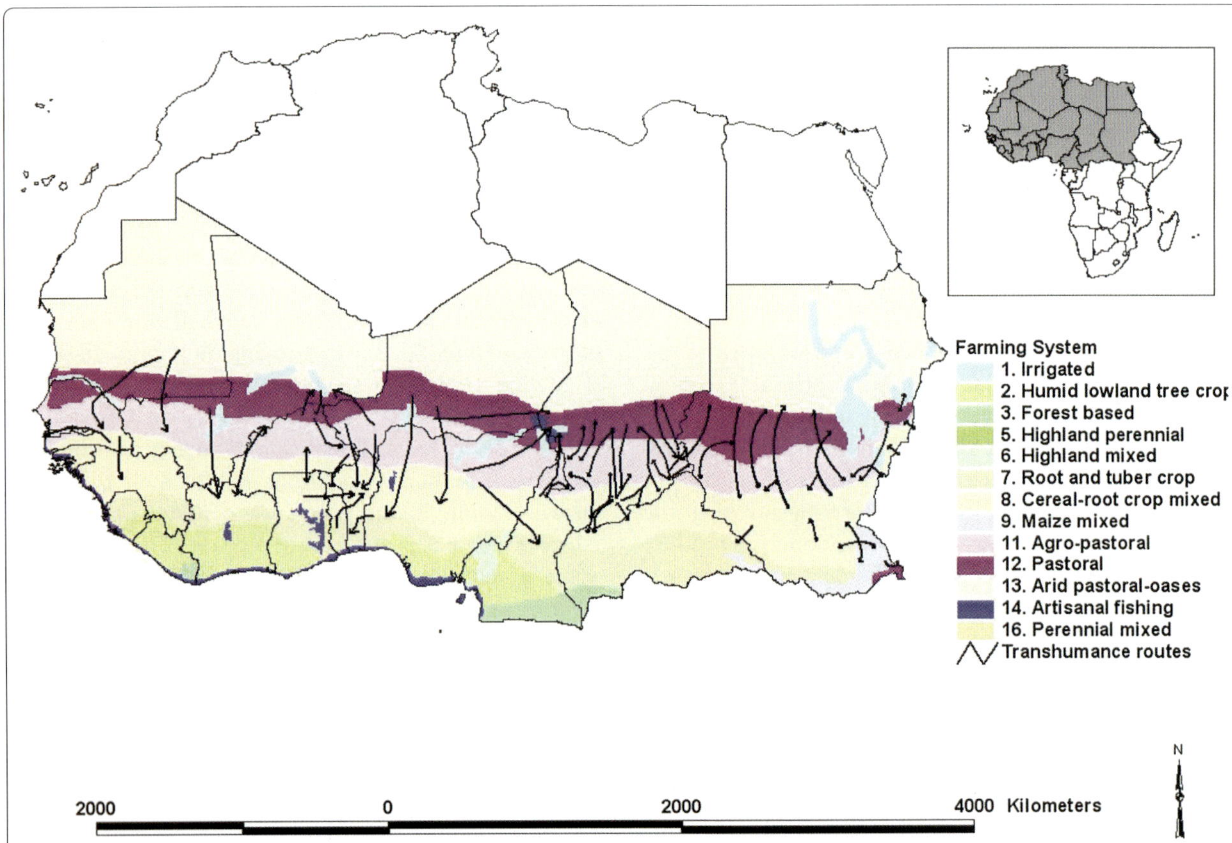

Fig. 2 Distribution of livestock movement across different farming systems in West Africa (*Source*: Zougmore et al. [27])

diseases (emerging and re-emerging diseases) [19]. A hotter and drier climate in the arid and semiarid zones of the region may favor livestock species and breeds that thrive well under heat stress and those with less water requirements such as small ruminants (sheep and goats) and camels [20]. This is already the case in West African Sahel with the shift in livestock species from cattle before the droughts of early 1970s and 1980s to sheep and goats [21] as the latter (small ruminants) are less costly, hardier, require lower feed and reproduce faster than cattle. A hotter and drier climate in subhumid and humid zones will modify the habitat of the endemic livestock breeds which are resistant to trypanosomiasis, the major animal disease in the zones, and will consequently alter the breeds that can be kept. However, climate change impacts on forage availability and quality may include: changes in herbage growth, changes in floristic composition of vegetation, changes in herbage quality and changes in the importance of crop residues as animal feed [22]. Generally, the impacts of climate change on herbage growth will depend on the plant species as increase in future CO_2 levels may favor different grass species than currently, while the opposite is expected under associated temperature increases [23]. The consequences of these impacts for livestock producing households depend on the development pathway taken, such as population growth, changes in income levels, growth in regional trade and degree of technological development [22]. Meanwhile, the current low adaptive capacity is expected to make the region particularly vulnerable to climatic shocks such as drought and flood. In general, the impacts of climate change on the poor livestock keepers will be context-specific, reflecting factors such as geographic location, socioeconomic profiles, prioritization and concerns of individual households, as well as institutional and political constraints [24].

A critical challenge that will face livestock keepers particularly pastoralists is the inevitable switch from their livelihoods that are intertwined with their customs and traditions over generations. Some climate models predict growing wetness in large areas of key cattle-producing countries like Niger, Nigeria and Mali. High rainfall could make these areas inappropriate for cattle production due to increase in disease pressure. The consequences will be very grave in acquiring much needed skills in exclusively growing crops particularly those that are adapted to wet conditions which the herders have not been accustomed to. In addition to threats to their cattle, the infrastructure with regard to houses and roads is adapted to dry conditions. Persistent rainfall will pull down the characteristic mud houses and fragile roads. In the meantime, however, livestock will remain an important asset to help households manage climate risks.

Adaptation strategies and mitigation options for the livestock production sector

In addressing climate change adaptation for livestock-based livelihoods, key questions to consider include: (1) Which types of livestock management are suited to climate change and where? (2) which animal species and breeds should be kept in which areas and what are the trade-offs? (3) which animal diseases should we focus on? (4) are there current livestock-based livelihood systems in the region that are best suited to climate change adaptation? (5) how can we add value to the existing livestock-based adaptation strategies? (6) are there policy and institutional mechanisms to enhance adaptation of livestock production systems to climate change and variability? (7) how could the capacity of rural institutions be strengthened to use appropriate tools and strategies to cope better with consequences of climate change? (8) how could we balance the need for short-term adaptation, which is often reactive, with long-term climate change adaptation planning? At community level, climate change adaptation should be considered in the context of other significant drivers of change (demographic change, economic development, market opportunities). Livestock production systems should be "climate-smart" by contributing to increasing food security, adaptation and mitigation in a sustainable way. Any livestock management practice that improves productivity or the efficient use of scarce resources can be considered climate-smart because of the potential benefits with regard to food security, even if no direct measures are taken to counter detrimental climate effects [25]. For livestock-based livelihood, adaptation options depend on the climatic risks, agro-ecological zones, the livestock production systems and the socioeconomic profiles of the household (Table 1). A community's capacity to adapt to climate change and the associated risks depends on its economic resources, geographic location, available technologies and information, infrastructures, institutions and networks [24]. Generally, poor infrastructures and weak institutions inhibit adaptive capacity and planning of a community. For most livestock keepers in West Africa, adaptation options are often not limited to livestock but a mixture of livelihood options including crop agriculture, non-agricultural activities and migration [14, 26].

Mitigation measures in the livestock sector could include technical and management options that promote reduction of greenhouse gas (GHG) emissions from livestock: efficient livestock feeding systems, balanced feed rations and efficient manure management. While improving feed resource use efficiencies would improve livestock productivity and reduce emissions per unit of product, one major challenge is finding the right behavioral incentives to encourage greater feed resource use

Table 1 Adaptation options for livestock-based livelihoods to major climatic risks in West Africa. *Source*: Zougmore et al. [27]

Agro-ecological zone	Dominant livestock system	Climatic risk	Adaptation option
Arid	Pastoral Agro-pastoral	Drought	Shift to small ruminants and camels Livestock mobility to semiarid/subhumid zones Commercial activities Growing of adapted crop varieties in the "oasis" Migration (local and regional)
Semiarid	Pastoral Agro-pastoral Peri-urban livestock	Drought Flood Bush fire	Shift to small ruminants Livestock mobility to subhumid zone Better integration of crop and livestock Commercial activities Growing of adapted crop varieties, e.g., drought-tolerant millet/sorghum Fodder conservation Migration (local and regional)
Subhumid/humid	Mixed crop livestock Peri-urban livestock	Flood Bush fire Vector-borne diseases, e.g., ticks, trypanosomiasis	Intensification of crop-livestock production Growing of cash crops, e.g., cotton Fodder production and conservation Use and conservation of endemic livestock breeds Commercial activities Moving out of agriculture to service industry
Coastal	Peri-urban livestock	Flood Water erosion Vector-borne diseases	Use and conservation of endemic livestock breeds Commercial activities Moving out of agriculture to service industry

efficiency, taking advantage of a win–win opportunity. It is now well understood that while many communities are highly adaptive and community-based approaches are critical, many governments and donors worry that a more concerted effort at higher levels of action and governance are needed to move beyond the so-called incremental adaptation and bring about the transformations that will allow for systemic adaptation [27]. This is even more necessary if synergies between adaptation and mitigation are to be found and taken advantage of to promote gains on both fronts. One challenge to this is the need for better integration between agriculture and livestock ministry staff and climate change units, so that climate change issues can be integrated into ongoing development planning. Another challenge is the lack of support for pastoral production systems and the failure to recognize that these are highly adaptive systems which resist conformation to the standard "intensification" model [14, 27]. In West Africa, livestock make tremendously important contributions to economies and food security. In terms of financing flows available, agriculture in general, and livestock in particular, suffers from inherent biases in the current mitigation financing mechanism. The Clean Development Mechanism excludes agriculture by and large, and there are a few voluntary carbon markets with interest in agriculture, particularly agroforestry and livestock production. Within these schemes, however, there is a bias toward "high potential" systems such as intensive dairy, where gains could be made from improving feeding practices. More extensive grazing-based systems are considered to be too difficult institutionally to manage. In terms of adaptation financing, many countries now have National Adaptation Programs of Action (NAPAs), but often agriculture is given short attention, and often the funds get "stuck" at the national level, failing to reach the local communities [3].

Fishery sector

Current status of fisheries in West Africa

West African fisheries encompass a wide range of ecological and socioeconomic components. Africa has huge potential for fish farming in terms of land availability with 31% of its surface area suitable for small-scale fish farming and 13% suitable for commercial fish production. African fisheries contribute significantly to food and nutritional security of an estimated 200 million people and provide a source of income for over 10 million who are engaged in production, processing and trade [28]. The artisanal fishery sector dominates employment in the fishing industry. The fishermen that operate the artisanal fishing industry use traditional wooden boats, sometimes motorized, with a variety of gear types, including nets, lines and seines. The industrial fishing is however operated by non-African trawlers and fleets with less direct economic and employment benefits. Total number of jobs created in the sector is over 5.6 million, while total number of direct jobs is over 1.8 million [29]. Women are responsible for artisanal processing and distribution of fish to urban centers and inland. Traditional processing methods include smoking, drying, salting and curing. Industrial processing exists in some countries. In many countries, inaccessibility of cold storage facilities inhibits the growth of a value-added industry [29].

West African population is estimated to increase to 430 million by 2025, and more than half of the region's population consumes fish products on a daily basis which accounts for up to 3–5% of total GDP [30]. As at 2012, total fish production in West Africa stood at about 2.9 million metric tonnes with the largest percentage contributions from Nigeria (32), Senegal (16), Mauritania (15) and Ghana (14) (Fig. 3) [12]. Although marine fish and invertebrates exported from West Africa amounted to only US$600 million annually [24] and contributed only about 2% to the total export value from West Africa countries, the fisheries sector in the region continues to play an important role in the local economy of countries like Mauritania and Senegal which are net exporters of fish [31]. However, Smith et al. [32] revealed that the low level of exports from West Africa relative to other regions reflects access agreements between West African countries and countries in Europe and Asia. Meanwhile, aquaculture, which is the world's fastest growing food production system, growing at 7% annually is still at developing stage in West Africa. It has, however, recently received higher levels of governmental and private support [32].

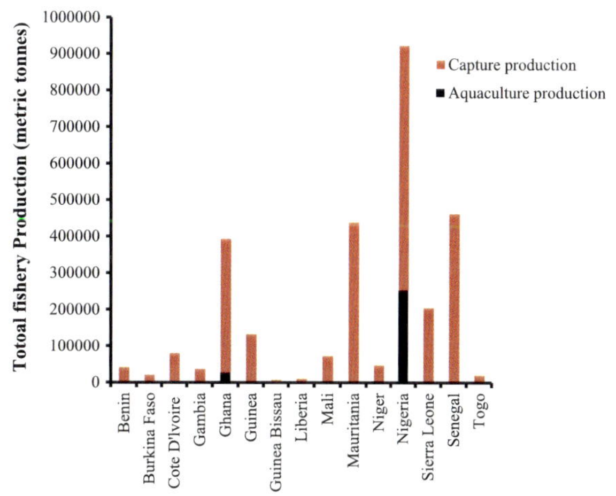

Fig. 3 The distribution of total fishery production among some West African countries in 2012 (*Source*: adapted from FAO [12])

Observed climate change impacts and projections for the fishery sector

Climate change is projected to impact the fisheries and aquaculture sector [33]. FAO [33] reported that the projected changes due to climate impact in West Africa coastal fisheries include changes in the composition, production and seasonality of plankton and fish populations. These projected impacts of climate change on fisheries will affect both the social and economics for fishing fleets and fishing communities. Many climate models have projected that the West Africa Inland fishery zones will be impacted by potential reduction in floodplain zones for seasonal inland fishing areas as a result of lowered precipitation. Increased demand for dam infrastructure for access to water or energy will also exacerbate the reduction in fish landings for seasonal inland fishing. Allison et al. [34] projected that in addition to precipitation and temperature changes, changes in sea level rise, land-based runoff and increasing frequency of storms and storm surges threaten coastal infrastructure, aquaculture located in riparian and coastal zones, and loss of harbors or homes. Omitoyin and Tosan [35] reported that climate change is already modifying the distribution of fish species, thus affecting habitat size, species diversity and productivity of Lagos Lagoon in Nigeria. The total landings of 14 West African countries were estimated to fall by about 8 and 26% percent from 2000 to 2050 due to low and high greenhouse gas emission scenario, respectively [31, 36]. In addition, the study indicated that the Exclusive Economic Zones (EEZs) of Ghana, Cote d'Ivoire, Liberia, Togo, Nigeria and Sierra Leone will experience up to and above 50% reductions in landings under a high emission scenario. The total landed value was estimated to drop from US$732 million to US$577 million between 2000 and 2050 for the high emission scenario [31, 36]. Associated Press [37] reported low fish catch in Lake Chad, while field observation on Lake Chad in 2012 shows that there is not only reduction of fish catch but also of total area covered by water. Rhodes et al. [36] reported that in Cote d'Ivoire, major fish species are being affected by changes in fresh water flows and greater intrusion of salt water into lagoons and lakes. Moreover, impact on aquaculture will include increasing seasonal and annual variability in rainfall leading to flooding and drought extremes. In Nigeria, Adebo and Ayelari [38] reported potential flood incidences which could wash away fish from small fish farms in Nigeria. Furthermore, Lam et al. [31] projected that climate change may lead to substantial reduction in marine fish production and decline in fish protein supply in West Africa by the 2050s under the Special Report on Emission Scenarios (SRES) A1B. Combining with economic parameters, they projected a 50% decline in fisheries-related jobs and a total annual loss of US$311 million in the whole economy of West Africa. These changes are expected to increase the vulnerability of the region (through economics and food security) to climate change [31].

Adaptation strategies and mitigation options for the fishery production sector

Small-scale fisheries and aquaculture have contributed little to the causes of climate change but will be among the first sectors to feel its impacts. For instance, surface water resources in West Africa are concentrated in a few watershed areas in the Niger, Lake Chad, Senegal, the Gambia and the Volta [39]. Following the decrease in rainfall since the 1970s, the main rivers have witnessed a drop in their stream flows. The Niger River's (Onitsha) stream flow fell by 30% between 1971 and 1989; those of the Senegal and Gambia Rivers fell by almost 60% [39]. According to studies by McCartney et al. [40], water resources development is vital for the well-being and livelihoods of the people living in the Volta River Basin and central to the economic development of the riparian countries. There remains great uncertainty about how climate change will affect water resources of the basin. However, it is shown that anticipated reductions in rainfall, and increases in temperature and potential evapotranspiration, would affect both river flow and groundwater recharge, which in turn will impact the performance of existing and planned reservoirs and hence irrigation and hydropower schemes [40]. On average, only 75% of annual irrigation water demand is expected to be supplied by 2050 and just 52% of potential hydroelectricity will be generated. The rise in sea levels has had a direct impact on submergence and coastal erosion, an increase in flood-prone areas and an increased salinity in estuaries and coastal water tables. Mangrove swamps, which occupy large surface areas in Nigeria, Guinea, Guinea Bissau, Cameroon and Senegal, are particularly sensitive [41]. The submergence of these mangrove swamps or coastal lagoons could lead to a loss in biodiversity. The cost of adaptation could amount to at least 5–10% of GDP. Changes in coastal ecosystems will have a direct bearing on settlements and productivity. Fish stocks and coastal dwellers will have no alternative but migrate. The resultant shortage of labor in source areas and pressure on limited natural resources in destination areas could result in many undesired outcomes. In the fisheries sector, CO_2 emissions from harvesting and shipping of fish and fish products are estimated at 0.05 Gt year^{-1} with Africa producing 3.6% of the world's CO_2 emissions. About three-fourth of the total emissions from agriculture and land use originate in developing countries [23]. At the scientific and technical levels, adaptation and mitigation options may include relevant research to enable fisheries and aquaculture to adapt to climate change with countries and regions streamlining on:

- reduction of GHG emissions from fishing activities by: (1) improving fuel efficiency through switching to more efficient gear types or vessels, (2) switching to sails or changing fishing practices; (3) use of bulk sea freight rather than air freight or non-bulk sea freight or (4) increasing consumption closer to the source (reducing travel distance);
- removal and storage of atmospheric carbon through coastal ecosystems management of mangroves, sea grass beds, salt marshes [42];
- use of more stable fishing vessels of all sizes to allow for fishing further away from the coastal area to follow targeted species and resist inclement weather;
- use of fish aggregating devices to lure fish back within the traditional fishing grounds;
- new diseases and preventive treatments;
- search for new and better adapted aquaculture species to confront sea level rise;
- better feeds and feeding practices that are more ecosystem friendly;
- strengthening of technology transfer mechanisms to share weather as well as market information with farmers;
- promoting insurance among fish producers and aquaculturist

Crop production sector

Overview of crop production in West Africa

Crop production in West Africa is mainly rainfed and remains vulnerable to climate change manifested in unpredictable rainfall patterns and high temperatures. The five leading crops by harvested area in West Africa are (in millions of hectares) millet (16.0), sorghum (14.3), cowpea (10.3), maize (7.8) and rice (5.7) [43]. Sustaining the yields of these major crops has been challenged (among other factors) by declining soil fertility [44] and expected climate change impacts. Farm sizes are usually small and involve the use of simple farm tools like hoes and cutlasses. Most farmers practice low-input agriculture with limited fertilizer and pesticide use. Diversity is the norm in West African farming systems. Even at the level of the individual farm unit, farmers typically cultivate 10 or more crops in diverse mixtures that vary across soil type, topographical position and distance from the household compound.

Observed climate change impact and projections for the crop production sector

Many projections on West Africa's future climate prognosticate adverse impacts that are likely to lead to productivity crises unless sustainable solutions are in place. While many CMIP5 models indicate a wetter core rainfall season with a small delay to rainy season by the end of the twenty-first century [45], many authors have argued that Regional Climate Models (RCMs) can alter the sign of rainfall change of the driving four General Circulation Models (GCM), especially in regions of high or complex topography [46–48]. The IPCC estimates that crop growing periods in West Africa may shorten by an average of 20% by 2050, causing a 40% decline in cereal yields and

Table 2 Projected changes (%) in crop yields from 2000 to 2050 owing to climate change in West Africa *Source* based on analysis in Jalloh et al. [26]

Water	Crop	Median	CNRM	CSIRO	ECHAM	MIROC
Rainfed	Groundnuts	−6.8	−5.8	−7.7	−9.2	0.3
Rainfed	Maize	−5.5	−2.3	−8.1	−6.0	−4.9
Irrigated	Rice	−19.0	−19.9	−12.4	−20.0	−18.2
Rainfed	Rice	0.9	4.4	0.5	0.9	1.0
Rainfed	Sorghum	−13.9	−15.9	−9.5	−14.8	−13.0
Rainfed	Soybeans	−5.0	−1.5	−8.4	−1.6	−14.2
Irrigated	Wheat	−21.4	−37.8	−10.9	−28.5	−14.3

All GCMs are from the AR4 and represent the A1B scenario

a reduction in cereal biomass for livestock [23]. This has been attributed to the expected 5% decline in rainfall and frequently long-lasting and intense droughts which is likely to increase the area of arid and semiarid land by 5–8% by 2050 [23].

However, the four climate models used by Jalloh et al. [26] all projected a rise in temperature, which is expected to hit hard in West Africa. The average rise in temperature for the four GCMs ranged from 1.5 to 2.3 °C for between 2000 and 2050, while changes in mean annual rainfall ranged from a drop of 23 mm per year to a rise of 22 mm per year. Table 2 shows the predicted change in yield of major food crops owing to climate change impacts in West Africa by 2050. The predicted yield changes in Table 2 have been based on the DSSAT crop model and the four GCM models [26, 49]. As reported in Table 2, only rainfed rice seems to show a positive response. Meanwhile, losses to wheat and irrigated rice are projected to be around 20%; sorghum, 14%; and maize, soybeans and groundnuts, between 5 and 7%. These projections have been similarly reported by Knox et al. [50] and Roudier et al. [10]. Figure 4 shows the detailed projected changes in yield of rainfed sorghum across West Africa for CSIRO, A1B and GCM models. For all the crops and GCMs analyzed, there is a band in the northern boundary of currently cultivable areas which might not be cultivated in future due to increase in temperature (Fig. 4). Some of the reductions in productivity along the southern portions of the coastal countries may be due to declining rainfall as predicted by some of the climate models. Table 3 reports the results of a different analysis which used a global partial equilibrium model of food and agriculture, called IMPACT. This model takes into account productivity changes from the crop model analyses, but includes projected technological growth [26]. As a result, there are large yield increases projected, even under climate change. Additionally, the model takes into consideration global demand for food, based on higher GDP projected for the future, and a large population. When equilibrium prices rise sufficiently, farmers intensify production, sometimes enough to increase yields under climate change above the yields of the no climate change scenario. However, Table 3 should not be read as suggesting that climate change will be generally positive for agriculture. Table 2 already shows that the direct effects of climate change on agriculture will be mostly negative. The large productivity increases in Table 3 emphasize the importance of continuing technological improvement and should suggest the value of continued and increased investment in agricultural research and extension, along with productivity-enhancing investments in irrigation, mechanization, fertilizer supply and climate-smart agriculture.

With climate change affecting crop yields and productivity, changes in the commodity price of major food crops are also inevitable. Table 4 shows the IMPACT model projections in world food prices for some major food crops in West Africa from 2010 to 2050. Median wheat price under climate change will more than triple, while without climate change, it would only increase by two-thirds of the price in 2010. This increase in the price of major food crops in the region is expected to reduce access to food among resource-poor households, a situation that will increase the incidences of malnutrition and undernourishment among children. Table 5 shows the IMPACT projections of climate change on malnutrition among children in West Africa based on two scenarios. While the reductions of numbers appear to be modest by 2050 (Table 5), with a growing population, the percentage of malnourished children might be reduced by almost half, even with climate change, due to projections in GDP and per capita growth outpacing the food price growth, especially under the optimistic economic–demographic scenario. In light of the recent comparisons between the Coupled Model Intercomparison Project (CMIP) and the Agricultural Model Intercomparison Project (AgMIP) [51], we have noted that there is much variation among climate models, crop models [51] and economic models. Scientists, unfortunately, have not been able to come to general agreement on future projections. With so many

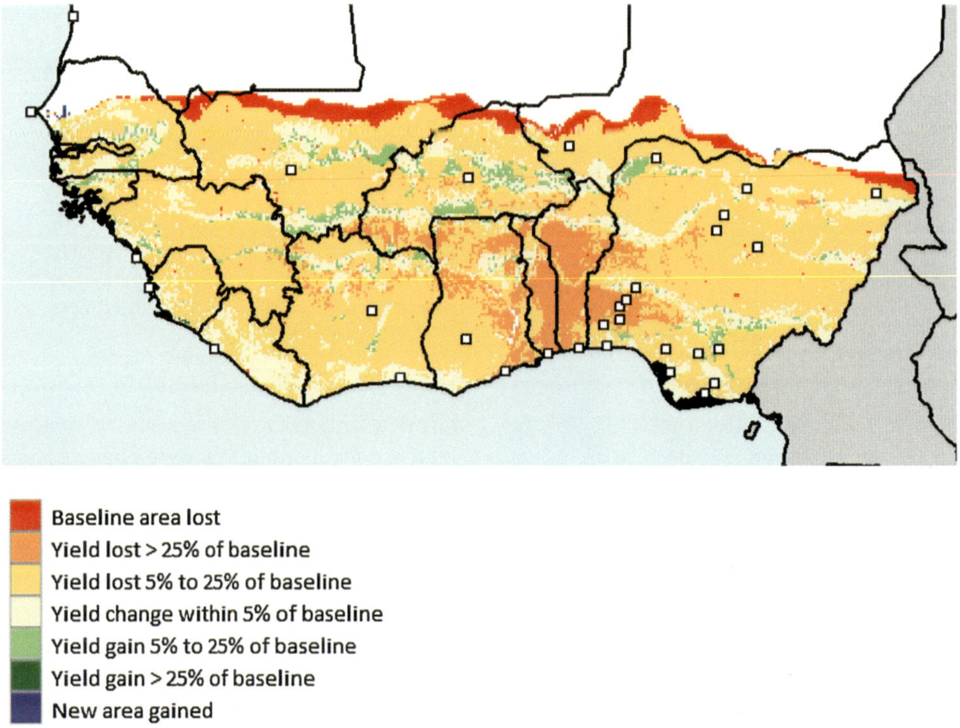

Fig. 4 Yield changes for rainfed sorghum based on CSIRO A1B GCM for the period 2000–2050. *Source*: Jalloh et al. [26]

Table 3 Projected changes (%) in crop productivity from 2000 to 2050 owing to climate change in West Africa based on the IMPACT model *Source* adapted from Nelson et al. [49]

Crop	No climate change	Median of 4 GCMs	MIROC	MIROCCSIRO		CSIRO A1B
				B1	A1B	B1
Cassava	49.5	46.3	37.2	62.5	35.5	55.5
Cotton	90.9	80.8	76.5	85.2	71.4	89.1
Groundnuts	42.0	42.5	43.9	47.3	35.4	41.1
Maize	57.4	57.3	59.8	58.7	53.0	55.9
Millet	147.2	154.0	176.2	156.2	151.9	147.5
Rice	89.3	89.1	89.1	89.7	87.5	89.1
Sorghum	94.1	97.4	106.3	99.4	95.5	95.2
Soybeans	81.5	79.2	77.7	78.5	80.0	84.6
Sweet potatoes and yams	73.5	60.7	49.1	84.0	48.1	72.3

Values are for the baseline economic–demographic scenario

diverse potential outcomes, it is unwise to put all investment into overcoming a single outcome. Rather, the approach should be multipronged and should be focused on giving farmers many options.

Adaptation strategies and mitigation options for the crop production sector

For the crop production sector, a number of CSA technologies including conservation agriculture, alternate wetting and drying approach (AWD) for rice production, agroforestry and "zaï" water conservation technology have been promoted to contribute to the development of sustainable cropping systems in the region. Important references including FAO's Climate-Smart Agriculture Sourcebook [52], and their earlier "Save and Grow" [53] publications outline a number of practices and technologies for building the resilience of crop production systems against climate change:

Table 4 Projected changes (%) in world prices of West African food crops owing to climate change impacts from 2010 to 2050 *Source* based on analysis conducted for Nelson et al. [49]

Crop	No climate change	Median of 4 GCMs	MIROC A1B	MIROC B1	CSIRO A1B	CSIRO B1
Rice	54	84	83	87	85	82
Wheat	66	202	121	106	99	93
Maize	103	160	209	165	156	145
Sweet potatoes and yams	60	130	141	96	156	120
Cassava	18	57	78	50	64	42
Sugarcane	77	110	125	113	108	103
Sorghum	82	107	115	104	110	104
Millet	8	10	8	8	14	13
Groundnuts	13	34	35	33	37	33

The price changes are from the baseline economic–demographic scenario

Table 5 Projected number and percentage of children (under 5 years) in West Africa expected to be malnourished due to climate change effects from 2010 and 2050 based on the IMPACT model *Source* based on analysis conducted by Nelson et al. [49]

Scenario	2010		2050			
			No climate change		Average of max. and min. of 4 GCM–SRES scenarios	
	Number	Percent	Number	Percent	Number	Percent
Baseline	15,157	31.0	12,415	20.9	13,913	23.4
Optimistic	14,733	30.2	7,615	15.1	8,949	17.1

(a) Cultivar development—in the realm of cultivar development, this would include developing varieties that withstand higher temperatures. But it would also include varieties that are resilient to drought, pest, weeds, salinity, flooding, etc. The best new varieties would be ones resilient to more than one threat considering it takes many years to develop and test new cultivars. A recent study on the economic impact of climate change on cereal crop production in Northern Ghana [54] indicated that early season precipitation was beneficial for sorghum, but harmful for maize. However, mid-season precipitation tended to promote maize production. Temperature levels for all seasons impacted negatively on net revenue for both crops, except during the mid-season, when temperature exerted a positive effect on net revenue for sorghum. Our findings suggest that appropriate adaptation strategies should be promoted to reduce the negative impacts of prevailing climate change on cereal crop production.

(b) Water management—along with cultivar development, supporting farming techniques should be developed. These might include irrigation and water harvesting. There may be benefits from shifting the focus from large-scale public irrigation to small-scale private irrigation, which may include more efficient management and distribution of water, related to the user actually bearing the costs of the water use, bypassing issues related to equitability and funding in large-scale schemes. Furthermore, one of the critical issues surrounding climate change is the variability in weather, with floods and droughts both becoming more frequent. Water conservation and supplementation are both important to develop when feasible, especially in marginal areas. For instance, research for development work to build climate-smart farming systems through integrated water storage and crop-livestock interventions has been conducted in Burkina Faso, Mali and Ghana [55]. Through this project, synergies that exist between water retention interventions (such as zaï, contour ridges, dugouts, small reservoirs) and crop-livestock interventions (such as trees and legumes, fodder production, crop residue management) have been identified to improve water availability for crops, livestock and humans throughout every year [55].

Although research studies on the economics of soil and water management on farmers' fields are still scarce, various scientific papers tend to report a cost-effectiveness of these practices. Reij and Smaling [56] estimated the costs of establishment and maintenance of zaï pits for soil water and fertility management at US$250 ha^{-1} $year^{-1}$ and US$65 ha^{-1} $year^{-1}$, respectively. Furthermore, Barro et al. [57] reported that manual zaï could be labor-intensive, requiring about 300 h·man ha^{-1}. Mechanized zaï using cattle traction has been studied on structurally degraded Lixisols in Burkina Faso [58]. Labor time was reduced to 22–36 h ha^{-1} with mechanized zaï, which also increased sorghum grain yield by 34 % compared to that obtained with manual zaï. The mechanization of the zaï generated a significant benefit of up to US$300 ha^{-1}) with sorghum cropping. It may therefore constitute an interesting alternative for increasing the income of the smallholders besides contributing to the preservation of the environment. Fox et al. [58] found that the combination of rainwater harvesting and surface irrigation yielded a net profit of US$151 ha^{-1} to US$622 ha^{-1} for smallholder irrigation in Burkina Faso [59]. Zougmoré et al. [60] also reported that in the semiarid West Africa, the combination of soil and water conservation measures (stone bunds, grass strips) with application of compost resulted in financial gains of 241–300 ha^{-1} $year^{-1}$ (in USD$ equivalent) under adequate rainfall condition. Without nutrient inputs, soil and water conservation (SWC) measures hardly affected sorghum yields, and without SWC, fertilizer inputs also had little effect. However, combining SWC and nutrient management caused an increase in sorghum yield.

(c) Agroforestry—integrating trees with crops can have several advantages when done properly. Trees can sometimes serve as "nutrient pumps," bringing nutrients that are too deep for crops. They can be used to enhance soil nitrogen, when nitrogen-fixing trees are planted. Their leaves can serve as mulch which might suppress some weed growth but would also help cool the soil, overcoming some of the impacts of temperature rise on crop growth. Furthermore, the litter would eventually be converted to soil organic matter (SOM), which has important properties in relation to soil nutrient retention. In Niger and the Sahel zones, an African alliance to combat desertification is improving food security, through tree planting and farmer-managed natural regeneration (FMNR). A synthesis report by Nyasimi et al. [61] and Reij et al. [56] showed that farmers have grown 200 million new trees on cultivated fields in West Africa. The natural regeneration and the improvements that it brings in soil fertility, fodder, food and fuelwood, have been valued at US$56 ha^{-1} $year^{-1}$ or a total annual value of US$280 million. These fields contribute an additional 500,000t of cereals, providing food for about 2.5 million people. The trees contribute to climate change adaption by reducing wind speed and decreasing damage to crops from windblown sand. In addition, a cost and benefit analysis showed FRMR has a low investment cost (about USD$20 ha^{-1}) with an internal rate of returns of 31% if managed for fuelwood production over 20 years [62] with cereal production within an initial five-year period before canopy closure. Taking into account all factors, including enhanced soil fertility and increased food, wood and fodder supply, FNMR can bring an estimated benefit of USD 56 ha^{-1} $year^{-1}$ [63, 64].

(d) Soil carbon sequestration—Soil organic matter has been known for years to be beneficial to cultivation, due to its abilities to improve soil structure and enhance water and nutrient retention. The challenge is to leave enough (or place enough) vegetation for the SOM to increase. Agroforestry is one option, but other possibilities include no-till agriculture, off-season cover crops, use of animal manure and biochar (a by-product of the pyrolysis of biomass under limited oxygen conditions) [65]. Although soil carbon sequestration has direct benefits to the farmer and mitigates climate change, increasing SOM may increase the emission of some greenhouse gases. Not all the science on this latter point is settled. However, it is reported that nitrous oxide, a greenhouse gas approximately 300 times more powerful than carbon dioxide, can be emitted during nitrification and denitrification from organic matter [66, 67].

(e) Seasonal weather and climate forecasting—With climate information services, farmers will be able to plan their planting and make projections about rainfall distribution patterns and temperature variations. Local ICT companies and meteorological institutions must be supported in providing the most accurate and reliable information to farmers. Recently, a sound approach was successfully implemented (1) to designing tailored climate information services and (2) to communicate them appropriately to farmers for their farm management decision making vis-à-vis climate variability in Senegal [68]. A collaboration between scientists and the national meteorological agencies of Senegal, Ghana and IT-based service providers allowed developing more accurate and specific seasonal rainfall forecasts and to raise capacity of partners to do longer-term analysis and provide more targeted information for farmers. The forecast information provided includes the total rainfall, the onset and end of the rainy season, plus a 10-day forecast across the rainy season. The information is conveyed to farmers as agro-meteorological advisories that are tailored to meet their local needs. In

Senegal for instance, through a partnership with 82 rural community-based radio stations promoting economic development through communication and local information exchange, the seasonal forecast is now reaching about 740,000 rural households across the 14 administrative regions [68]. In Ghana, through a private ICT-based platform, market price alerts, climate-smart agricultural advice, weather forecast and voice messages on climate-smart agricultural practices are sent out to farmers from Northern Ghana in the language of their (farmers) choice. This platform has so far trained and improved about 835 farmers' (of which 33% are females) access to and use of downscaled seasonal forecast and climate-smart agriculture technologies and practices (agro-advisories) through mobile phones [69]. Furthermore, the agricultural value chain programs in Burkina Faso and Senegal also disseminated seasonal forecast information and climate-smart agricultural options to farmers from various agricultural sectors [70]. A cost–benefit analysis in Burkina Faso by Ouedraogo et al. [70] showed that farmers exposed to climate information have used less local seed and more improved seed for cowpea and sesame production. They also used less organic manure and more fertilizers for sesame production. Cowpea producers exposed to climate information obtained higher yields while covering lower inputs costs and their gross margins were therefore higher compared to non-exposed farmers.

(f) Fertilizer efficiency—In addition to the use of inappropriate types of fertilizer depending on the soil type and native soil fertility, too much fertilizer, or fertilizer applied at less than optimal levels and times can be wasted, going beyond the reach of the crop. Not only is this economically inefficient, but the fertilizer can be converted to nitrous oxide and emitted to the atmosphere. However, the right type of fertilizer applied at proper times and amounts can be used efficiently by the crops while minimizing emissions. This may involve promoting integrated soil fertility management that seeks, among other things, to enhance the soil organic matter content of the soils, which improves nutrient retention [71].

(g) Rice water management—With optimal management of water in a rice system, such as alternate wet and dry (AWD), methane emissions can be reduced without adversely impacting yield and potentially increasing yields. It may also prove to be a more efficient use of water in many locales. The danger of such a system is that if done incorrectly, the nitrous oxide emissions will increase so much that they will negate any gains in methane emission reduction [72].

Socioeconomic consequences of climate change and the economic implications for climate-smart agriculture adoption in West Africa

With almost 70% of West African populace dependent on agriculture, climate change will have far socioeconomic consequences for national economies and individuals. Climate change impacts on production are expected to translate into economic impacts at various scales: (1) At the farm level, climate change will cause reduced income for households which will limit the capacity to acquire physical assets and meet the cost of child education and health [73]. In Northern Ghana for instance, estimates from the Ricardian regression models showed that precipitation and temperature impacted significantly on net revenue per hectare of maize and sorghum. Indeed, increasing precipitation (by 1 mm), while decreasing temperature (by 1 °C), would affect net revenue, but differently across seasons and among various crops [54]. This will worsen existing poverty, exacerbate inequalities and trigger both new vulnerabilities and some opportunities for affluent individuals and communities who can take advantage of shocks and crises, given their flexible assets and power status [74]. This confirms the important link between climate and crop revenue and the need to take action to reinforce existing adaptation options and to even develop new ones. (2) At the national level, climate change may have macroeconomic consequences for countries whose national economies are dependent on agriculture. Interacting with other non-climatic stressors such as environmental degradation, market volatility and declining soil fertility, climate change can hinder agricultural development by discouraging investments [73]; (3) at the regional level, climate change impact on production will have (i) quantity and price effects, with increased tension on markets; and (ii) impacts on bilateral contracts and/or import/export behavior, with disruption of trade patterns [73].

The quest to promote climate-smart agriculture in West Africa will not come at a low cost. This necessitates regional-level large-scale investments that require innovation, cooperative action and political will to urgently and adequately address current and projected shortfalls for adaptation and mitigation [3]. Generally, information relating to the investment needs for agriculture and climate finance is limited and may not include all related investment needs [75]. Earlier projections have shown the entire Africa will require about USD 48.5 billion per year to meet the investment needs of the agriculture sector and about USD 3 billion per year investments toward climate change adaptation over the period 2005/7–2050 [75, 76]. At least a quarter of these investment needs will be required for West Africa which has a high dependence on agriculture and its related activities. Further, there will be extra costs toward research, capacity building and planning for climate change adaptation. These investments would have to come from national budgets (aligned with national goals and priorities relevant for CSA), private sector funds and bonds, concessionary mechanisms, bilateral and multilateral funding, development banks, global adaptation funds (e.g., Green Climate Funds), etc.

Regional initiatives and policy development efforts for promoting climate-smart agriculture

Developing or reinforcing adaptive mechanisms to deal with the negative effects of climate change is considered a priority by West Africa policy makers, but this requires considerable changes in national and local governance, legislation, policies and financial mechanisms. Policy initiatives and developments that promote CSA in West Africa do exist, but studies monitoring and evaluating their effectiveness are limited. Before the African Union Summit in 2014 (which led to the development of the African Climate Smart Agriculture Coordination Platform, NEPAD Planning and Coordinating Agency, (NPCA)), most countries in the Committee for Drought Control in the Sahel (CILSS) in the ECOWAS region had between 2007 and 2009 formulated and adopted their National Adaptation Programs of Action (NAPA) which included various interventions under CSA for climate change mitigation and adaptation [77, 78]. In addition, some projects under the NAPA covered more specific areas such as (i) mainstreaming climate change-related issues in agricultural policies, programs, plans and projects; (ii) implementing good adaptation practices in the sectors most vulnerable to climate change; (iii) preventing and mitigating food crises; (iv) providing information, education and communication to stakeholders; (vi) strengthening early warning systems on climate change, etc. [79]. To improve and sustain the aforementioned, ECOWAS adopted in 2005 a common agricultural policy, the ECOWAP/CAADP as the regional version of the NEPAD's Comprehensive Africa Agriculture Development Program (CAADP) [79]. The implementation of ECOWAP/CAADP was based on coordinated interventions at national and regional levels through the development of National Agricultural Investment Programs (NAIPs) and a Regional Agricultural Investment Programme (RAIP) at community level [79, 80]. A synthesis report from the ECOWAS communications indicates ECOWAS intensions to improve the shortcomings of the NAIPS and RAIPs in the quest to sustainable food security in the face of climate change challenges by (i) integrating CSA in the NAIPs of the 17 ECOWAS/CILSS countries and in the RAIP at regional level; (ii) assisting the countries in increasing climate funding mobilization for CSA within the framework of their NAIPs and developing a CSA monitoring-evaluation framework in the context of ECOWAP/CAADP monitoring-evaluation system; and (iii) strengthening interinstitutional dialogue and cross-sector cohesion around CSA between agricultural policies and programs for climate adaptation and water management both at national level and at regional level [79]. Furthermore, ECOWAS and regional stakeholders developed in 2015 an intervention framework for CSA in West Africa coupled with the creation of a West Africa Climate-Smart Agriculture Alliance for proper implementation. This was presented during COP 21 in Paris as a regional initiative (dubbed: "Promotion of Smart Agriculture towards Climate Change and agroecology transition in West Africa") that aims to support the transition toward agro-ecology in West Africa, dwelling on CSA principles to reinforce the resilience of vulnerable populations [81]. Protocols in the implementation of the initiative involve the (1) creation of a platform to share knowledge: capitalization of technics and ecology-intensive practices, data on investments and data on carbon sequestration; (2) promotion of these practices through agro-meteorological support: climate modeling and of its impacts on agriculture, strengthening of the production and dissemination of information systems; (3) production and dissemination of best practices; (4) supporting the scale-up of best practices: the use of agroforestry species with carbon sequestration capacities, storm water management, soil regeneration and fertilization; (5) reinforcement of national and regional capacity building in policies and strategies: promotion of best practices in programs and projects, trainings; and (6) mobilizing financial and technical resources: access to the International Climate Fund (primary channel of climate change finance), creation of expert pools.

Within the ECOWAS region, some countries have now mainstreamed climate change and CSA into national action plans and policies. In 2011, the Government of Nigeria and civil society organizations developed a National Adaptation Strategy and Plan of Action on Climate Change for Nigeria (NASPA-CCN). The document was developed through multistakeholder consultations [79]. Ghana also launched the National Climate-Smart Agriculture Action Plan (2016–2020) in 2015 under the technical and scientific auspices of the Ministry of Food and Agriculture and the Ghana science-policy dialogue platform on climate change, agriculture and food security [82]. Mali has also formulated a National Policy on Climate Change (PNCC) accompanied by a National Implementation Strategy consisting of 147 actions for the 2012–2017 periods. The sectorial orientations on which CSA is based within the framework of the PNCC are yet to be identified [79]. However, a CSA prioritization exercise has been undertaken with key national actors to define cost-effective climate-smart technological options that could be considered for implementation in each agro-ecological zone [83].

Conclusions

There is substantial evidence that climate change is already impacting West African livestock, fishery and crop production sectors and would continue to have disastrous effects in the future if appropriate mitigation and adaptive measures are not in place. For instance, current cereal production is expected to reduce by 40% in 2050, while an expected 5% decline in rainfall and frequently long-lasting and intense droughts will predictably increase the area of arid and semiarid land by 5–8%. Intense droughts will have consequential effects on floristic composition of vegetation, changes in herbage quality and changes in the importance of crop residues as animal feed. In addition, climate change is expected to modify the distribution of fish species, thus affecting habitat size, species diversity and productivity of lagoons in the region. The mounting consensus of the vulnerabilities of the crop, livestock and fishery production sectors to climate change means that every day of inaction could make the consequences more catastrophic and irreversible. Although the Kyoto Protocol places great emphasis on mitigation efforts (reducing greenhouse gas emissions and creation of sinks), its impact on climate change will not be seen immediately even if the most effective reductive measures are implemented. Therefore, developing or reinforcing adaptive mechanisms to deal with the negative effects of climate change must be a high priority. The promotion of climate-smart agricultural (CSA) practices is one mainstream opportunity to mitigate climate change while sustaining the productivity of agricultural systems. In addition, CSA can help build adaptive capacity, so that farmers, service providers to farmers and key institutions have the ability to respond effectively to longer-term climate change as well as being able to manage the risks associated with increased climate variability. Recent developments at the political level give hopes for large-scale CSA adoption for improved resilience to climate change and food security in the region. CSA is already endorsed for inclusion in the NEPAD program on agriculture and climate change by the African Union, while in the framework of the formulation of the ECOWAP + 10, the new common agricultural policy for the region, ECOWAS seeks for a focus on the mainstreaming of climate change and CSA into local plans and policies of member countries. While policy developments are advancing, governments would have to raise the levels of national agricultural investments and create adequate and effective financial mechanisms to achieve large-scale landscape adoption of CSA. Further, research for development and dissemination of CSA technologies has to be intensified in the region.

Abbreviations

AgMIP: Agricultural Model Intercomparison Project; AU: African Union; AWD: alternate wet and dry; CAADP: Comprehensive Africa Agriculture Development Program; CMIP: Coupled Model Intercomparison Project; CILSS: Committee for Drought Control in the Sahel; CSA: climate-smart agriculture; CSIRO: Commonwealth Scientific and Industrial Research Organization; DSSAT: Decision Support System for Agrotechnology Transfer; ECOWAS: Economic Community of West African States; NAIP: National Agricultural Investment Programs; NASPA-CCN: National Adaptation Strategy and Plan of Action on Climate Change for Nigeria; NAPA: National Adaptation Programs of Action; NEST: Nigerian Environmental Study/Action Team; NPCA: NEPAD Planning and Coordinating Agency; PNCC: National Policy on Climate Change (translated from French); RAIP: Regional Agricultural Investment Programme.

Authors' contributions

RZ and AJ wrote the review on the crop production sector; RZ again coordinated the entire literature review process. STP worked on combining the various reviews, improving the coherence of content, writing a background to the review, conclusions and abstract. BO wrote the review on the fishery sector. MO and TT edited and reviewed the policy context of the paper. AA, PE and MS wrote the review on the livestock sector. All authors contributed to the final editing and reviewing of the entire manuscript. All authors read and approved the final manuscript.

Author details

CGIAR Research Program on Climate Change, Agriculture and Food Security (CCAFS), International Crops Research Institute for the Semi-Arid Tropics (ICRISAT), BP 320, Bamako, Mali. [2] Department of Aquaculture and Fisheries Management, University of Ibadan, Ibadan, Nigeria. [3] Environment and Production Technology Division (EPTD), International Food Policy Research Institute (IFPRI), 2033 K Street NW, Washington, DC 20006, USA. [4] International Livestock Research Institute, Ouagadougou, Burkina Faso. [5] International Livestock Research Institute, Nairobi, Kenya. [6] West and Central African Council for Agricultural Research and Development (CORAF/WECARD), BP 48, Dakar RP, Senegal.

Acknowledgements

This work was implemented as part of the CGIAR Research Program on Climate Change, Agriculture and Food Security (CCAFS), a strategic partnership of CGIAR and Future Earth, led by the International Center for Tropical Agriculture (CIAT). We acknowledge the CGIAR Fund Council, Australia (ACIAR), European Union, International Fund for Agricultural Development (IFAD), Ireland, New Zealand, Netherlands, Switzerland, USAID, UK and Thailand for funding to CCAFS.

Competing interests

The authors declare that they have no competing interests.

Funding

CGIAR Fund Donors.

References

1. CCAFS. West Africa: our work. https://ccafs.cgiar.org/west-africa-our-work#.VwKVfPmLRD8 (2013). Accessed 26 Feb 2016.
2. FAO. Regional overview of food insecurity: African food insecurity prospects brighter than ever. Accra: FAO; 2015.
3. FAO. Climate smart agriculture: policies, practices and financing for food security, adaptation and mitigation. Rome: Food and Agriculture Organization; 2010.
4. Campbell BM, Thornton P, Zougmoré R, van Asten P, Lipper L. Sustainable intensification: what is its role in climate smart agriculture? Curr Opin Environ Sustain. 2014;8:39–43.
5. Kristjanson P, Waters-Bayer A, Johnson N, Tipilda A, Njuki J, Baltenweck I, Grace D, MacMillan S. Livestock and women's livelihoods. In: Gender in agriculture 2014 (pp. 209–233). Springer Netherlands. http://libcatalog.cimmyt.org/download/general/98958.pdf#page=220 (2014). Accessed 13 Mar 2016.
6. Nyberg G, Knutsson P, Ostwald M, Öborn I, Wredle E, Otieno DJ, Mureithi S, Mwangi P, Said MY, Jirström M, Grönvall A. Enclosures in West Pokot, Kenya: transforming land, livestock and livelihoods in drylands. Pastoralism. 2015; 3,5(1):1.
7. SWAC-OECD/ECOWAS. Livestock and regional market in the Sahel and West Africa—potentials and challenges. Paris: Sahel and West Africa Club/OECD; 2008.
8. Martin R, Linstädter A, Frank K, Müller B. Livelihood security in face of drought–assessing the vulnerability of pastoral households. Environ Modell Softw. 2016;75:414–23.
9. Zaibet L, Traore S, Ayantunde A, Marshall K, Johnson N, Siegmund-Schultze M. Livelihood strategies in endemic livestock production systems in sub-humid zone of West Africa: trends, trade-offs and implications. Environ Dev Sustain. 2011;13(1):87–105.
10. Roudier P, Sultan B, Quirion P, Berg A. The impact of future climate change on West African crop yields: what does the recent literature say? Glob Environ Change. 2011;21:1073–83.
11. Baudron F, Jaleta M, Okitoi O, Tegegn A. Conservation agriculture in African mixed crop-livestock systems: expanding the niche. Agr Ecosys Environ. 2014;187:171–82.
12. FAO. FAO statistical year book 2014 Africa. Accra: FAO; 2014.
13. Hiernaux P, Ayantunde A, Kalilou A, Mougin E, Gerard B, Baup F, Grippa M, Djaby B. Trends in productivity of crops, fallow and rangelands in Southwest Niger: impact of land use, management and variable rainfall. J Hydrol. 2009;375(1–2):65–77.
14. Turner MD, McPeak JG, Ayantunde A. The role of livestock mobility in the livelihood strategies of rural peoples in semi-arid West Africa. Hum Ecol. 2014;42(2):231–47.
15. Niang I, Ruppel OC, Abdrabo MA, Essel A, Lennard C, Padgham J, Urquhart P. Africa. In: Barros VR, Field CB, Dokken DJ, Mastrandrea MD, Mach KJ, Bilir TE, Chatterjee M, Ebi KL, Estrada YO, Genova RC, Girma B, Kissel ES, Levy AN, MacCracken S, Mastrandrea PR, White LL, editors. Climate change 2014: impacts, adaptation, and vulnerability. Cambridge: Cambridge University Press; 2014. p. 1199–265.
16. Freier KP, Bruggemann R, Scheffran J, Finckh M, Schneider UA. Assessing the predictability of future livelihood strategies of pastoralists in semi-arid Morocco under climate change. Technol Forecast Soc. 2012;79(2):371–82.
17. Schilling J, Freier KP, Hertig E, Scheffran J. Climate change, vulnerability and adaptation in North Africa with focus on Morocco. Agr Ecosyst Environ. 2012;156:12–26.
18. Mortimore MJ, Adams WM. Farmer adaptation, change and "crisis" in the Sahel. Global Environ Chang. 2001;11:49–57.
19. Thornton PK, Ericksen PJ, Herrero M, Challinor AJ. Climate variability and vulnerability to climate change: a review. Glob Change Biol. 2014;20:3313–28.
20. Rust JM, Rust T. Climate change and livestock production: a review with emphasis on Africa. S Afr J Anim Sci. 2013;43(3):256–67.
21. Dossa LH, Sangaré M, Buerkert A, Schlecht E. Production objectives and breeding practices of urban goat and sheep keepers in West Africa: regional analysis and implications for the development of supportive breeding programs. SpringerPlus. 2015;4:1–12.
22. Thornton PK, van de Steeg J, Notenbaert A, Herrero M. The impacts of climate change on livestock and livestock systems in developing countries: a review of what we know and what we need to know. Agr Syst. 2009;101:113–27.
23. Intergovernmental Panel on Climate Change (IPCC). Climate change 2014: impacts, adaptation and vulnerability. IPCCWGIIAR5 technical summary. http://ipccwg2.gov/AR5/images/uploads/WGIIAR5-TS_FGDall.pdf (2014). Accessed Aug 19 2014.
24. FAO. Adaptation to climate change in agriculture, forestry and fisheries: perspective, framework and priorities. Rome: FAO, Interdepartmental Working Group on Climate Change; 2007.
25. Rosenstock TS, Lamanna C, Arslan A, Richards M. The scientific basis of climate-smart agriculture: a systematic review protocol. CCAFS working paper no. 138; 2016.
26. Jalloh A, Nelson GC, Thomas TS, Zougmoré R, Roy-Macauley H, editors. West African agriculture and climate change. Washington: International Food Policy Research Institute; 2013.
27. Zougmoré R, Traoré AS, Mbodj Y. Overview of the scientific, political and financial landscape of climate-smart agriculture in West Africa (no. 118). Working paper; 2015.
28. Belhabib D, Sumaila UR, Pauly D. Feeding the poor: contribution of West African fisheries to employment and food security. Ocean Coas Manag. 2015;111:72–81.
29. USAID West Africa. West African fisheries profile. USAID. 2008.
30. FAO/SFLP. Building adaptive capacity to climate change. Policies to sustain livelihoods and fisheries. New directions in fisheries—a series of policy briefs on development issues. no.8. Sustainable Fisheries Livelihoods Programme. Rome: FAO; 2007.
31. Lam V, Cheung WW, Swartz W, Sumaila UR. Climate change impacts on fisheries in West Africa: implications for economic, food and nutritional security. Afr J Mar Sci. 2012;34:103–17.
32. Smith MD, Roheim CA, Crowder LB, Halpern BS, Turnipseed M, Anderson JL, Asche F, Bourillón L, Guttormsen AG, Khan A, Liguori LA. Sustainability and global seafood. Science. 2010;327:784–6.
33. Cochrane K, De Young C, Soto D, Bahri T, editors. Climate change implications for fisheries and aquaculture: overview of current scientific knowledge. FAO Fisheries and Aquaculture Technical Paper. No. 530. Rome: FAO; 2009.
34. Allison EH, Perry AL, Badjeck MC. Vulnerability of national economies to the impacts of climate change on fisheries. Fish Fish. 2009;10:173–96.
35. Omitoyin SA, Tosan FB. Potential impacts of climate change on livelihood and food security of Artisanal Fisherfolks in Lagos State, Nigeria. J Agr Sci. 2012;9(4):20–30.
36. Rhodes ER, Jalloh A, Diouf A. Africa interact: enabling research-to-policy dialogue for adaptation to climate change in Africa. Review of research and policies for climate change adaptation in the agriculture sector in West Africa. Working paper 090; 2014.
37. Associated Press. Shrinking of Lake Chad. www.globalpolicy.org/component/content/article/198/40377.html (2006). Accessed 14 Dec 2015.
38. Adebo GM, Ayelari TA. Climate change and vulnerability of fish farmers in South Western Nigeria. Afr J Aggr Res. 2011;6(4230):4238.
39. Niasse M, Afouda A, Amani A (ed). Réduire la vulnérabilité de l'Afrique de l'Ouest aux impacts du climat sur les ressources en eau, les zones humides et la désertification. https://portals.iucn.org/library/efiles/documents/Climate-impactsF-prelims.pdf (2004). Accessed 18 Feb 2016.
40. McCartney M, Forkuor G, Sood A, Amisigo B, Hattermann F, Muthuwatta L. The water resource implications of changing climate in the Volta River Basin. Colombo, Sri Lanka: International Water Management Institute (IWMI). IWMI research report 146. http://www.iwmi.cgiar.org/Publications/IWMI_Research_Reports/PDF/PUB146/RR146.pdf (2012). Accessed 18 June 2015.
41. ECOWAS-SWAC/OECD. Atlas on regional integration in West Africa. http://www.atls.westafrica.org (2015). Accessed 18 Dec 2015.
42. McLeod E, Chmura GL, Bouillon S, Salm R, Björk M, Duarte CM, Lovelock C, Schlesinger W, Silliman B. A blueprint for blue carbon: toward an improved understanding of the role of vegetated coastal habitats in sequestering CO_2. Front Ecol Environ. 2011;9:552–60.
43. FAOSTAT. Food and Agriculture Organization of the United Nations Statistics Division. Rome: FAO; 2015.
44. Partey ST. Effect of pruning frequency and pruning height on the biomass production of *Tithonia diversifolia*. Agroforest Syst. 2011;83:181–7.
45. Biasutti M. Forced Sahel rainfall trends in the CMIP5 archive. J Geophys Res. 2013;118:1613–23.
46. Cook KH, Vizy EK. Projected changes in East African rainy seasons. J Clim. 2013;26:5931–48.
47. Saeed F, Haensler A, Weber T, Hagemann S, Jacob D. Representation of extreme precipitation events leading to opposite climate change signals over the Congo Basin. Atmosphere. 2013;4:254–71.

48. Sylla MB, Gaye AT, Jenkins GS. On the fine-scale topography regulating changes in atmospheric hydrological cycle and extreme rainfall over West Africa in a regional climate model projections. Int J Geophys. 2012;. doi:10.1155/2012/981649.
49. Nelson GC, Rosegrant MW, Palazzo A, Gray I, Ingersoll C, Robertson R, Tokgoz S, Zhu T, Sulser TB, Ringler C, Msangi S. Food security, farming, and climate change to 2050: scenarios, results, policy options. Washington DC, USA: International Food Policy Research Institute; 2010.
50. Knox J, Hess T, Daccache A, Wheeler T. Climate change impacts on crop productivity in Africa and South Asia. Environ Res Lett. 2012;7:034032.
51. Rosenzweig C, Elliott J, Deryng D, Ruane AC, Müller C, Arneth A, Boote KJ, Folberth C, Glotter M, Khabarov N, Neumann K. Assessing agricultural risks of climate change in the 21st century in a global gridded crop model intercomparison. Proc Natl Acad Sci. 2014;111(9):3268–73.
52. FAO. Climate-smart agriculture sourcebook. Rome: Food and Agricultural Organization of the United Nations; 2013.
53. FAO. Save and grow: a policymaker's guide to the sustainable intensification of smallholder crop production. Rome: Food and Agricultural Organization of the United Nations; 2011.
54. Bawayelaazaa Nyuor A, Donkor E, Aidoo R, Saaka Buah S, Naab JB, Nutsugah SK, Bayala J, Zougmoré R. Economic impacts of climate change on cereal production: implications for sustainable agriculture in Northern Ghana. Sustainability 2016; 08 (08):1–17. ISSN 2071-1050.
55. Amole TA, Ayantunde AA. Climate-smart livestock interventions in West Africa: a review. CCAFS working paper no. 178. Copenhagen, Denmark: CGIAR Research Program on Climate Change, Agriculture and Food Security (CCAFS); 2016.
56. Reij C, Tappan G, and Smale M. Re-greening the Sahel farmer led innovation in Burkina Faso and Niger. In: Reij C, Tappan G, Smale M. Agroenvironmental transformation in the Sahel: another kind of "Green Revolution". International Food Policy Research Institute discussion paper 00914; 2009.
57. Barro A, Zougmoré R, Taonda SJB. Mécanisation de la technique du zaï manuel en zone semi-aride. Cahiers Agricultures. 2005;14:549–59.
58. Fox P, Rockstrom J, Barron J. Risk analysis and economic viability of water harvesting for supplementary irrigation in semi-arid Burkina Faso and Kenya. Agri Syst. 2005;83:231–50.
59. Rhodes ER, Jalloh A, Diouf A. Review of research and policies for climate change adaptation in the agriculture sector in West Africa. Future Agricultures. Working paper.090; 2014.
60. Zougmoré R, Mando A, Stroosnijder L, Ouédraogo E. Economic benefits of combining soil and water conservation measures with nutrient management in semiarid Burkina Faso. Nutr Cycl Agroecosys. 2005;70:261–9.
61. Nyasimi M, Amwata D, Hove L, Kinyangi J, Wamukoya G. 'Evidence of impact: Climate-smart agriculture in Africa'. CCAFS working paper no. 86. Copenhagen, Denmark: CGIAR Research Program on Climate Change, Agriculture and Food Security (CCAFS). http://ccafs.cgiar.org/publications/evidence-impact-climate-smart-agriculture-africa (2014).
62. Reij C, Tappan G, Smale M. Agroenvironmental transformation in the Sahel: another kind of "Green Revolution". International Food Policy Research Institute discussion paper 00914; 2009.
63. Cooper PJM, Cappiello S, Vermeulen SJ, Campbell BM, Zougmoré R, Kinyangi J. Large scale implementation of adaptation and mitigation actions in agriculture. CCAFS working paper no. 50. Copenhagen: CGIAR Research Program on Climate Change, Agriculture and Food Security (CCAFS); 2013.
64. Dinesh D, Frid-Nielsen S, Norman J, Mutamba M, Loboguerrero Rodriguez AM, and Campbell B. Is climate-smart agriculture effective? A review of selected cases. CCAFS working paper no. 129. Copenhagen, Denmark: CGIAR Research Program on Climate Change, Agriculture and Food Security (CCAFS). www.ccafs.cgiar.org (2015).
65. Partey ST, Saito K, Preziosi RF, Robson GD. Biochar use in a legume-rice rotation system: effects on soil fertility and crop performance. Arch Agron Soil Sci. 2016;62:199–215.
66. Corsi S, Friedrich T, Kassam A, Pisante M, de Moraes Sà J. Soil organic carbon accumulation and greenhouse gas emission reductions from conservation agriculture: a literature review. Integrated Crop Management, vol. 16. Plant Production and Protection Division. Rome: FAO; 2012.
67. Duxbury JM. Soil carbon sequestration and nitrogen management for greenhouse gas mitigation in the natural farmer archives. http://tnfarchives.nofa.org/?q=article/soil-carbon-sequestration-and-nitrogen-management-greenhouse-gas-mitigation (2012). Accessed 2 Jan 2016.
68. CCAFS. The impact of climate information services in Senegal. CCAFS outcome case no. 3. Copenhagen, Denmark: CGIAR Research Program on Climate Change, Agriculture and Food Security (CCAFS); 2015.
69. ICRISAT. Building climate-smart farming communities. ICRISAT annual report 2015. ISSN 1017-9933; 2015.
70. Ouédraogo M, Zougmoré R, Barry S, Somé L, Baki G. The value and benefits of using seasonal climate forecasts in agriculture: evidence from cowpea and sesame sectors in climate-smart villages of Burkina Faso. CCAFS Info Note. Copenhagen, Denmark: CGIAR Research Program on Climate Change, Agriculture and Food Security (CCAFS); 2015.
71. Zougmoré R, Jalloh A, Tioro A. Climate-smart soil water and nutrient management options in semiarid West Africa: a review of evidence and analysis of stone bunds and zaï techniques. Agricult Food Secur. 2014;3:16.
72. Weller S, Kraus D, Ayag KRP, Wassmann R, Alberto MCR, Butterbach-Bahl K, Kiese R. Methane and nitrous oxide emissions from rice and maize production in diversified rice cropping systems. Nutr Cycl Agroecosys. 2015;101:37–53.
73. FAO. Climate change and food security: risks and responses. Rome: Food and Agriculture Organization of the United Nations; 2016.
74. Olsson L, Opondo M, Tschakert P, Agrawal A, Eriksen SH, Ma S, Perch LN, Zakieldeen SA. Livelihoods and poverty. In: Field CB, Barros VR, Dokken DJ, Mach KJ, Mastrandrea MD, Bilir TE, Chatterjee M, Ebi KL, Estrada YO, Genova RC, Girma B, Kissel ES, Levy AN, MacCracken S, Mastrandrea PR, White LL, editors. Climate change 2014: impacts, adaptation, and vulnerability. Cambridge: Cambridge University Press; 2014. p. 793–832.
75. Branca G, Tennigkeit T, Mann W, Lipper L. Identifying opportunities for climate-smart agriculture investment in Africa. Rome: Food and Agriculture Organization of the United Nations; 2012.
76. Schmidhuber J, Bruinsma J, Boedeker G. Capital requirements for agriculture in developing countries to 2050. In FAO expert meeting on "How to Feed the World in vol. 2050, p. 24–6; 2009.
77. African Union. Report of the chairperson of the NEPAD heads of state and government orientation committee. Assembly of the Union. Twenty-Third Ordinary Session. 26–27 June 2014. Malabo, Equatorial Guinea. http://africacsa.org/introducing-the-africa-csa-alliance-ascaa/ (2014). Accessed 9th Oct 2015.
78. ECOWAS. Sub-regional action program to reduce vulnerability to climate change in West Africa. Part1: overview of West Africa vulnerability to climate change and of response strategies, Abuja, Nigeria: Economic Community of West African States; 2009.
79. ECOWAS. Synthesis of the national communications from the ECOWAS/UEMOA/CILSS countries to mainstream Climate-Smart Agriculture into National Agricultural Investment Programmes. http://www.hubrural.org/IMG/pdf/note_agenda_ecowas_high_level_forum_of_csa_stakeholders_in_west_africa_.pdf (2015). Accessed 23 Jan 2016.
80. ECOWAS. Sub-regional action program to reduce vulnerability to climate change in West Africa. Part 2: The Strategic Action Plan, Abuja, Nigeria: Economic Community of West African States; 2009.
81. WACSAA, West Africa Climate-Smart Agriculture Alliance: Framework document, WACSAA, High Level Forum of Climate-Smart Agriculture Stakeholders in West Africa (Bamako, Mai, 15–18 June 2015), ECOWAS, UEMOA, CILSS, Hub Rural, USAID, ASDI, European Union, Africa Lead, UNOPS, June 2015.
82. Essegbey GO, Nutsukpo D, Karbo N, Zougmoré R. National climate-smart agriculture and food security action plan of Ghana (2016–2020). Working paper no.139. Copenhagen, Denmark: CGIAR Research Program on Climate Change, Agriculture and Food Security (CCAFS); 2015.
83. Sogoba B, Andrieu N, Howland F, Samake O, Corner-Dolloff C, Bonilla-Findji O, Zougmore R. Climate-smart solutions for Mali. CCAFS Info Note. Copenhagen, Denmark: CGIAR Research Program on Climate Change, Agriculture and Food Security (CCAFS); 2016.

Small ruminant fattening practices in Amhara region, Ethiopia

Shewangzaw Addisu Mekuria[1*], Aschalew Assefa Teshager[1], Addis Getu Endeshaw[1], Malede Birhan Atinaw[1] and Assemu Tesfa Sendeku[2]

Abstract

Background: In Ethiopia, small ruminant had a great role in the economic development for farmers or producers, food-insecure areas and one of the main sources of meat production. The study was conducted in five small ruminant potential zones of Amhara region from November to June, 2017, with the objective of assessing small ruminant fattening practices in Amhara region, Ethiopia. Twenty representative kebeles were selected purposively. From each of selected kebeles, 10 small ruminant fatteners were selected purposively based on their fattening experience. A total of 200 households were selected for interviewing. Data were analyzed by using SPSS version, 20.

Results: Majority of (47%) small ruminant fatteners can read and write. Mean and SD of family size per household was shown to be 5.22 ± 2.00. Mean and SD of private and communal grazing lands and total land for traction was 0.30 ± 0.60, 0.81 ± 1.81 and 1.84 ± 2.90, respectively. From the total of 100% respondents only 46.6% had there owen private grazing land this mean the remaining 53.4% were utilizing communal grazing land. Among the 46.6% of respondents only 74.5% were primarily grazed fattening animals. About 59% of respondents responded that the status of grazing land was decreased from the previous status, but the remaining 31.5 and 9.5% responded that there was no change and that it was increasing, respectively. Mean and SD of sheep and goat fattening per household was 2.86 ± 2.90 and 1.65 ± 4.67, respectively. Majority (71.5%) of producers preferred fattening sheep species to goat. In Amhara region, the purpose of small ruminant fattening was mainly income source (81.5%) and the remaining 12, 2, 1.5, 1, 1, and 1% were risk/benefit, meat, social or cultural function, sacrifices/rituals, saving and others, respectively. Natural pasture grazing and stubble grazing were the major feed sources in the wet and dry seasons, respectively. Fatteners were provided with supplementary feeds for small ruminant fattening twice a day (36%), once in a day (31%) and whenever available (12%). Majority of respondents (77.6%) were not provided with supplementary feed scientifically recommended due to the low supply and high cost of concentrate feed in the market. The main water source was pipe water. The average distance of water from their home was less than one km, and watering frequency was twice a day.

Conclusions: In general, the present study showed that fatteners in the region did not fatten small ruminants scientifically recommended, because there were high cost of supplementary feed in the market and lack of enfaces on the development of small ruminant fattening in Amhara region. So, there should be given more enfaces by the government and researches should be done on alternative and non-conventional feeds with low price by considering the export standard of meat production.

Keywords: Amhara region, Fattening, Feed, Purpose, Small ruminant

*Correspondence: shewakibr12@gmail.com
[1] University of Gondar, P.O. Box 196, Gondar, Ethiopia
Full list of author information is available at the end of the article

Background

Small ruminants are important components of the livestock subsector and are sources of cash income and play a vital role as sources of meat, milk and wool for smallholder keepers in different farming systems and agroecological zones of the country. They are also sources of foreign currency [2]. Sheep and goat in Ethiopia and most developing regions are kept under traditional extensive systems. Sheep and goats are largely produced in mixed crop–livestock, specialized pastoral and agro-pastoral systems. Livestock production is of subsistence nature. Market-oriented or commercial production is almost nonexistent [10].

According to the report of [8], the total annual meat production comes from cattle (63%), sheep (25%) and goats (12%). At the national level, sheep and goat account for about 90% of the live animal/meat and 92% of skin and hide [8] export trade value. In the lowlands, sheep with other livestock are the mainstay of the pastoral livelihoods.

In traditional production systems, small ruminants provide both tangible benefits such as cash income from animal sales, meat for home consumption, manure, fiber and skins and intangible benefits such as savings and insurance against emergencies, employment, cultural and ceremonial purposes [6]. Thus, small ruminants contribute their share in fundamental issues related to reducing under-nutrition, enhancing food security, combating rural poverty and achieving rates and patterns of agricultural growth that would contribute to the overall economic development and environmental protection [9].

Ethiopia has some important comparative advantages in the Middle Eastern livestock and meat markets. Live animal exports are high, as an estimated 1.6 million livestock [1]. However, feed shortages are often highlighted as a constraint to Ethiopia's livestock and meat industry [12].

Live animal exports contributed 70% of the earnings, while 30% was obtained from meat exports. However, the lack of exporting routes and ports, illegal live animal trade, the shortage of live animals and the lack of appropriate breeding programs are some of the main challenges faced by the sector. The presence of large livestock population with diverse and adaptable genotypes, and diverse agroecologies for production of different types of livestock; the expansion of agro-industries and the increase of by-product feedstuffs allowing for enhanced productivity; proximity to Middle East countries; high demand for meat and live animals in the market including the domestic market are some of the opportunities that the sectors have [7]. However, based on the above facts there was no documented information in order to improve and know the status of small ruminant fattening in Amhara region. Thus, on the basis of this background this study was initiated with the objective of small ruminant fattening practices in Amhara region.

Methods

Description of the study area

The study was conducted in five parts of Amhara regional zones of (N/Gondar, S/Gondar, S/Wollo, W/Gojjam and N/Shewa). The Amhara region has 105 woredas. According to figures of the Central Statistical Agency [3], the region has an estimated total human population of 19,120,000, consisting of 9,555,000 males and 9,565,000 females. About 16,925,000 or 88.5% of the population are rural inhabitants, while 2,195,000 or 11.5% are urban. The annual population growth rate is 2.7%. With an estimated area of 159,173.66 km^2, the region has an estimated density of 120 persons/km^2. The Amhara region is the second most densely populated region in Ethiopia. Some 18–20% of the region's population is critically food insecure. The livestock population in Amhara region includes cattle (14,710,911), sheep (10,024,277), goats (6,064,944), horses (420,760), mules (157,213), donkeys (2,677,429), camels (66,364), poultry (18,031,121) and beehives (1,361,329). Mutton and chevon household consumption, sale, wages in kind and others utilization-private peasant holders in Amhara region were 94.15, 1.87, 0.23 and 3.74%, respectively [4] (Fig. 1).

Sampling design, techniques and procedures

From the total of the selected five zones of Amhara region, a total of 20 kebeles were selected. Each zone was stratified by midland and low land agroecology zones and then from each midland and low land agroecology zones one potential woreda was selected purposively. From the selected district, two kebeles were selected randomly. Then, from each of selected kebeles, ten small ruminant fatteners were selected purposively for data collection. Therefore, a total of two hundred households were selected.

Data requirements, sources and methods of data collection

Both primary and secondary sources of data were used to obtain qualitative and quantitative data by structured and semi-structured questioner. Primary data such as grazing land, feed sources and feeding, water sources and watering of small ruminant were generated from interviewed respondents, and secondary sources that covered households' demographic characteristics, institutional settings were obtained from District Agricultural Rural Development and Livestock Agency office.

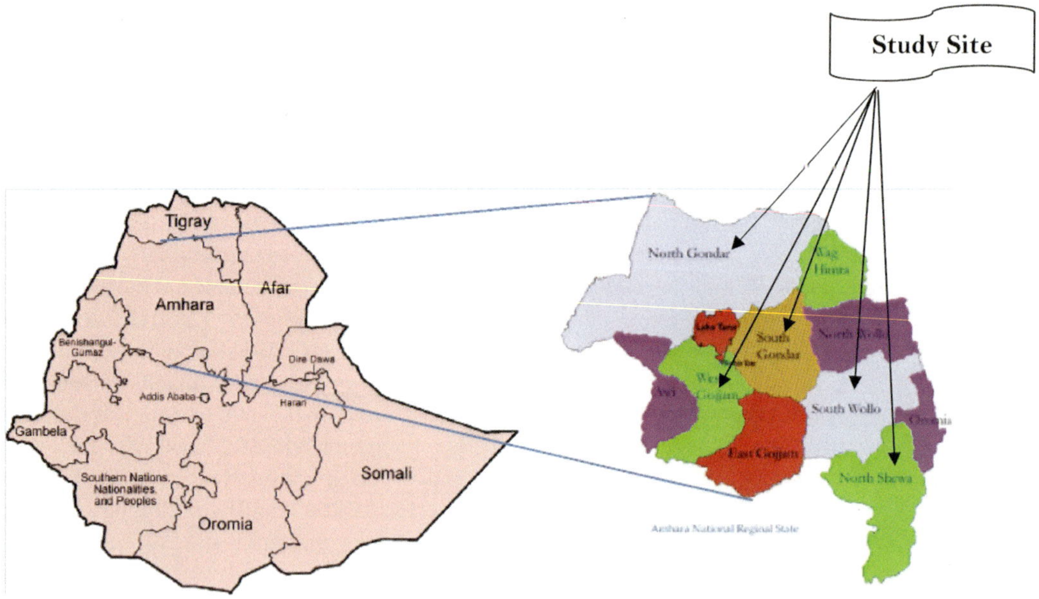

Fig. 1 Study map of Amhara region

Data analysis

The collected data were directly entered and analyzed by Statistical Package for Social Sciences (SPSS) version 20 software.

Results and discussion

Socioeconomic characteristics of households in the study area

The characteristics of households in the study area of Amhara region are presented in Fig. 2. From the total of

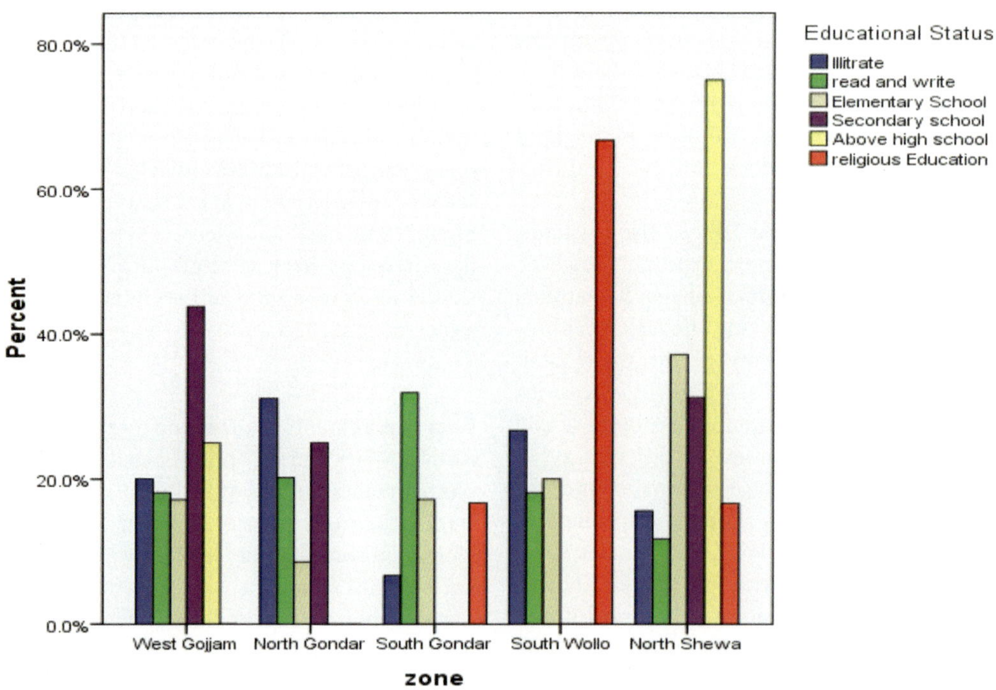

Fig. 2 Educational status of household in the study area

Table 1 Average household family size in the study area

Household	Mean	SD
Total family size	5.22	2.00
Male	2.65	1.38
Female	2.55	1.26

Table 2 Mean and SD of private, communal and total lands for grazing and traction purpose in hectare in the study area

Land description per ha	*N*	Min	Max	Mean ± SD
Private grazing land	200	0.00	4.00	0.30±0.61
Communal grazing land	200	0.00	20.00	0.81±1.81
Total land for traction	200	0.00	21.00	1.84±2.90

(N=200) respondents, majority of small ruminant fatteners can read and write (47%), while the remaining 22.5, 17.5, 8.0, 2.0 and 3.0% were illiterate, elementary school, secondary school, above high school and religious education, respectively. Educational status of zones showed that in West Gojjam majority of fatteners were secondary school, whereas in North Gondar, South Gondar, South Wollo and North Shewa they are illiterate, can read and write, are religiously educated and had above high school, respectively. Mean and standard deviation (SD) of family size per household in the study area was 5.22±2.00, and that of females and males was 2.65±1.38 and 2.55±1.26, respectively (Table 1).

Grazing land utilization

Mean and SD of private, communal and total lands for grazing and traction purpose in the study area is shown in Table 2. In the region, the mean and SD of private grazing land for fattening purpose was 0.30±0.61 ha and that of communal grazing land of its mean and SD were 0.81±1.81 ha. Generally, the mean and SD of total land for traction was shown to be 1.84±2.90 ha.

The mean and standard deviation of private grazing land in the present study was almost in line with that in the study of [5], i.e., 0.22±0.01. However, the mean and standard deviation of land for traction did not agree with that in this study, i.e., 1.08±0.04. This may be due to the presence of large size of land per household in the low land area of Amhara region.

Private grazing land utilization

Utilization of private grazing land for fattening in the study area is indicated in Fig. 3. From the total of 100% respondents only 46.6% had there owen private grazing land this mean the remaining 53.4% were utilizing communal grazing land. Among the 46.6% of respondents only 74.5% were primarily grazed fattening animals

The results were in line with the study of [13] who reported that the land-owning people who have private grazing lands (29%) to feed and supplement their animals during feed scarcity periods (during the rainy season). However, the average area of private grazing lands was very small as compared to the present study which was 0.04 ha per household.

Status of private and communal grazing land

Status of grazing land in the study area is indicated in Table 3. Even though the respondents had private and communal grazing lands in the study zones, 59% responded that the status of grazing land was decreasing from time to time, whereas 9.5% responded that it was

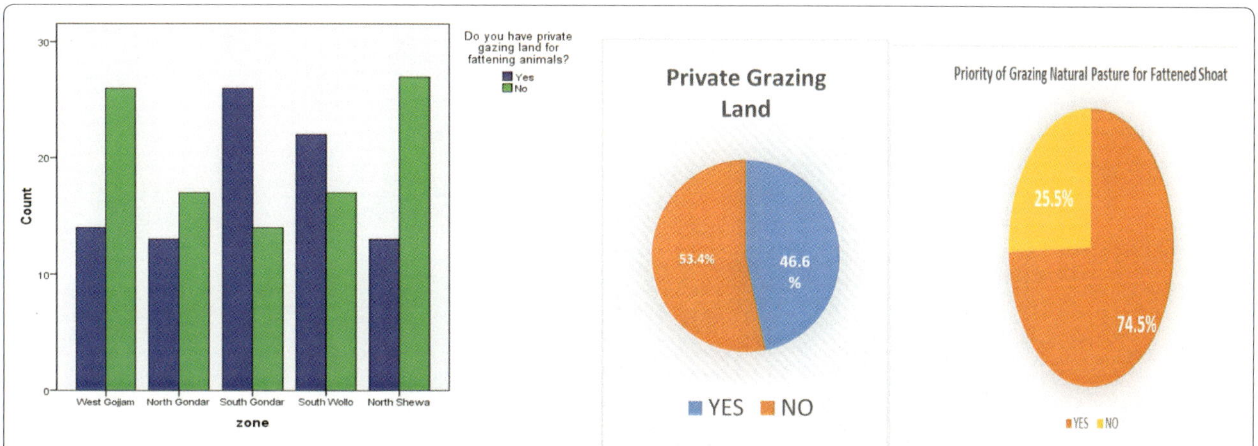

Fig. 3 Utilization of private grazing land for fattening

Table 3 Status of grazing land in the study area

Grazing land status	Frequency	%
Decreasing	118	59.0
Increasing	19	9.5
No change	63	31.5
Total	200	100.0

Table 4 Maximum, mean and SD of small ruminant fattening per individual in the study area

Sheep and goat population	Max	Mean	SD
Sheep Ewe	69.00	5.1800	8.57703
Sheep Ram	17.00	1.6050	2.53002
Sheep Lamb	28.00	2.9200	3.93331
Sheep fattening	19.00	2.8600	2.90042
Goat Doe	80.00	3.5050	9.72176
Goat Buck	22.00	1.0450	2.76063
Goat Kid	32.00	2.0200	4.82764
Goat fattening	50.00	1.6450	4.66129

increasing and the remaining 31.5% responded that there was no change.

The size of land decrease may be due to an increment of human population from time to time which was the land utilized for human food crops, expansion of urban and cities, utilization of communal grazing land for public service such as school, health center, and construction of offices for different sectors and for investment purpose.

Small ruminant fattening per household

Maximum, mean and SD of small ruminant fattening per household in the study area are indicated in Table 4. In Amhara region, there was a good potential of keeping small ruminant population per individual or per household. The mean and SD of Sheep Ewe, Ram and Lamb population was shown to be 5.18 ± 8.58, 1.61 ± 2.53 and 2.92 ± 3.93, respectively. From these an average of 2.86 sheep were used for the purpose of fattening. In addition to this, the average population of Goat Doe, Goat Buck and Goat Kid was shown to be 3.51 ± 9.72, 1.05 ± 2.76 and 2.02 ± 4.83, respectively, of which an average of 1.65 were used for the purpose of fattening. Generally, according to the report of the respondents in the study areas there were involved in sheep keeping than goat per household for fattening. This might be due to the presence of suitable climatic condition and more preference of the community for mutton in the study region than chevon.

Standard deviation of sheep fattening in the study area was almost in line with that in the study of [13], which was 2.46.

Preferred small ruminant species for fattening

Small ruminants selected for fattening purpose in the study zones are indicated in Fig. 4. The result showed that in West Gojjam the households selected sheep for fattening than goats, in north Gondar both sheep and goat were preferred or selected for fattening, but the figure shows that the fatteners were mostly involved in goat fattening than sheep fattening, which might be due to the climatic condition of the area which was suitable for goat production than sheep, in south Gondar producers were involved in both goat and sheep fattening, in South Wollo

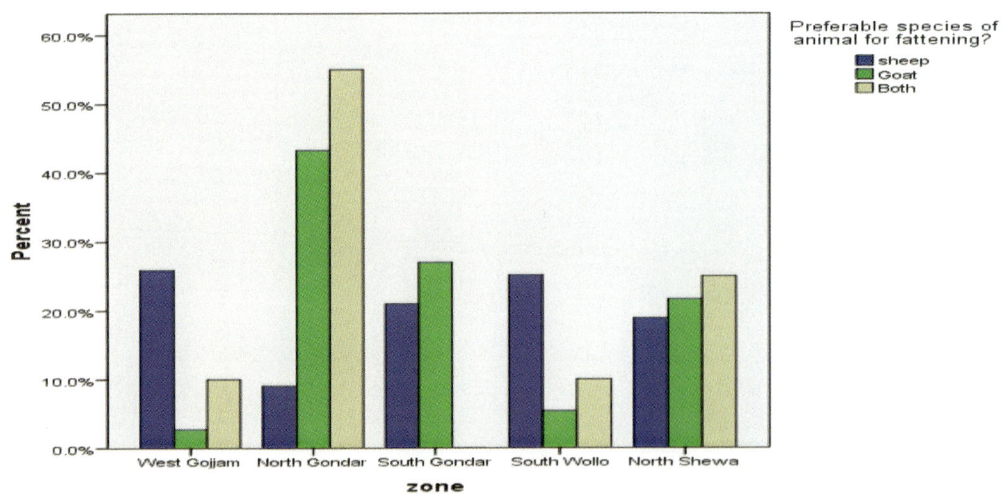

Fig. 4 Small ruminant selected for fattening in the study zones of Amhara region

majority of producers were involved in sheep fattening and in North Shewa they were almost similarly involved in sheep and goat fattening. Generally, small ruminants selected for fattening in the region (Table 5) showed that 71.5% of producers were involved in sheep fattening than goat fattening.

Purpose of sheep and goat fattening in the study areas

The purpose of keeping sheep and goat (small ruminant) fattening in the study area is indicated in Fig. 5. According to the respondents of west Gojjam zone, the purpose of fattening small ruminants was income sources, meat, sacrifice/rituals, saving and risk benefit, whereas in North Gondar, South Gondar, South Wollo and North Shewa the major purpose of fattening small ruminants was meat and saving, all, saving, sacrifice/rituals, respectively. However, generally in Amhara region producers fattened small ruminants for the purpose of income. In addition to this, the purpose of small ruminant fattening in the Amhara region (Table 6) indicated that in Amhara region producers were involved in small ruminant fattening for the purpose of income source, meat, sacrifice/rituals, social/cultural function, saving, risk/benefit and others which were 81.5, 2.0, 1.0, 1.5, 1.0, 12.0 and 1.0%, respectively.

The present study agreed with the study of [10]. Farmers were keeping sheep and goats for the purpose of income source in the area of Fogera, Alaba, Metema, Mieso, Goma and Alamata. However, according to their study in Dale and Bure they kept small ruminants primarily for saving and meat (home consumption), respectively. It was also in line with the finding of [5] in Oromia region of Ilu Aba bora zone who pointed out that the purpose of keeping small ruminants was primarily income generation.

Small ruminants feed sources and supplementary feeds

The major feed sources in the dry and wet seasons for small ruminants in the study area are shown in Table 7. According to the respondents, the first feed source which was utilized for fattening small ruminant was natural pasture, whereas in the dry season stubble grazing. The feeding practices of fattening small ruminant in the study

Table 5 Small ruminants selected for fattening in Amhara region

Species of animals	Frequency	%
Sheep	143	71.5
Goat	37	18.5
Both	20	10.0
Total	200	100.0

Table 6 Purpose of sheep and goat fattening in the Amhara region

Purpose/reasons	Frequency	%
Income source	163	81.5
Meat	4	2.0
Sacrifice/rituals	2	1.0
Social/cultural function	3	1.5
Saving	2	1.0
Risk/benefit	24	12.0
Others	2	1.0
Total	200	100.0

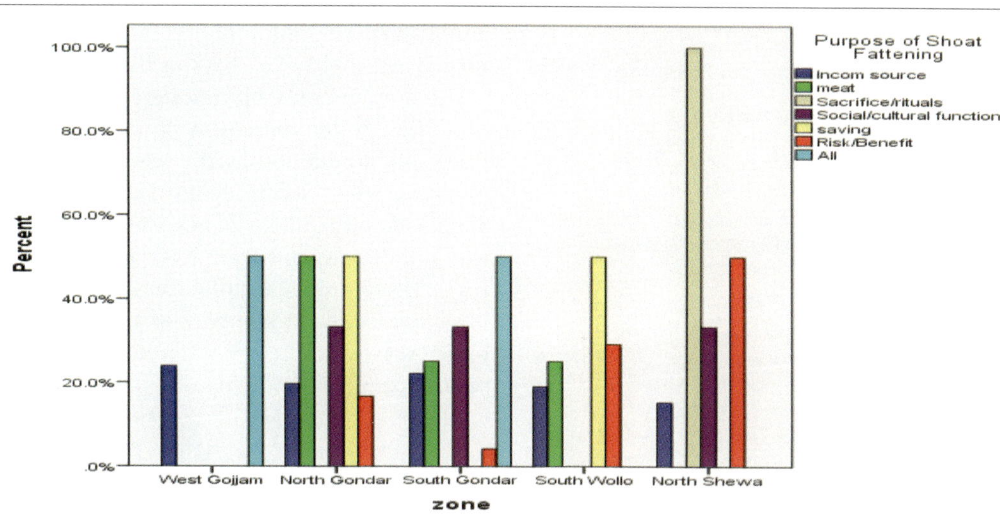

Fig. 5 Purpose of sheep and goat fattening in the study zones

Table 7 The major feed sources in the dry and wet seasons for small ruminants in the study area

No.	Feed sources	Ranking	
		Wet season	Dry season
1	Natural pasture	1	2
2	Crop residue	2	3
3	Grass hay	5	5
4	Fodder tree	3	6
5	Agro-industrial by-product	4	4
6	Stubble grazing	6	1
7	By-product of Millhouse	7	7

area showed that 48% of fatteners utilized crop residue and grass hay with supplements.

The study was in line with the result of [5]. In Ilu Aba Bora zone, the major sources of feed for small ruminants were natural pasture grazing and browsing on communal grazing lands, road and riverside and aftermath grazing.

Supplementary feed utilized for fattening is shown in Table 8. According to the respondents and direct observed of the farms in the study zones showed that fatteners formulate the rations without the background of scientific ration formulation or without the assistant of expertise. This might affect the fatteners exposed to extra feed cost and labor, over- and under-feeding and delay time of fattening.

Supplementary feeding practices of households in the study area are indicated in Table 9. Majority of fatteners (36%) were provided with the formulated ration twice a day. However, the remaining were provided with the supplemented feed once in a day (31%), whenever available (21%) and three times a day (12%) because according to the respondents it was because of expensive (77.6%), not available in the market (19.4%), no need of (2%) and others (1%).

Table 8 Supplementary feed utilized for fattening

No.	Zones	Feed resources and ingredients
1	West Gojjam	Grass, straw, crop residue, surplus kitchen food, *Frusika*, *Fagulo*, *Atela* (*Birinte*), salt
2	North Gondar	*Frushica*, *Mashila*, *Fagulo*, salt, surplus kitchen food, corn, grinded meal by-product (*Yewofcho Tiragi*), salt, Dashen brewery by-product
3	South Gondar	Crop residue, leaves, fodder trees, urea-treated feed, salt
4	South Wollo	Alfalfa, Susbania, pigeon pea (Ye-ergib Ater), straw, salt, grinded meal by-product (*Yewofcho Tiragi*), straw, *Fagulo*, *Frushica* and Tree Lucerne
5	North Showa	*Frushica*, salt, local brewery by-product (*Atela*), Vetch, grinded bean, bean bran, lentil bran, barley bran, *Fagulo*, wilted and grinded *Germia*, straw, salt, grinded barley, salt, wilted bean, wilted *Gerima*

Table 9 Supplementary feeding in the study area

Supplementary feeding	Frequency	%
Frequency of supplementary feeding		
Once in a day	62	31.0
Twice a day	72	36.0
Whenever available	42	21.0
Three times	24	12.0
Total	200	100.0
Reasons not providing supplementary feed scientifically recommended		
Not available in the market	38	19.4
Expensive	152	77.6
No need of	4	2.0
Others	2	1.0
Total	200	100.0

According to the report of [5] in Ilu Aba Bora zone, most of small ruminants were supplemented with concentrate feed whenever available. However, in the case of the present study the report was agreed. Fatteners were given supplementary feed twice a day. The reason supplementary feed was not given sufficiently for their small ruminants was its expensiveness in the market.

Water sources and watering

Water sources and watering for small ruminant fattening in the region are indicated in Table 10. In Amhara region, majority of small ruminant fatteners utilized water sources from pipe water (62.5%), and the remaining 26.5, 5.5 and 5.5% got water from river, pond and spring water, respectively. During dry season, the average distance from home to the water source was less than 1 km (45%) and 23 and 32% of the respondents got water from with no distance and 1–5 km, respectively.

The present study did not agree with the study of [5]. The major water sources in Ilu Aba bora zone for small ruminant production in the wet and dry seasons were rain and river. According to the study of [11] in West Shoa and South Wollo Zones, the major water sources in wet and dry seasons were river and river and tap water, respectively. Frequency of water in West Shoa and South Wollo was shown to be once in a day and once in 3 days in dry season, and twice a day and once in a day in wet season, respectively.

Table 10 Water sources and watering for small ruminant fattening in the region

Water sources	Frequency	%
River	53	26.5
Pond	11	5.5
Spring water	11	5.5
Pipe water	125	62.5
Average distance of water from home to water sources during dry season		
Water at home	46	23
<1 km	90	45
1–5 km	64	32
Watering frequency		
Add libitum	25	12.5
Once in a day	34	17
Twice a day	136	68
Sometimes	5	2.5

Conclusions

In general, small ruminant fattening practices in Amhara region showed that majority of fatteners did not have a private grazing land for fattening purpose. The status of private and communal grazing lands shows a decreasing trend from time to time. This might be due to the increment of human population, the decline in the productivity of land, shifting of grazing land into crop land, the expansion of settlement, the increment of animal population and so on. In the region, producers were mostly involved in sheep fattening than goat fattening. The main purpose of fattening small ruminants in the region was used for income generation. The main feed sources in the wet and dry seasons were natural pasture and stubble grazing, respectively. In addition to this, the source of supplementary feeds was surplus kitchen food, wheat bran, nuge seed cake, maize or corn, Mashila, grinded meal by-product, pigeon pea, *alfalfa*, Susbania, Tree Lucerne, brewery by-product, local brewery by-product (Atela and Brinte), Vetch, grinded bean, wilted and grinded *Gerima* bean and lentil bran, salt, etc. Some of fatteners were also provided with these supplementary feeds twice a day. This was due to an expensiveness of the supplementary feeds in the market. The major water sources for small ruminants were rivers, and the average distance was less than one km. Water frequency was twice a day. Based on the above information, the livestock and fishery ministry should implement the meat and feed improvement strategic plan, and universities and research institutes should conduct a research on feed improvement and fattening trial with low cost. Further research on alternative feeds for fattening of the above-listed supplementary feeds on feed intake, body weight change and carcass yield as well as digestibility on sheep and goats in the region should be done. A policy for small ruminant fattening program should be given attention.

Authors' contributions

The corresponding author SA contributed to the conceptualization and methodology; formal analysis contributed to investigation and writing of the original draft; and SAM and AAT, AGE, MBA and ATS contributed to data curation, validation, writing, review and editing. All authors read and approved the final manuscript.

Author details

[1] University of Gondar, P.O. Box 196, Gondar, Ethiopia. [2] Andassa Livestock Research Center, P.O. Box 77, Bahir Dar, Ethiopia.

Acknowledgements

We would like to thank University of Gondar for supporting this research project. We would like to thank the Amhara regional district and kebele experts for their collaboration of data collection and information. We would like to also thank small ruminant fatteners in the Amhara region for their great collaboration of information sharing.

Competing interests

The authors declare that we have no competing interests.

Funding

This research was supported by funding from University of Gondar.

References

1. ACDI/VOCA. Djibouti Study Tour—Trip Report. ACDI/VOCA Pastoralist Livelihoods Initiative Livestock Marketing Program. 26–28 Aug 2008. p. 12. https://www.guidestar.org/ViewEdoc.aspx?eDocId=3194250&approved=True.
2. Berhanu G, Hoekstra D, Azege T. Improving the competitiveness of agricultural input markets in Ethiopia: experiences since 1991. Paper presented at the symposium on seed-fertilizer technology, cereal productivity and pro-poor triennial conference of the International Association of Agricultural Economics (IAAE), Gold Coast, Australia; 2006. https://cgspace.cgiar.org/bitstream/handle/10568/.../improvingCompetitiveness.pdf?.
3. CSA (Central Statistics Agency). Agricultural Sample. Addis Ababa, Ethiopia; 2005. www.csa.gov.et/monthly-retail-price/category/121-eth-agss-2005?download...2005.
4. CSA (Central Statistics Agency). Agricultural sample survey 2014/15. Volume II report on livestock and livestock characteristics (Private Peasant Holdings), Central Statistical Agency (CSA), Addis Ababa, Ethiopia; 2015. p. 188. www.csa.gov.et/ehioinfo-internal/category/131-eth-agss-2015?download=397.
5. Dhaba U, Belay D, Solomon D and Taye T. Breeding practices and reproductive performance of traditionally managed indigenous sheep and goat in Ilu Abba Bora Zone of Oromia Regional State, Ethiopia; 2013. p. 676–80. https://doi.org/10.5829/idosi.gv.2013.10.6.64163.
6. Jainter J, Sowe J, Secka NE, Dempfle L. Ownership pattern and manage-

ment practices of small ruminants in the Gambia-implications for a breeding programme. Small Rumin Res. 2001;40:101–8.
7. Kefyalew A, Tarekegn A. Meat and live animal export in Ethiopia: status, challenges and opportunities; 2013. p. 054–9.
8. Mueller B, Acero F, Estruch E. Creating employment potential in small-ruminant value chains in the Ethiopian Highlands. FAO Animal production and health working paper no. 16, Rome, FAO; 2017. http://www.fao.org/ag/againfo/resources/en/publications.html.
9. Pelant RK, Chandra B, Pu JB, Lohani M, Suknaphasawat N, Xu G. Small ruminants in development. Small Rumin Res. 1999;34(3):249–57. https://doi.org/10.1016/s0921-4488(99)00077-2.
10. Solomon G, Azage T, Berhanu G and Dirk H. Sheep and goat production and marketing systems in Ethiopia: characteristics and strategies for improvement. IPMS (improving productivity and market success) of Ethiopian farmers project working paper 23. ILRI (International Livestock Research Institute), Nairobi, Kenya; 2010. p. 58. https://cgspace.cgiar.org/bitstream/handle/10568/2238/IPMS_Working_Paper_23.pdf.
11. Trevor AC. Advancing Ethiopian small ruminant production systems. Borlaug-Ruan International Intern World Food Prize International Livestock Research Institute Addis Ababa, Ethiopia. https://www.worldfoodprize.org/documents/filelibrary/images/youth_programs/2015_interns/2015_br_research_papers/Clemens_ILRI_Ethiopia_FinalReport_7C63C2512D16D.pdf. Accessed 17–18 Dec 2015.
12. USAID. End market analysis of Ethiopian livestock and meat. A desk study micro report#164. https://www.bdsknowledge.org/dyn/bds/docs/801/USAID. Accessed May 2010.
13. Yenesew A. Assessment of small ruminant production systems and on-farm evaluation of urea treated wheat straw and concentrate feeding on sheep body weight change in Burie Woreda, West Gojjam; 2010. https://cgspace.cgiar.org/handle/10568/792.

12

Evaluation of livestock feed balance under mixed crop–livestock production system in the central highlands of Ethiopia

Mekete Bekele Tahir[1,2*], Ashenafi Mengistu Wossen[2] and Berhan Tamir Mersso[2]

Abstract

Background: Livestock production under smallholder mixed crop–livestock (MCL) production systems is constrained by feed shortage and scarcity of land on which to grow feed. Livestock feeds are obtained from different sources including crop residues (CR), grazing lands (GL), crop aftermath, fallow land and purchased. But the contribution of these feed resources and the extent of feed balance at farm level are not quantitatively examined. The study was conducted to assess the major feed resources available and evaluate feed balance for the prevailing livestock in MCL system.

Methods: Feed types and amount available, livestock holding size and feed demand were estimated for 159 smallholder farmers stratified into wealth status, which were selected following a multistage sampling procedure. The feed balance was evaluated as the difference between requirements of livestock (feed demand) and amount of utilizable feed (supply) per year in terms of dry matter (DM), metabolizable energy (ME) and digestible crude protein (DCP).

Results: Regardless of farmers' wealth status, CR mainly sourced from cereals, particularly barley, contributed more than half of the annual feed supply, followed by GL. The contribution of CR to total feed supply sourced on-farm and purchased combined was 55%. Significant differences in the supply, demand and balance of livestock feed were observed across wealth group of farmers. The wealthier have higher quantities of feed supply and demand, but suffered more in feed insufficiency. Overall, about 51, 19 and 38% annual feed deficit in DM, ME and DCP were observed, respectively. But when the rate is considered based on feed produced on-farm only, the deficit worsened and goes up to 60, 34 and 52% in DM, ME and DCP, respectively.

Conclusions: Expansion of grazing land is not a practical option to increase feed supply. Therefore, increasing food-feed crops production per unit area, conservation of surplus forages, strategic feeding based on productivity and traction services of livestock, purchase of feeds, and increasing livestock off-take during time of scarcity would help to correct the observed feed shortage. Moreover, refinement of the feed balance analysis at specific nutrients level would be compulsory for effective strategic interventions.

Keywords: Debre Berhan, Dry matter, Metabolizable energy, Digestible crude protein, Feed balance, Feed supply and demand, Wealth status

Background

Ethiopia is an agrarian country known for possession of huge livestock numbers. The total livestock population estimated in millions during the 2014/2015 annual livestock sample survey in the sedentary areas of the country was about 56.7 cattle, 29.3 sheep, 29.1 goats, 2.0 horses, 7.4 donkeys, 0.4 mules, 1.2 camels and 56.9 poultry with total tropical livestock unit (TLU) of 52.9, excluding the non-sedentary three zones of Afar and six zones of Somali Region of the country [1]. Livestock perform economic and social functions both at the national and

*Correspondence: meketebekele@gmail.com
[2] Department of Animal Production Studies, College of Veterinary Medicine and Agriculture, Addis Ababa University, PO Box 34, Debre Zeit, Ethiopia
Full list of author information is available at the end of the article

household levels. Albeit variations among data sources, livestock contribute 15–17% of national gross domestic product (GDP), 35–40% of agricultural GDP and 37–87% of the household incomes [2]. Moreover, livestock contribute to improve the nutritional status and income gain of the people by providing meat, milk, eggs, cheese, butter, etc. and commodities, such as live animals, hides and skins for home use and export, and avert risks in times of crop failures [1, 2].

Despite the huge numbers of possession, the current contribution of livestock to the producers and to the national economy is dismal to its size. It has increasingly been unable to meet the demands for the rapidly growing population [3–5]. Among the many factors that could explain this disproportionate role of the sector often mentioned are the inadequate quantity and quality of feed supply throughout the year to satisfy the annual demand of livestock [6, 7].

Smallholder farmers in the mixed crop–livestock (MCL) systems keep some form of livestock in conjunction with crop production. However, better soils are allocated for food-feed crops, whereas mountainous, sloppy and less fertile marginal lands, which are naturally vulnerable for soil erosion and land degradation, are left for livestock grazing. Eventually, livestock became more dependent on common feed resources derived from low biomass producing food-feed crops and poorly managed grazing lands (GL). In order livestock to express their productive potentials and increase productivity, the available feed resources should match their demands for dry matter (DM) and nutrients. However, frequently livestock are exposed to seasonal feed shortages both in quantity and quality, especially during the dry season [8]. Moreover, the nutritive quality of native pasture is low especially in dry season and it is much worse for crop residues (CR) owing to the lower content of digestible nutrients [9]. Furthermore, the continuing trends of expansion in cropland cultivation at the expense of GL in the MCL systems have resulted in shrinkage of the area and productivity of GL and reversed the proportional contribution of CR for feed upward.

The need to increase both crop and livestock production from the existing resources requires responsive action throughout the production system. Upon examining the livestock feed resources and feed balance at smallholder farm level, Kassa et al. [10] suggested the possible use of livestock feed balance as potential indicator to assess sustainability of the farms. However, the extent and persistency of feed deficit in rainfed MCL systems in the highlands of Ethiopia described in several previous reports [6, 9–12] are a formidable challenge for sustained livestock production. A negative feed balance in the MCL system disrupts the interactions between the system components, impairs livestock performance and compromises the potential roles of livestock in driving the economic development. For instance, Kassa et al. [10] reported that better-off and medium wealth groups of smallholder mixed farmers did not produce enough feed more than the poor group to support their livestock in the *Harar* highlands of eastern Ethiopia. However, it is anticipated that the analysis at farm scale that constitutes the major available feeds and the livestock resources would unveil the prevailing status and hint strategies to address feed shortage problems encountered in the mixed farming systems. In the study area like most other MCL systems areas of Ethiopia, the ongoing land use change from grazing land to other land uses mainly to cropland resulted in shortage of grazing lands. On the other hand, feed obtained as a by-product from crop production on converted land from grazing to cropland is not likely to compensate the feed supply due to the inherently low feed value of crop residues. Moreover, feed from other alternative feed sources is limited. The present study was aimed at assessment of the potential feed resources supply for the prevailing livestock feed demand and evaluate the annual feed balance in the MCL system.

Methods

Description of the study area

The study was conducted in Debre Berhan milkshed in the Amhara National Regional State, central highlands of Ethiopia, 130 km away to the Northeast of Addis Ababa. The area is located between 9°30′ and 9°50′ latitudes and 39°20′ and 39°44′ longitudes (Fig. 1). The elevation ranges from 2840 to 2943 masl [13].

The mean annual minimum and maximum temperatures averaged between 2000 and 2014 years are 6.7 and 19.9 °C, respectively. The mean annual rainfall within the span of the same years is 1026 mm with a potential evapotranspiration of 1396 mm. Rainfall distribution is bimodal, usually the long rains last from June to the beginning of September, and the period of the short rains falls between February and May. About 85.5% of the rain falls between June and September which is the main cropping season (Fig. 2). Most of the area is covered by moderately and poorly drained soils, predominantly black Vertisol [13, 14].

Rainfed MCL farming is the dominant system carried out primarily to meet the subsistence requirements for most of the farmers with their families in the study area. Small-scale irrigated farming is limited to few farmers with small patchy areas along Beressa River mainly for vegetables. The principal rainfed food-feed crops include barley, wheat, faba bean and field pea. The cereals which covered the largest portion of cropped areas are the major sources of CR for livestock feeding. Cultivations

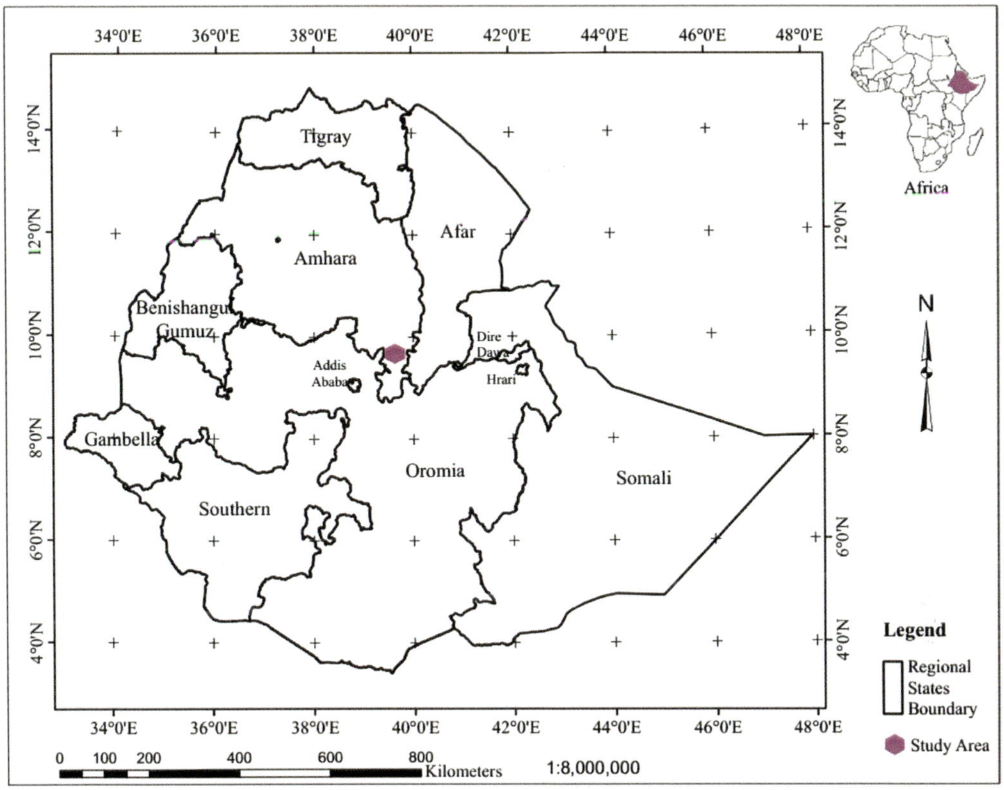

Fig. 1 Map of Ethiopia showing the location of the study area

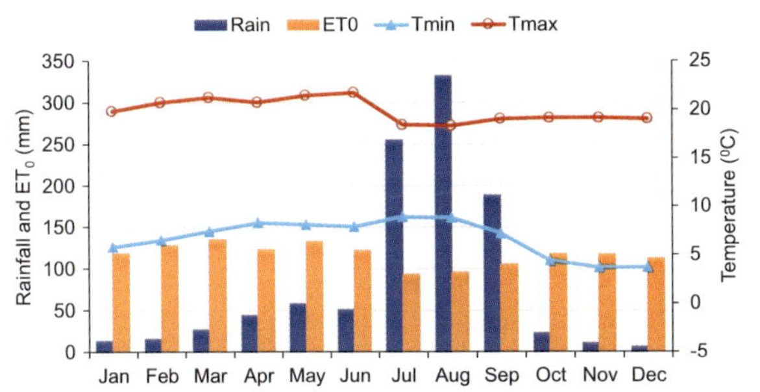

Fig. 2 Average monthly rainfall, potential evapotranspiration (ET_0), maximum (T_{max}) and minimum (T_{min}) temperatures at Debre Berhan weather station (monthly data averaged from 2000 to 2014)

of *teff* (*Eragrostis teff*), lentils, chickpea, oats, linseed and vegetables are intermittent on small plots of cropland. Natural GL and CR are the major livestock feed resources, while fallow land, crop aftermath grazing and concentrates are occasionally used feeds [8, 15]. The livestock species reared include cattle, sheep, goats, donkeys, horses, mules and poultry. Cattle production with indigenous and cross-bred animals predominates the livestock production followed by sheep production [8].

Wealth status classification criteria

The setting of criteria for wealth status was made in consultation with district experts, development agents and confirmed based on the perspectives of local farmers during group discussions. Multiple criteria focused on physical ownership of key assets and their anticipated values at the time of the study were used rather than precarious annual cash income. Ownership of houses with corrugated iron or thatched grass roofs, number and types of livestock, area of land and the capacity of the farmers to satisfy annual household basic needs were the major focuses. Nonetheless, setting an absolute cut-off point to each criterion was not possible, and an overlap in the range of values for the set criterion was evident. Instead of fixing the judgment based on the value of a single criterion, the contribution of the whole was assessed together to group a farmer under one of the three wealth categories (better-off, medium and poor). The descriptions of each criterion are summarized in Table 1.

Survey design and data collection

Sampling procedures and sample size

A multistage sampling procedure was employed to select sample farmers engaged in MCL farming and deliver fluid milk to the nearby milk collection centres (MCC).

Table 1 Description of wealth status grouping criteria generalized based on farmers' perspectives in the mixed crop–livestock system, central highlands of Ethiopia

No.	Criteria	Wealth categories		
		Better-off	Medium	Poor
1	Cropland holding (ha)	>1.50	1–2.5	0.25–1.5
2	Number of ploughing oxen	>2	≥2	≤1
3	Number of milking cows (local and crossbreds)	>2 most of crossbreds	At least 2, few crossbreds	<2 and usually local breeds
4	Number of sheep (adults)	10–50 or more	5–20	≤10
5	Number of equines (adults)	A mule, a horse and donkeys	At least horse and donkeys	No mule & horse may be donkey
6	Roof of houses			
	Corrugated iron sheet	Yes	Yes (often)	No
	Grass thatched	Yes	Yes	Yes
7	Annual food production and supply to sustain household	Enough with more surplus	Enough but meagre surplus	Not enough, in need of support

Debre Berhan milkshed was purposely selected based on its accessibility and potential representativeness of the MCL system in central highlands of Ethiopia. From 48 villages in the milkshed, six representative MCC (*Wush-awushign, Angolela, Kormargefia, Kebele01, Kebele07* and *Kebele09*) were selected using simple random sampling procedure. Farmers sample size was determined using G * power 3.1.7 software considering the farmer wealth group as fixed effect and assuming 0.25 effect size, $\alpha = 0.05$ and 80% power of the statistical test [16]. List of farmers obtained from the selected MCC was categorized into three wealth groups to form the sampling frame. The number of sample farmers in each wealth group was based on the 'probability proportional to size' sampling technique [17]. Finally, a total of 159 farmers (50 in the poor, 58 in the medium and 51 in the better-off wealth groups) were selected using systematic random sampling with whom the questionnaires were administered.

Data collection

A socio-economic survey using pretested semi-structured questionnaires was carried out using a face to face interview in 2014 and 2015. The questionnaires covered data on: household demographic characteristics, land and livestock ownership, livestock species composition and herd structure, feed types, area of private and communal GL, types of food-feed crops produced, area cultivated, input used and crop yield at farm level. Focus group discussions were conducted with farmers at each of the selected MCC. A total of 40 farmers (6–8 per session) representing wealth groups have participated in six sessions. Knowledgeable farmers from all wealth groups were selected anticipating an effective communication between the moderator and within themselves. The secondary data were extracted from previous studies and information documented at agricultural development and research offices. Local climate data were obtained from Debre Berhan agricultural research centre. Data enumerators recruited from agricultural development agents were trained to assist on primary data collection during face to face interview and group discussions.

Estimation of annual feed availability (supply)

Quantity of feed DM available per year was estimated from the major feed resources including CR, crop aftermath (stubble that remain after harvest) and GL. Crop residues DM obtained was derived from grain yield, harvest indices and area of cropland cultivated. Conversion factors derived from harvest indices, 1.5 t/ha for wheat and barley and 1.2 t/ha for faba bean and field pea [18], were used to estimate crop residues from grain yields. The amount of CR DM collected per year by an individual farmer was quantified based on the size of cropland plot allocated for growing a particular crop type during the cropping season. Given the feed shortage and farmers' priority to the use of CR in the highland MCL systems, it is assumed that about 90% of the CR used as feed and 10% for other purposes and wastage [19]. Available DM from crop aftermath grazing on cropland was estimated using conversion factor of 0.5 ton DM/ha per year [20] and the area cultivated. Available feed DM from GL is estimated by taking the private and communal ownership pattern into account. Farmers are eligible to use whole of available feed on their entitled private GL, but can only share certain amount of feed from openly accessible communal GL. It was assumed that the amount of share from communal grazing is a function of livestock density, which is ascribed to the size of livestock ownership that had access to use this communal resource. A livestock density of 14.8 TLU/ha derived based on data on the size of livestock and GL area was used to allocate the communal grazing to each livestock owner relative to livestock possession. The total area of privately owned and part of the communal GL allocated was considered to estimate the available feed DM from GL per household per year. The feed DM productivity on GL was estimated based on multiplier of 2 t/ha established from previous GL condition and productivity studies [20]. Utilization factor of 75% as suggested by [21], for extensive grassland, is used to quantify the DM that would be utilized by livestock. The quantities of metabolizable energy (ME) and digestible crude protein (DCP) of feed resources were calculated based on the in vitro digestibility of organic matter in dry matter (IVDOMD) and crude protein (CP) contents of each feed type reported by [9] in the study area. The following equations were used to estimate the annual energy and protein supply at farm level in relation to the type of feed resource and amount obtained per year [22, 23].

$$\text{ME}\left(\text{MJ/kg DM}\right) = 0.015 * \text{IVDOMD}\left(\text{g/kg}\right);$$
$$\text{DCP}\left(\text{g}\right) = 0.929 * \text{CP}\left(\text{g}\right) - 3.48.$$

The seasonal availability of feed resources was assessed based on farmers' judgment and scores given for a particular feed type in each month throughout the year. Availability of feed over the year was scored on a scale of 0–10, where 10 = excess feed available, 5 = adequate feed available and 0 = no feed available [24].

Estimation of livestock feed requirements (demand)

Livestock holdings per household were aggregated into TLU considering the annual average livestock ownership derived based on the number of animals at the beginning and end of the study year. This was done to take into account the annual inflow and outflow dynamics of livestock at famer level. Species-specific TLU conversion factors of 0.7 for cattle, 0.1 for sheep and goats, 0.5 for donkeys and 0.8 for horses were used [25]. The dry matter demand (DMD) was estimated based on the expected daily dry matter intake (DMI) suggested for the standard TLU of 250 kg at 2.5% of the body weight, which is equivalent to 6.25 kg/day or 2280 kg/year [25]. Comparable rate of DMI was also suggested by [26]. The ME and DCP requirements for maintenance were calculated according to the daily average recommendations given by [26]. Based on metabolic body weight, 118.0, 93.0 and 103.45 kcal of ME/$W_{kg}^{0.75}$ and 2.86, 1.72 and 2.51 g of DCP/$W_{kg}^{0.75}$ per day for cattle, sheep and goats, respectively, were used for maintenance [26]. Accordingly, the daily nutrient requirement tables for maintenance recommended for ruminants in developing countries by [26] were used to estimate the annual ME and DCP requirements for cattle, sheep and goats. In the case of donkeys and horses, the daily ME and DCP maintenance requirements of 14.9 and 27.6 MJ and 0.18 and 0.37 kg recommended by McCarthy as cited in [9] were used, respectively. Then, farm level DM, ME and DCP requirements for maintenance per year were extrapolated relative to the livestock ownership per household.

Livestock feed balance

Livestock feed balance at individual farmer level over the entire production year was determined as the difference between the annual feed DM, ME and DCP supply estimated from major feed resources and the annual feed DM, ME and DCP demands for the annual average livestock holding of farmers.

Statistical analysis

Means, standard deviations and percentages were used to describe variables observed among farmers stratified into wealth status. To compare the differences across farmer wealth groups in terms of harvested grain yields, net feed supply, net demand and feed balance, a one-way analysis of variance (ANOVA) was used. The one-way ANOVA model is given by:

$$Y_{ij} = \mu_i + F_i + \varepsilon_{ij},$$

where Y_{ij} is the jth observation in the ith wealth group, μ_i is the common effect for the whole wealth group, F_i is the effect of the ith wealth group and ε_{ij} is the random error associated with the jth observation in the ith wealth group assumed to be normally and independently distributed, with mean zero and variance $\sigma^2\varepsilon$, i designates the wealth group, and j denotes a specific observation. The tests were done at 95% level of confidence ($\alpha = 0.05$). Tukeys' HSD mean comparison procedure was used to test mean differences. The analysis was carried out using SPSS version 23.0 statistical software [27].

Results

Farm households and key farm resources characteristics

Descriptions of farm households and key farm resources characteristics with respect to farmer wealth groups are summarized in Table 2. Average family size of the better-off and medium wealth group households was somehow comparable, but for both wealth groups it was higher than the poor farmers. The family labour force followed the same trend of trajectory as that of average family size. In contrast, age dependency ratio was higher in the poor than both the medium and better-off wealth groups. The magnitude of livestock and land holdings matched the wealth status gradient of farmers, exhibiting a decreasing trend from better-off to poor wealth groups. Livestock holding aggregated in terms of TLU for the whole

Table 2 Households demographic and key farm resources characteristics in the mixed crop–livestock system, central highlands of Ethiopia

Households and farm resources characteristics	Wealth groups			Total ($n = 159$)
	Poor ($n = 50$)	Medium ($n = 58$)	Better-off ($n = 51$)	
Family size	4.9 (2.0)	6.2 (1.5)	6.4 (1.6)	5.9 (1.8)
Labour force (in adult equivalent)	2.5 (1.2)	3.5 (1.1)	3.8 (1.1)	3.3 (1.3)
Age dependency ratio (ADR)	1.0 (0.6)	0.7 (0.5)	0.6 (0.5)	0.8 (0.6)
Livestock holding (TLU)	5.1 (1.2)	8.2 (2.0)	12.0 (2.4)	8.5 (3.4)
Cattle	3.0 (0.9)	4.6 (1.2)	6.4 (1.4)	4.7 (1.2)
Sheep and goats	0.9 (0.5)	1.7 (0.8)	2.7 (1.2)	1.8 (1.1)
Equines	1.2 (0.7)	1.9 (1.0)	2.9 (1.0)	2.0 (1.1)
Cropland entitled (ha)	1.1 (0.5)	1.3 (0.6)	1.8 (0.6)	1.4 (0.6)
Fallow land (ha)	0.1 (0.1)	0.1 (0.2)	0.3 (0.3)	0.2 (0.3)
Cultivated and grazing land (ha)	2.3 (0.4)	3.2 (0.7)	4.1 (0.7)	3.2 (1.0)
Cropland cultivated (ha)	1.6 (0.4)	2.2 (0.6)	2.5 (0.6)	2.1 (0.6)
Grazing land (ha)	0.7 (0.2)	1.0 (0.4)	1.6 (0.5)	1.1 (0.5)

N number of respondents; numbers in parenthesis are standard deviations (SD); *ADR* age dependency ratio; *TLU* tropical livestock unit; *1 TLU* 250 kg live weight; *ADR* the dependency ratio relates the number of children (0–14 years old) and older persons (65 years or over) to the working-age population (15–64 years old)

and grouped/disaggregated into species was higher in the better-off followed by medium and poor farmers, respectively. The size of cultivated cropland that includes cropland righteously entitled and cropland temporarily acquired based on local lease agreements between farmers, GL and the total land holdings similarly reflected the image of farmers' wealth status consistently being higher for the wealthier.

Food-feed crops production

Descriptions of cultivated crops which are the staple human food and major sources of CR for feeding livestock are presented in Table 3. The same table also presents the comparisons of mean grain yields (ton) of the major crops cultivated per household between farmer wealth groups. The trends in the size of plot of cropland allocation to a particular crop type and proportion of farmers who have grown the crop during the cropping season were similar irrespective of their wealth status. Barely was the most extensively grown food-feed crop followed by faba bean, wheat and field pea, respectively, which is manifested by the larger size of cropland plot allocated for barley growing than for the other crop types and it was cropped by every wealth group of farmers. The quantity of both barley and wheat grain yield harvested per household was significantly higher for better-off than both medium and poor farmer wealth groups ($P < 0.001$). The grain yield of faba bean harvested per household by the better-off and medium wealth groups was significantly higher than the poor group ($P < 0.001$). However, no significant difference in field pea grain harvested was observed among the wealth groups.

Types and availability of feed resources

Estimated farm level annual feed supply from different sources in terms of DM, ME and DCP is presented in Table 4. On-farm produced feed resources were derived from food-feed crops grown, GL, crop aftermath and fallow land. In addition, livestock feeds were sourced off-farm, though the quantities procured were dependent on the capability of farmers to afford. The wealthier were relatively more capable to afford considerable amount of additional purchased feed sources (hay and some concentrates) off-farm. Corresponding to the amount of grain yield harvested, the quantity of CR obtained from individual crop types and aggregated were higher for the wealthier group of farmers. Moreover, the quantities of DM, ME and DCP obtained from on-farm feed sources including GL, fallow land and aftermath grazing were lower compared to the amount attained from total CR; however, the trend and extent of disparities between farmer groups have mirrored their wealth status.

Livestock feed supply, demand and feed balance

Comparisons of the annual feed demand, supply and balance quantified on DM, ME and DCP basis per farm household across wealth groups of farmers are presented in Table 5. The demand of livestock estimated for expected annual intake of DM, ME and DCP was significantly different ($P < 0.001$) across wealth group of farmers, consistently increasing along with the wealth gradient of farmers. The aggregated annual supply obtained on-farm from different sources, as well as the total annual supply (on-farm obtained combined with purchased feeds), differs significantly ($P < 0.001$) across

Table 3 Area of plots of cropland allocated, proportion of farmers who grow the specific food-feed crop types and grain yields per household in the mixed crop–livestock system, central highlands of Ethiopia

Area cultivated, grain yield and proportion of farmers who grow	Wealth groups			Total ($n = 159$)
	Poor ($n = 50$)	Medium ($n = 58$)	Better-off ($n = 51$)	
Land allocated (ha)				
Barley	0.9 (0.3)	1.1 (0.6)	1.2 (0.4)	1.1 (0.5)
Wheat	0.3 (0.2)	0.3 (0.2)	0.5 (0.3)	0.4 (0.2)
Faba bean	0.4 (0.2)	0.6 (0.3)	0.7 (0.4)	0.6 (0.3)
Field pea	0.1 (0.1)	0.1 (0.1)	0.1 (0.1)	0.1 (0.1)
Farmers who grow (%)				
Barley	100.0	100.0	100.0	100.0
Wheat	78.0	93.1	100.0	90.6
Faba bean	100.0	100.0	92.2	97.5
Field pea	20.0	43.1	49.0	37.7
Grain yield (t/hh)				
Barley	1.7 (0.7)^{a}	2.1 (1.0)^{ab}	2.5 (0.9)^{c}	2.1 (0.9)***
Wheat	0.5 (0.4)^{a}	0.7 (0.4)^{ab}	1.2 (0.7)^{c}	0.8 (0.6)***
Faba bean	0.6 (0.3)^{a}	0.9 (0.5)^{bc}	1.0 (0.5)^{c}	0.8 (0.5)***
Field pea	0.1 (0.2)	0.2 (0.2)	0.2 (0.2)	0.1 (0.2)^{NS}

Values in parentheses indicate standard deviations (SD)

n number of respondents, *ha* hectare, *t* ton, *hh* household

^{a,b,c} Means in a row with different superscripts differ significantly ($P < 0.05$), *NS* not significant at the 0.05 probability level

***Significant at 0.001 probability level

wealth group of farmers. Similar to the demands, the wealthier obtained more quantities of DM, ME and DCP from feeds sourced on-farm and from the total feed supply including the additional purchased feeds. The annual balance of DM, ME and DCP at farm household level assessed based on the on-farm feed supply and total feed supply that included purchased feeds was significantly different ($P < 0.001$) across wealth groups of farmers, except that the balance in ME was not significantly different between the poor and medium wealth groups. However, in all cases the wealthier encountered more deficit than the poor group of farmers in the quantities of DM, ME and DCP balance attained.

The degree at which the demands for DM, ME and DCP were fulfilled or not with respect to the estimated quantities for maintenance requirement of livestock at farm household level is illustrated in Fig. 3. The magnitude of insufficiency in feed DM, ME and DCP supply for livestock was unlike between the stratified wealth groups of farmers. The better-off group had more inadequate supply followed by the medium and poor in order of importance. When the supply of DM, ME and DCP was observed at farmer level in a particular wealth status group, the DM supply shortage was more pronounced followed by DCP and ME shortages in order of the magnitude of insufficiency. It was also observed that acquiring additional purchased feeds helped to relief part of the deficiencies in the DM, ME and DCP supply differently between the poor and wealthier group of farmers. The poor farmer group was able to reduce the shortage in the DM, ME and DCP supply by 6, 9 and 9%, respectively, whereas the medium and better-off wealth groups have reduced the DM and ME shortage equally by 10 and 15%, respectively. The deficiency in DCP was reduced by 15% among the medium and by 16% among the better-off wealth groups. The overall supply shortage in DM, ME and DCP in the farming system was mitigated by 9, 14 and 14%, respectively, due to additional purchased feeds acquired.

The proportion of farmers under wealth status groups who were in a positive balance are presented in Table 6. Overall, very few farmers were able to source sufficient feed to satisfy the maintenance requirements of livestock throughout the year. Relatively more number of farmers from the poor wealth group were in positive state of feed balance in the DM, ME and DCP supply. The situation was better for ME irrespective of farmers' wealth status, especially when farmers acquire additional purchased feeds. Based on the feed obtained from on-farm only, none in any wealth group were in a positive DM balance, but less than 2%, all of them from the poor farmers group when purchased feeds were considered. Generally, the proportion of farmers in a positive feed balance for DM, ME and DCP were contrary to their wealth status, the wealthier suffered more than the poor in feed insufficiency.

Table 4 Feed resource types and estimated quantities of DM, ME and DCP obtained per year per farm household in the mixed crop–livestock system, central highlands of Ethiopia

Feed resource types	Wealth groups			Total ($n = 159$)
	Poor ($n = 50$)	Medium ($n = 58$)	Better-off ($n = 51$)	
DM (t/hh)				
Crop residues	3.8 (1.2)	5.1 (1.4)	6.6 (1.9)	5.2 (1.9)
Barley straw	2.3 (0.9)	2.8 (1.3)	3.4 (1.2)	2.8 (1.2)
Wheat straw	0.7 (0.6)	0.9 (0.5)	1.6 (1.0)	1.1 (0.8)
Faba bean straw	0.7 (0.4)	1.2 (0.7)	1.5 (1.0)	1.1 (0.8)
Field pea straw	0.1 (0.2)	0.2 (0.2)	0.2 (0.2)	0.2 (0.2)
Grazing land	1.1 (0.3)	1.6 (0.5)	2.1 (0.5)	1.6 (0.6)
Fallow land	0.1 (0.2)	0.1 (0.3)	0.4 (0.4)	0.2 (0.4)
Crop aftermath	0.6 (0.2)	0.8 (0.2)	0.9 (0.2)	0.8 (0.2)
Purchased	0.7 (0.5)	1.9 (1.7)	2.5 (2.0)	1.7 (1.7)
ME ('000 MJ/hh)				
Crop residues	28.4 (9.3)	38.1 (10.8)	48.8 (14)	38.5 (14)
Barley straw	18.3 (7.4)	22.2 (10.4)	26.7 (9.5)	22.4 (9.8)
Wheat straw	4.6 (3.8)	5.9 (3.2)	10.3 (6.5)	6.9 (5.2)
Faba bean straw	4.7 (2.5)	8.7 (4.9)	10.4 (7.1)	8.0 (5.7)
Field pea straw	0.8 (1.7)	1.2 (1.6)	1.3 (1.5)	1.1 (1.6)
Grazing land	9.0 (2.6)	12.6 (3.7)	17.2 (4.4)	12.9 (4.9)
Fallow land	0.7 (1.5)	1.1 (2.6)	3.4 (3.4)	1.7 (2.9)
Crop aftermath	4.4 (1.2)	5.7 (1.8)	6.5 (1.6)	5.6 (1.7)
Purchased	5.0 (3.9)	13.6 (12.3)	18.6 (14.8)	12.5 (12.6)
DCP (kg/hh)				
Crop residues	128.8 (42.0)	182.7 (51.2)	230 (70.8)	180.9 (68.7)
Barley straw	68.3 (27.7)	83.0 (38.8)	100 (35.4)	83.8 (36.6)
Wheat straw	18.4 (15.1)	23.6 (12.5)	41.0 (25.7)	27.5 (20.7)
Faba bean straw	35.8 (18.6)	66.1 (37.1)	78.5 (53.7)	60.5 (42.7)
Field pea straw	6.4 (13.7)	10.1 (12.8)	10.5 (11.9)	9.1 (12.9)
Grazing land	70.1 (20.4)	98.0 (28.7)	134 (34.2)	100.8 (38.1)
Fallow land	5.4 (11.5)	8.2 (20.4)	26.5 (26.3)	13.2 (22.2)
Crop aftermath	11.4 (3.1)	15.0 (4.2)	16.9 (4.0)	14.5 (4.4)
Purchased	36.4 (28.6)	99.3 (90.2)	136.0 (108.5)	91.3 (92.4)

Values in parentheses indicate standard deviations (SD)

n number of respondents, *DM* dry matter, *ME* metabolizable energy, *MJ* mega joules, *DCP* digestible crude protein, *kg* kilogram, *t* ton, *hh* household

Relative contribution of feed resources

Proportional contributions of livestock feed resources produced on-farm only (Fig. 4a) and combined with purchased off-farm sources (Fig. 4b) to the total feed DM obtained per year at farmer level are illustrated among different wealth groups of farmers. Regardless of farmers' wealth status, the foremost on-farm feed source that covered more than half of the annual feed DM supply was CR derived from major food-feed crops (Fig. 4a). Even when the contribution of purchased feed resources was considered, share of CR prevailed (Fig. 4b). Following CR, natural GL, crop aftermath and fallow land contribute to the bulk of on-farm annual feed supply in order of importance (Fig. 4a), whereas purchased feed resources take the second level of contribution except for the poor farmers group where it takes the fourth level, when the proportional contribution of feed resources was considered based on the on-farm produced and off-farm procured sources combined on aggregated at farm level of the different wealth groups (Fig. 4b). The contribution of fallow land was very limited as there was scarcity of cropland. Among CR that make up the largest part of the livestock feed resource base, cereals (barley and wheat) contribute more than legumes (faba bean and field pea). Moreover, the proportional contribution of barley straw was the highest followed by faba bean and wheat which have comparable contributions, whereas the contribution of field pea was the least, when the share of individual crops was considered irrespective of farmers' wealth status (Fig. 4a, b).

Seasonal feed resources availability

The seasonal availability of feed resources throughout the year is illustrated in Fig. 5. Crop residues obtained from on-farm crop production were the major feed sources, which were abundantly available from months of December to March and steadily decline from April to November. Grazing lands even though utilized throughout the year, feed availability is limited to the months of August to December. During the rest of the months, feeds available from grazing land were meagre. Availability of fallow land grazing coincides with the trend of grazing lands, but with very low contribution from February to August. However, the contribution of fallow land and natural grazing land is increasing from September to November. Availability of crop aftermath grazing follows harvesting periods of crops grown and it is short lived from December to February. Months from April to August are critical periods of feed shortage when farmers used most of purchased feed resources. During these critical feed shortage periods, conserved hay and crop residues were used, preferentially for feeding lactating cows and oxen used for cropland preparation.

Discussion

Household characteristics and farm resources: implications on livestock feed availability

The overall average family size of 5.86 in the present study is higher than the national average of 5.14 reported

Table 5 Mean annual livestock feed supply, demand and balance per household in the mixed crop–livestock system, central highlands of Ethiopia

Feed supply, livestock demand and feed balance	Wealth groups			Total ($n = 159$)
	Poor ($n = 50$)	Medium ($n = 58$)	Better-off ($n = 51$)	
DM demand (t/hh)	11.6 (0.4)^{a}	18.7 (0.6)^{b}	27.4 (0.8)^{c}	19.3 (0.6)***
DM supply (t/hh)				
On-farm	5.6 (0.2)^{a}	7.6 (0.2)^{b}	10.0 (0.3)^{c}	7.7 (0.2)***
On-farm and purchased	6.3 (0.2)^{a}	9.4 (0.4)^{b}	12.6 (0.4)^{c}	9.5 (0.3)***
Balance (t/hh)				
On-farm	– 6.0 (0.5)^{c}	– 11.2 (0.7)^{b}	– 17.4 (0.7)^{a}	– 11.5 (0.5)***
On-farm & purchased	– 5.3 (0.5)^{c}	– 9.3 (0.7)^{b}	– 14.8 (0.7)^{a}	– 9.8 (0.5)***
ME demand ('000 MJ/hh)	54.9 (1.8)^{a}	86.4 (2.7)^{b}	122.8 (2.9)^{c}	88.1 (2.6)***
ME supply ('000 MJ/hh)				
On-farm	42.5 (1.5)^{a}	57.5 (1.9)^{b}	75.9 (2.2)^{c}	58.7 (1.5)***
On-farm and purchased	47.5 (1.7)^{a}	71.0 (2.7)^{b}	94.4 (3.3)^{c}	71.1 (2.1)***
Balance ('000 MJ/hh)				
On-farm	– 12.3 (2.5)^{c}	– 28.9 (3.4)^{b}	– 46.9 (3.1)^{a}	– 29.5 (2.1)***
On-farm and purchased	– 7.3 (2.6)^{cb}	– 15.4 (3.8)^{b}	– 28.3 (3.5)^{a}	– 17.0 (2.1)***
DCP demand (kg/hh)	392.2 (14.6)^{a}	630.4 (23.5)^{b}	907.1 (21.8)^{c}	644.2 (20.2)***
DCP supply (kg/hh)				
On-farm	215.7 (6.8)^{a}	303.9 (9.3)^{b}	407.4 (11.8)^{c}	309.4 (8.2)***
On-farm and purchased	252.1 (8.3)^{a}	403.1 (16.3)^{b}	543.4 (21.9)^{c}	400.6 (13.3)***
Balance (kg/hh)				
On-farm	– 176.4 (16)^{c}	– 326.5 (24.7)^{b}	– 499.7 (17.3)^{a}	– 334.9 (15.5)***
On-farm and purchased	– 140.0 (16.9)^{c}	– 227.2 (26.4)^{b}	– 363.7 (20.2)^{a}	– 243.6 (14.6)***

Values in parentheses indicate standard errors (SE)

n number of respondents, *DM* dry matter, *ME* metabolizable energy, *MJ* mega joules, *DCP* digestible crude protein, *kg* kilogram, *t* ton, *hh* household

[a,b,c] Means in a row with different superscripts differ significantly ($P < 0.05$), *NS* not significant at the 0.05 probability level

***Significant at 0.001 probability level

by [28], which is closer to the poor but lower than the wealthier group of farmers in the study area. It is still much higher compared to the average family size for both the Amhara Region of 4.62 and the North Shewa Zone of 4.45, where the study site is located [28]. The lower family labour force availability among the poor farming families than the other wealth groups could be explained by the observed small number of family size coupled with higher age dependency ratio attributed to the presence of more productively inactive dependent family members, especially children under 15 years of age. The age distribution of family members and the consequent state of age dependency ratio, particularly among the poor farming families, are in agreement with the national demographic characteristics reported by [1], which explains that about 45% of the population are under 15 years of age, reflecting the dominance of young section of the population with less contribution to the farming labour force requirement.

Availability of farm resources such as family labour, land and livestock has an influence on the level of production and availability of CR for livestock feeding. The limited availability of family labour force in the poor family group may have restrained their capacity to intensify human labour use in the farming operation, which may negatively contribute to the cause for lower farm productivity. It influences farmers to apply adequate labour input for optimum agronomic activities like tillage frequency, timely land cultivation and weeding that are required to increase crop biomass productivity. In addition, farming families with small size of cropland ownership are hardly able to diversify and increase the crop grain yield and associated CR biomass production due to limitation of available cropland and subsequent allocation of smaller plots for the intended crop types. Moreover, the small number of livestock ownership, especially oxen among poor farming families, has a limiting influence on the use of needed animal labour for traction services, such as timely cultivation of cropland with optimum tillage frequency required for the crop type grown. On the contrary, wealthier farming families with relatively better availability of ploughing oxen

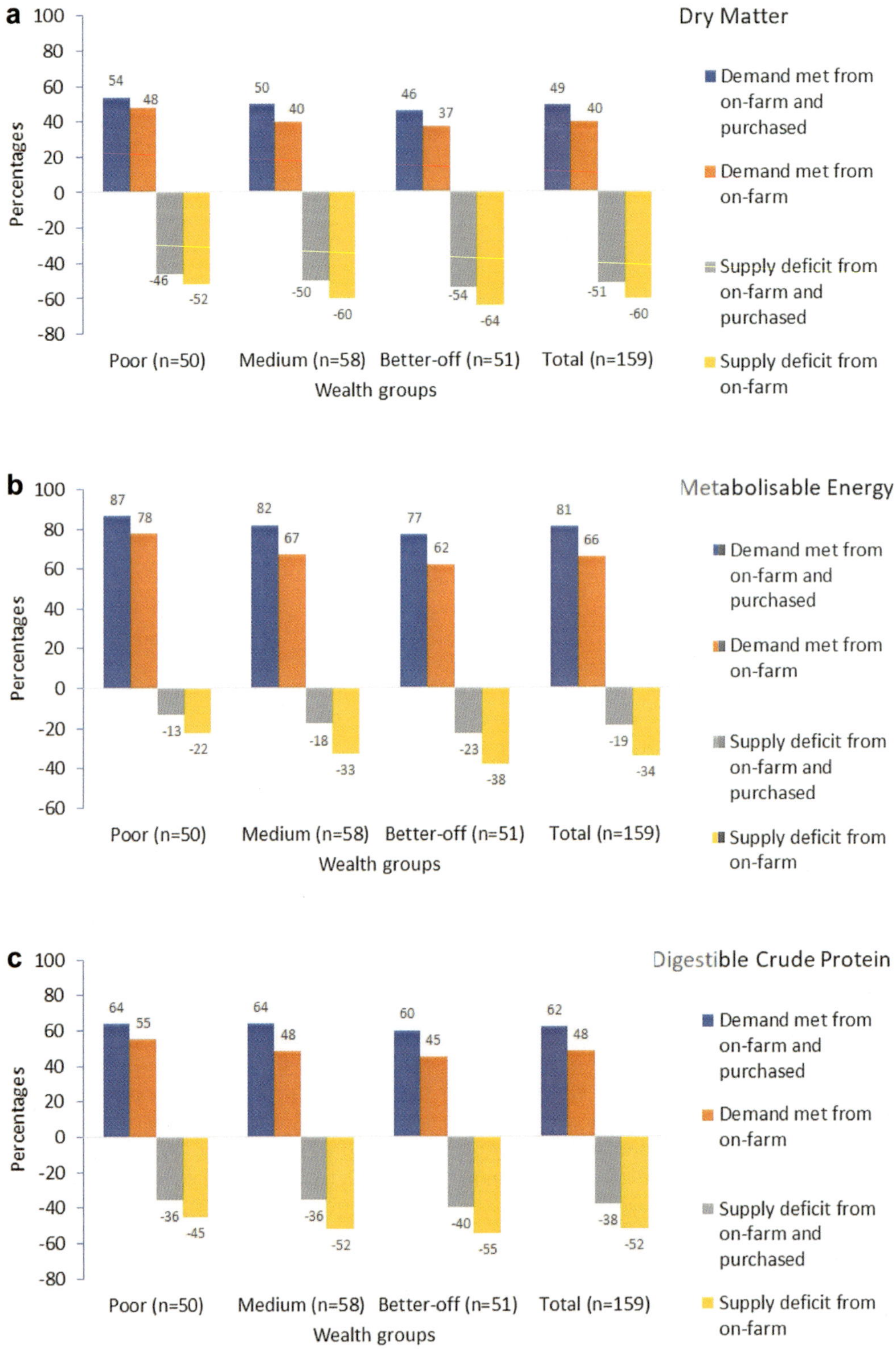

Fig. 3 Percentages of dry matter (**a**), metabolizable energy (**b**) and digestible crude protein (**c**) supply and demand balances of farm households categorized into wealth groups in the mixed crop–livestock system, central highlands of Ethiopia

Table 6 Numbers and proportions of farmers with positive livestock feed balance in the mixed crop–livestock system, central highlands of Ethiopia

Feed nutrients	Feed sourced from	Wealth groups			Total ($n = 159$)
		Poor ($n = 50$)	Medium ($n = 58$)	Better-off ($n = 51$)	
DM	On-farm	0 (0.0)	0 (0.0)	0 (0.0)	0 (0.0)
	On-farm and purchased	2 (4.0)	0 (0.0)	0 (0.0)	2 (1.3)
ME	On-farm	15 (30.0)	3 (5.2)	0 (0.0)	18 (11.3)
	On-farm and purchased	23 (46.0)	17 (29.3)	7 (13.7)	47 (29.6)
DCP	On-farm	2 (4.0)	0 (0.0)	0 (0.0)	2 (1.3)
	On-farm and purchased	7 (14.0)	7 (12.1)	0 (0.0)	14 (8.8)

Values outside and inside parentheses indicate frequencies (*n*) and percentages (%), respectively

n number of respondents

are able to cultivate their cropland adequately in time and acquire additional cropland for share cropping from other farmers through local lease agreement and attain more feed from the share of CR attained in part or as a whole depending on the agreement with the cropland owners. In support of the present study, several studies reported that inadequate cropland preparation due to limited access to production factors such as land, human and animal labour affect the food-feed crop biomass production [29–32]. The wealthier are also relatively more efficient in collecting hay from private GL and store for later use in times of scarcity or practice strategic feeding by providing for the more productive animals such as milking cows and working oxen [8, 9].

Types and availability of feed resources

The types and amount of feed obtained per household depend on land use, size of land ownership and cropping pattern during the production season. Irrespective of wealth status, large proportion of land is allocated for crop cultivation than other land use types including grazing, which is in agreement with reports of [14, 6]. The increasing expansion of land cultivation for crop production at the expense of GL has resulted in shrinkage of the area and productivity of GL [14]. Consequently, CR generated from production of food-feed crops comprise the largest proportion of livestock feed resources. In agreement, [33] reported the progressive decline of GL and the use of CR as the major source of livestock feed particularly during the dry season in the mixed farming system of the *Bale* highlands of Ethiopia. In addition, [9] also observed the heavy reliance of farmers at *Debre Derhan* in the central highlands of Ethiopia on CR for feeding their livestock.

Besides the suitability of the local agro-climate for barley growing [8, 14], the higher quantity of barley straw collected is associated with the preference of farmers and the consequent allocation of larger plots of cropland for the barley crop cultivation. Apparently, the remaining smaller plots of land are shared for other crop types, which correspondingly leads to lesser quantity of straw obtained from crop types such as wheat, faba bean and field pea. In line with the present study, [33] explained the variation in the quantity and quality of CR produced could be due to the cropping intensity besides the variations in altitude, soil type and rainfall patterns where the crop has been grown.

In relation to the size of cropland ownership and parcel allocated to a specific crop type, higher volume of biomass is harvested by better-off and medium wealth groups than poor farmers. This crop biomass productivity is accordingly translated into the higher amount of CR harvest which reflected the gradient of farmers' wealth status. In addition to the plot of cropland allocation, the availability of family labour force, size of livestock owned particularly oxen, natural fertility of the cropland and other variable factors associated with the capacity of farmers to afford various inputs and manage timely agronomic practices influence the volume of CR produce per household. Moreover, the amount of feed derived from the crop field in the form of crop aftermath grazing is intrinsically influenced by the cropping pattern and plot of cropland allocation, as part of the feed recovered, is the remainder of the parent crop obtained after harvest in addition to the different weeds and other herbaceous plants grown on the crop field. In line with this, [34] reported the quantity and quality of stubble left varies considerably depending on the harvesting techniques applied for the various crops and the grass weeds grown underneath the main crop. Therefore, it is predictable that the wealthier could be able to benefit more than the poor farmer groups from crop aftermath grazing which is attributed to the larger size of cropland ownership.

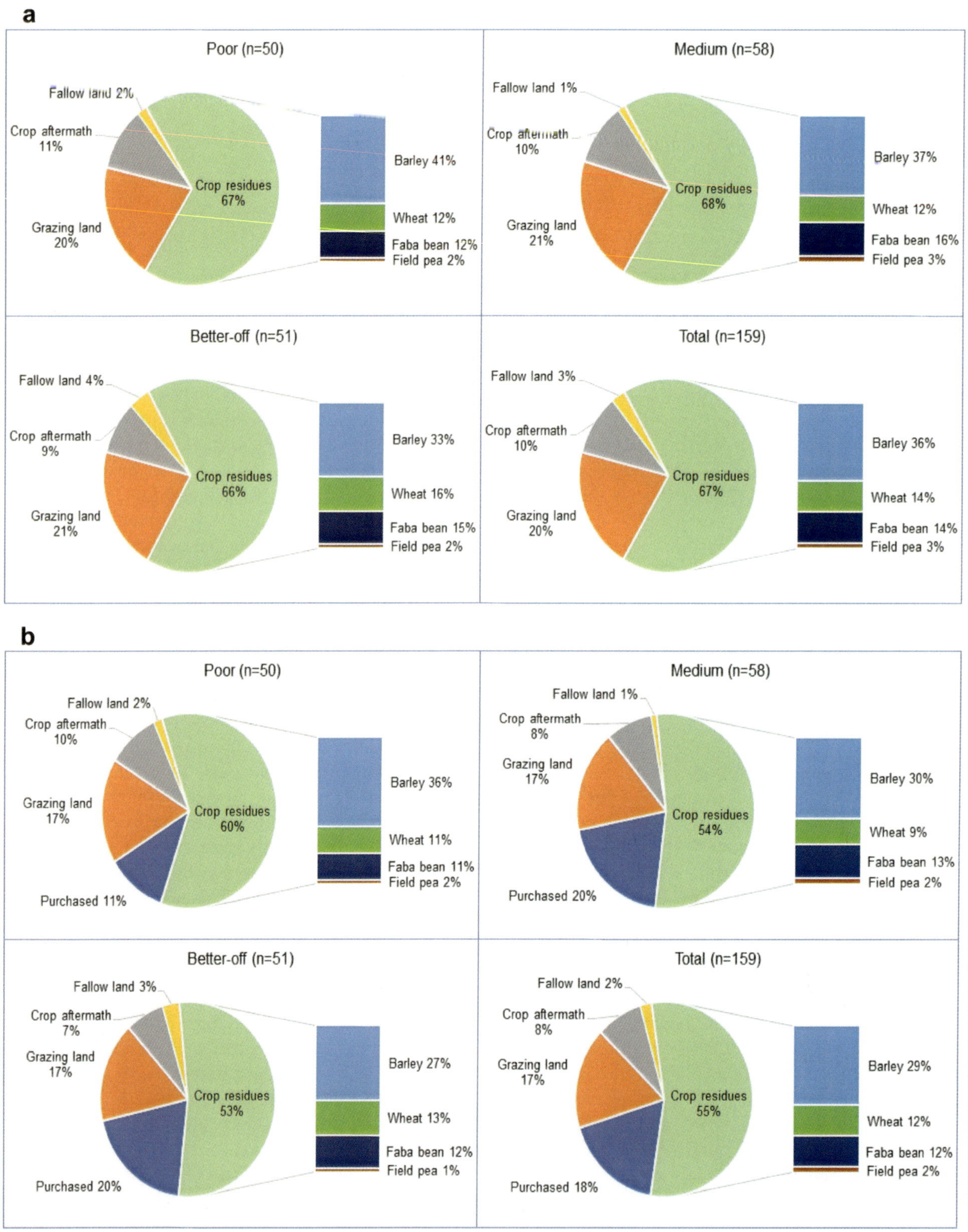

Fig. 4 Relative contributions of feed resources **a** obtained on-farm and **b** obtained on-farm combined with purchased feeds to the total annual livestock feed dry matter supply in the mixed crop–livestock system, central highlands of Ethiopia

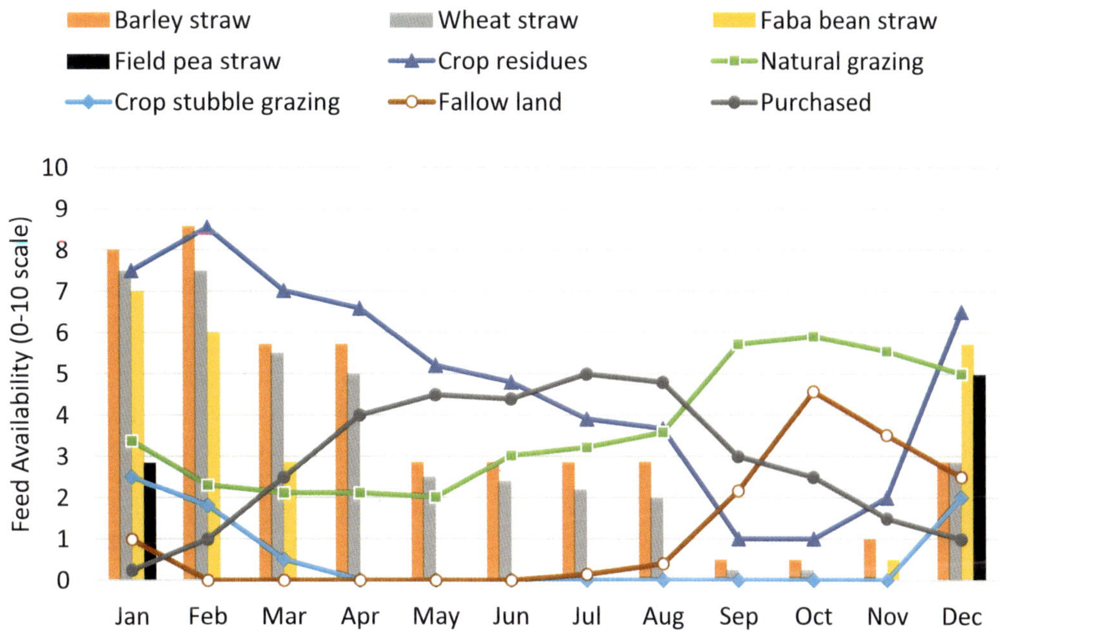

Fig. 5 Seasonal availability of feed resources as perceived by farmers in the mixed crop–livestock system, central highlands of Ethiopia. Availability of feed over the year on a scale of 0–10, where 10 = excess feed available, 5 = adequate feed available and 0 = no feed available

The size of available GL in the study area is small irrespective of wealth status, though the wealthier are more favoured to exploit available feed resource from freely grazing the communal land due to ownership of relatively larger livestock herd size. Moreover, fallow lands as livestock feed source were very rare, as there is critical land shortage for crop cultivation as well. In agreement, [34] reported that it was in the past that GL, hay and fallow land grazing were the main sources of feed; however, at present the situation has changed due to the rapidly increasing human population and expansion of crop production in the grazing areas. Same author asserted that there is very little that livestock can forage from the limited communal grazing due to continuous grazing and the mismatch between the feed supply potential of the land and the number of animals kept. Consequently, crop residues are increasingly becoming the main sources of livestock feed resources in the MCL systems [34–36]. Other several studies have also disclosed that, in recent years, GL in the MCL systems are devastatingly overstocked and that the herbage biomass productivity is declining both in quantity and quality [8, 11, 37–40]. Therefore, expansion of GL to increase feed availability is not a practical option as the increasing trends of both the human and livestock population are going to claim more additional land further for crop production, settlement and feed sourcing.

Hence, strategic options to improve the availability of CR and other alternative and/or additional feed resources should be intervened. From evidences captured in the present study which are supported by several other findings, most of smallholder farms in the MCL systems are dependent on CR as the main source of feed for livestock. Moreover, nowadays straws have also become an important source of revenue [34] and used for other purposes that compete livestock feed supply. References [41, 34] suggested close collaboration of crop and livestock scientists aimed at multidimensional crop improvement to increase the quantity and feeding value of CR without negatively affecting the grain yield and quality. In addition, Tolera [34] further suggested the introduction of cover crops that could be grazed after short rainy season, high yielding forage crops with possibility of multiple harvests per year and multipurpose leguminous trees used as supplemental green fodder could be explored to alleviate the feed shortage pertaining to the existing situation. In addition, interventions to increase feed biomass per unit area of GL, proper grazing management and purchase of supplemental feeds whenever possible would be among the options that help to reduce the influence of feed shortage on livestock health and productivity.

The fluctuations in the seasonal availability of major feed resources are mainly associated with the rainfall distribution and cropping pattern of food-feed crops. Crop residues are plenty during the dry season, which is the period of threshing for harvested crops. Availability and utilization of crop aftermath grazing is aligned with harvesting periods of crops in the dry season. Crop residues and hay are also conserved for mitigating feed shortage, especially during peak periods of cropland ploughing in the wet season. Forages from grazing lands and fallow lands are available following the main/long and short rainy seasons, when the moisture is sufficient enough to support plant growth.

Livestock feed demand and feed balance

In the present study, livestock feed balance evaluated is generally found to be negative irrespective of farmers' wealth status, indicating the critical feed shortage in the system to fulfil the DM, ME and DCP requirement for the prevailing livestock. Overall, only about 49, 81 and 62% of DM, ME and DCP requirements are fulfilled, respectively, with the current amount of feed supply from on-farm production combined with purchased feeds. Several previous study findings support the present report in that in the MCL system where land is limited, the contribution of CR is prominent and GL are deteriorating in size and productivity, and the livestock production is performed under annual feed deficit. For instance, Wondatir et al. [9] reported that around Debre Birhan the annual feed supply only satisfies 64% of the maintenance DM 81% ME and 66% DCP requirements of the animals per farm. Study by Amsalu et al. [11] in Gummara-Rib watershed of the Lake Tana sub-basin in Amhara Region shows that the available feed sources address only 72% of the annual DM requirement per household. These studies confirmed the long-standing impediments for increasing livestock productivity consistently claimed by smallholder farmers and development practitioners in the MCL systems, where seasonal feed shortage both in quality and quality coupled with inadequate successful interventions, which are remained unabated.

This study and others mentioned here used same amount of average daily DMI for a standard TLU of 250 kg at 2.5% of live body weight suggested by [25], which is 6.25 kg/day, to arrive at an aggregated demand for feed DM at household level per year. However, the daily DM feed intake is affected by various animal and feed-related factors. FAO [42] described where there is good access to feed, and health and welfare are optimum, the DM intake for young ruminant animals is likely to be 4% of body weight and decline to 2–2.5% as animals are getting mature. Moreover, where feeds of low digestibility are the main or only sources, voluntary intake will be significantly influenced due to the longer time it takes to progress through the digestive tract, and as a result low digestibility is reflected in lower intakes.

Livestock feed balance reported under smallholders in the MCL system is frequently negative [10, 43, 44], and even so many times declared as below the maintenance requirements [6, 9, 11]; however, in fact livestock continued to survive, reproduce, provide and perform their vital functions, whereas the different products and services obtained cannot be imagined at maintenance requirement level let alone when an overwhelming annual feed deficit below maintenance is revealed. Hence, series of critical questions could arise on the assumptions, methods and validations of livestock feed balance estimations, which were also iterated by Kassa et al. [10]. For instance, is the suggested daily DMI amount at 2.5% of live body weight is limited to only satisfy the maintenance requirement, as described in various livestock feed balance studies? To what extent the suggested daily allowance fulfils the animal requirement? Do feed types of any quality and form can be ingested at the suggested rate by any type and/or class of livestock throughout all seasons of the production year? Is that amount of feed could be affordable daily by smallholders in the MCL system consistently throughout the production year? etc.

The feed balance results should be cautiously interpreted, in that the livestock production system is running in an annual feed deficit does not mean that the livestock are currently without any benefit to offer. It is not unimportant either, since it quantitatively unveils the picture of what farmers and experts claim to be the most important determining factors of livestock production, which are feed shortage and the increasing trend of land scarcity on which to produce livestock feed. First of all, the suggested rate for the DMI of feed does not indicate the quality of the feed supplied but the average quantity that the animals could potentially consume daily. In addition, it does not mean that whatever is ingested is equally available to all types of animals. This is because the nutritional requirement of an animal depends on several factors: species, size, age, weight, physiological status (pregnancy, lactation, etc.), level of production (rate of gain, amount of milk produced, etc.), general health, amount of work, weather condition and season [19, 42, 45].

Secondly, there is a need to closely observe the dynamics of farm elements and situation analysis throughout the production year. Actually, the feed balance estimate showed the situations aggregated on annual basis, as if the feed availability and livestock DM, ME and DCP requirements per day are distributed evenly throughout the production year. However, a lot of dynamics can happen in the seasonal feed sources availability, nutritional composition and quality, and in the individual animal body and in the livestock herd in the course of the annual production process. It is an established fact that quantity and quality of feed available from natural pasture are affected by season [10, 35, 43–46]. Feed sourced from crop production is also affected by type of crop, cropping pattern, amount of input used like fertilizers and agronomic practices applied [36, 47, 48]. Different feed sources of varied quantity and quality may be available to livestock across different seasons, which could be produced on-farm or purchased depending on the capacity of producers to afford. The animals may undergo the phenomenon of compensatory growth, i.e. lose weight during periods of feed scarcity and put on weight (gain) during times of good feed availability. In the meantime, livestock could deliver any forms of products and services during periods of feed availability and/or mobilize what is deposited during good times of the year. At herd level, an inflow from natural perpetuation and purchase of animals, and an outflow due to death, sell and slaughter, and shared rearing of animal in either case could all happen within the year following the scarcity and availability of feed at household level. Hence, the seasonal feed availability coupled with farmers feeding practice and coping mechanism of feed shortage, the productivity gain or loss and the inflow and outflow dynamics of livestock at household level should be considered, while interpreting the livestock feed balance.

Conclusions

The available annual feed dry matter, metabolizable energy and digestible crude protein supplies and requirements of livestock did not match at present, in the study area characterized by smallholder mixed crop–livestock production system. Livestock feed insufficiency unanimously affects the whole group of smallholder farmers irrespective of wealth status. However, the wealthier group of farmers which owned larger size of livestock collected more quantity of feed on-farm and acquired additional feed through purchase found to suffer more in terms of feed deficit than the poor farmers, when livestock feed balance was assessed. Scarcity of GL due to expansion of cultivation on GL coupled with low productivity of biomass from grazing and food-feed crops production further exacerbate the mismatch between feed supply and demand. The trend of land scarcity testifies that smallholder livestock production in the MCL system is likely to continue to rely upon CR. Crop residues which are claimed to have suboptimal feeding value due to the high fibre content and low digestibility comprised the major portion of annual feed supply, and thus the inherent limitations of CR should not be ignored. Farmers who promoted cross-breeding of local animals with breeds of temperate origin to increase milk production are found unable to supply sufficient quantity and quality of feed to achieve the anticipated level of productivity. The present evidences implied that unless the feed deficit is relieved through increasing the availability and quality of feeds and practicing strategic feeding the productivity of livestock will be more profoundly affected.

Abbreviations

ADR: age dependency ratio; CR: crop residues; DM: dry matter; DMD: dry matter demand; DMI: dry matter intake; ME: metabolizable energy; MJ: mega joules; DCP: digestible crude protein; kg: kilogram; t: ton; hh: household; GDP: gross domestic product; GL: grazing lands; masl: metres above sea level; MCC: milk collection centres; MCL: mixed crop–livestock; *n*: number of respondents; SD: standard deviations; SE: standard errors; TLU: tropical livestock unit.

Authors' contributions

MBT designed the study, collected and analysed the data and drafted the manuscript. AMW and BTM supervised the design of the study and data analysis and revised the manuscript. All authors read and approved the final manuscript.

Authors' information

MBT is a lecturer (M.Sc. in animal production) in Debre Berhan University and a Ph.D. candidate in animal production in Addis Ababa University. AMW is an associate professor (Ph.D. in animal nutrition) in the Department of Animal Production, College of Veterinary Medicine and Agriculture, Addis Ababa University. BTM is a professor (Ph.D. in animal nutrition) and head of Animal Production Department in the College of Veterinary Medicine and Agriculture, Addis Ababa University.

Author details

[1] Department of Animal Science, College of Agriculture and Natural Resource Sciences, Debre Berhan University, PO Box 445, Debre Berhan, Ethiopia.
[2] Department of Animal Production Studies, College of Veterinary Medicine and Agriculture, Addis Ababa University, PO Box 34, Debre Zeit, Ethiopia.

Acknowledgements

The authors like to acknowledge Addis Ababa University and Debre Berhan University for financial support. The authors like to thank staff of the Bosona Worana and Angolela ena Tera district agriculture and rural development offices, development agents who assisted in data collection and farmers of the study area for their cooperation.

Competing interests

The authors declare that they have no competing interests.

Funding

Funding for this study was provided by Addis Ababa University and Debre Berhan University. Both Universities had no role in the study design, collection, analysis or interpretation of the data, writing of the manuscript, or the decision to submit the paper for publication.

References

1. Central Statistical Agency of the Federal Democratic Republic of Ethiopia (CSA). Agricultural sample survey in 2014/15. Report on livestock and livestock characteristics (Private Peasant Holdings). Addis Ababa, Ethiopia; 2015.
2. GebreMariam S, Amare S, Baker D, Solomon A, Davies R. Study of the Ethiopian live cattle and beef value chain. ILRI Discussion paper 23. Nairobi, Kenya; Addis Ababa, Ethiopia: International Livestock Research Institute (ILRI); 2013.
3. Swanepoel F, Stroebel A, Moyo S. The role of livestock in developing communities: Enhancing multifunctionality. Bloemfontein, South Africa: The Technical Centre for Agricultural and Rural Cooperation (CTA); 2010.
4. Shapiro BI, Gebru G, Desta S, Negassa A, Nigussie K, Aboset G, et al. Ethiopia livestock master plan, roadmaps for growth and transformation, a contribution to the growth and transformation plan II (2015–2020). Ethiopia: Addis Ababa; 2015.
5. Asresie A. Contribution of livestock sector in Ethiopian economy: a review. Adv Life Sci Technol. 2015;29:79–91.
6. Assefa D, Nurfeta A, Banerjee S. Assessment of feed resource availability and livestock production constraints in selected Kebeles of Adami. Afr J Agric Res. 2013;8:4067–73. https://doi.org/10.5897/AJAR2013.7096.
7. Yadessa E, Ebro A, Fita L, Asefa G. Livestock feed production and feed balance in Meta-Robi District, West Shewa Zone, Oromiya Regional State, Ethiopia. Aca Res J Agric Sci Res. 2016;4:45–54. https://doi.org/10.14662/ARJASR2016.072.
8. Hassen A, Ebro A, Kurtu M, Treydte AC. Livestock feed resources utilization and management as influenced by altitude in the Central Highlands of Ethiopia. Livest Res Rural Dev. 2010;22:1–19.
9. Wondatir Z, Mekasha Y, Wouters B. Assessment of productive and reproductive performance of dairy cattle nexus with feed availability in selected peri-urban areas of Ethiopia. J Cell Anim Biol. 2011;5:308–15. https://doi.org/10.5897/JCAB11.037.
10. Kassa H, Gibbon D, Tamir B. Use of livestock feed balance as a potential indicator of sustainability of tropical smallholder mixed farms—prevailing knowledge gaps: a case study from the Harar highlands of eastern Ethiopia. J Sustain Agric Lond. 2003;22:37–41.
11. Amsalu T, Addisu S. Assessment of grazing land and livestock feed balance in Gummara-Rib watershed, Ethiopia. Curr Agric Res J. 2014;2:114–22. https://doi.org/10.12944/CARJ.2.2.08.
12. Tegene N, Dinku G, Mohammed B. Assessment of potential of natural pasture and other feed resources in sweet potato production system of Shebedino District, Sidama Zone, SNNPRS, Ethiopia. Int J Livest Prod. 2015;6:91–8. https://doi.org/10.5897/ijlp2015.0264.
13. Molla A. Farmers' knowledge helps develop site specific fertilizer rate recommendations, central highlands of Ethiopia. World Appl Sci J. 2013;22:555–63. https://doi.org/10.5829/idosi.wasj.2013.22.04.1196.
14. Amsalu A, Stroosnijder L, Graaff JD. Long-term dynamics in land resource use and the driving forces in the Beressa watershed, highlands of Ethiopia. J Environ Manag. 2007;83:448–59. https://doi.org/10.1016/j.jenvman.2006.04.010.
15. Bekele M, Mengistu A, Tamir B. Livestock and feed water productivity in the mixed crop–livestock system. Animal. 2017;11:1852–60. https://doi.org/10.1017/S1751731117000416.
16. Faul F, Erdfelder E, Buchner A, Lang A-G. G * Power 3: a flexible statistical power analysis program for the social, behavioral, and biomedical sciences. Behav Res Methods. 2007;39:175–91.
17. Alam M, Sumy SA, Parh YA. Selection of the samples with probability proportional to size. Sci J Appl Math Stat. 2015;3:230–3. https://doi.org/10.11648/j.sjams.20150305.13.
18. Kossila V. The availability of crop residues in developing countries in relation to livestock populations. In: Reed JD, Capper BS, Neate PJH, editors. Proceedings of plant breeding and the nutritive value of crop residues, 7–10 December 1987. Addis Ababa, Ethiopia: International Livestock Centre for Africa (ILCA); 1988. p. 29–39.
19. Tolera A, Said A. Prospects for integrating food and feed production in Welayita Sodo, Ethiopia. In: Stares JES, Said AN, Kategile JA, editors. The complementarity of feed resources for animal production in Africa. Proceedings of the joint feed resources networks workshop held in Gaborone, Botswana, 4–8 March 1991. Addis Ababa, Ethiopia: ILCA (International Livestock Centre for Africa); 1992. p. 309–18.
20. Food and Agriculture Organization of the United Nations (FAO). Master Land Use Plan, Ethiopian Range Livestock Consultancy Technical Report, AG/ETH/82/020/FAO. Rome, Italy; 1987.
21. Woody Biomass Inventory and Strategic Planning Project (WBISPP). A strategic plan for the sustainable development, conservation, and management of the woody biomass resources. Amhara Regional State, Final report. Addis Ababa, Ethiopia; 2002.
22. Ministry of Agriculture Fisheries and Food (MAFF). Energy allowances and feeding systems for ruminants. Reference Book 413. HMSO, London, UK; 1984. p. 85.
23. Church D, Pond W. Basic animal nutrition and feeding. New York: Wiley; 1982.
24. International Livestock Research Institute (ILRI). Feed Assessment Tool (FEAST) individual farmer interview questionnaire [Internet]. Nairobi, Kenya: ILRI; 2014. p. 21. https://www.ilri.org/feastmethodology. Accessed 27 Dec 2014.
25. Jahnke HE. Livestock production systems and livestock development in Tropical Africa. Kiel: Kieler Wissenschaftsverlag Vauk; 1982.
26. Kearl LC. Nutrient requirements of ruminants in developing countries. Logan: Utah State University; 1982.
27. International Business Machines Corporation (IBM Corp. Released). IBM SPSS Statistics for Windows, Version 23.0. Armonk, NY, USA: IBM Corp.; 2015.
28. Central Statistical Agency of the Federal Democratic Republic of Ethiopia (CSA). Agricultural sample survey in 2015/2016. Report on land utilization (private peasant holdings, *Meher* season). Addis Ababa, Ethiopia; 2016.
29. Blümmel M, Samad M, Singh OP, Amede T. Opportunities and limitations of food–feed crops for livestock feeding and implications for livestock–water productivity. Rangel J. 2009;31:207–12. https://doi.org/10.1071/rj09005.
30. Haileslassie A, Peden D, Gebreselassie S, Amede T, Wagnew A, Taddesse G. Livestock water productivity in the Blue Nile Basin: assessment of farm scale heterogeneity. Rangel J. 2009;31:213–22. https://doi.org/10.1071/RJ09006.
31. Kebebe EG, Oosting SJ, Haileslassie A, Duncan AJ, de Boer IJM. Strategies for improving water use efficiency of livestock production in rain-fed systems. Animal. 2015;9:908–16. https://doi.org/10.1017/S1751731114003115.
32. Mcdermott JJ, Staal SJ, Freeman HA, Herrero M, Van de Steeg JA, Van De Steeg JA. Sustaining intensification of smallholder livestock systems in the tropics. Livest Sci. 2010;130:95–109. https://doi.org/10.1016/j.livsci.2010.02.014.
33. Bogale S, Melaku S, Yami A. Matching livestock systems with available feed resources in the Bale Highlands of Ethiopia. Outlook Agric. 2008;37:105–10. https://doi.org/10.5367/000000008784648960.
34. Tolera A. Feed resources for producing export quality meat and livestock in Ethiopia: examples from selected Woredas in Oromia and SNNP regional states. Ethiopia: Addis Ababa; 2007.
35. Bogale S, Melaku S, Yami A. Potential use of crop residues as livestock feed resources under smallholder farmers conditions in Bale highlands of Ethiopia. Trop Subtrop Agroecosyst. 2008;8:107–14.
36. Duncan AJ, Bachewe F, Mekonnen K, Valbuena D, Rachier G, Lule D, et al. Crop residue allocation to livestock feed, soil improvement and other uses along a productivity gradient in Eastern Africa. Agric Ecosyst Environ. 2016;228:101–10. https://doi.org/10.1016/j.agee.2016.05.011.
37. Asfaw MA. Livestock–water interactions: the case of Gumara watershed in the Upper Blue Nile Basin, Ethiopia. Ph.D. dissertation, Der Humboldt-Universität zu Berlin, Germany; 2013; p. 1–141.
38. Haileslassie A, Peden D, Gebreselassie S, Amede T, Descheemaeker K. Livestock water productivity in mixed crop–livestock farming systems of the Blue Nile basin: assessing variability and prospects for improvement. Agric Syst. 2009;102:33–40. https://doi.org/10.1016/j.agsy.2009.06.006.
39. Hussen M, Kechero Y, Molla M. Availability, yield and utilization practices of livestock feed resources in Gilgel Gibe catchments of Jimma Zone, Southwestern Ethiopia. Glob Vet. 2016;17:78–94. https://doi.org/10.5829/idosi.gv.2016.17.01.1046.
40. Tschopp R, Aseffa A, Schelling E, Zinsstag J. Farmers' perceptions of livestock, agriculture, and natural resources in the rural Ethiopian highlands. Mt Res Dev. 2010;4:381–90. https://doi.org/10.1659/MRD-JOURNAL-D-09-00072.1.

41. Sharma K, Pattanaik AK, Anandan S, Blümmel M. Food-feed crops research: a synthesis. Anim Nutr Feed Technol. 2010;10S:1–10.
42. Food and Agriculture Organization of the United Nations (FAO). Conducting national feed assessments—FAO animal production and health manual no. 15. Coughenour MB, Makkar HPS, editors. Rome: FAO; 2012.
43. Wondatir Z, Mekasha Y. Feed resources availability and livestock production in the central rift valley of Ethiopia. Int J Livest Prod. 2014;5:30–5. https://doi.org/10.5897/IJLP2013.0158.
44. Yisehak K, Janssens GPJ. The impacts of imbalances of feed supply and requirement on productivity of free-ranging tropical livestock units: links of multiple factors. Afr J Basic Appl Sci. 2014;6:187–97. https://doi.org/10.5829/idosi.ajbas.2014.6.6.9168.
45. Moore AD, Bell LW, Revell DK. Feed gaps in mixed-farming systems: insights from the Grain and Graze program. Anim Prod Sci. 2009;49:736–48. https://doi.org/10.1071/AN09010.
46. Abegaz A, Van Keulen H, Oosting SJ. Feed resources, livestock production and soil carbon dynamics in Teghane, Northern Highlands of Ethiopia Teghane, Northern Highlands of Ethiopia. Agric Syst. 2007;94(94):391–404. https://doi.org/10.1016/j.agsy.2006.11.001.
47. Valbuena D, Tui SHK, Erenstein O, Teufel N, Duncan A, Abdoulaye T, et al. Identifying determinants, pressures and trade-offs of crop residue use in mixed smallholder farms in Sub-Saharan Africa and South Asia. Agric Syst. 2015;134:107–18. https://doi.org/10.1016/j.agsy.2014.05.013.
48. Lenne JM, Fernandez-Rivera S, Blümmel M. Approaches to improve the utilization of food-feed crops-synthesis. Food Crop Res. 2003;84:213–22. https://doi.org/10.1016/s0378-4290(03)00152-7.

13

Fishing for food: piloting an exploration of the invisible subsistence harvest of coastal resources in Connecticut

Syma Alexi Ebbin*

Abstract

Background: In Connecticut, the subsistence harvest of coastal and marine resources, with the explicit goal of consuming what is collected or caught, is counted and regulated as recreational fishing. Little information exists regarding the harvest and use of these resources. It is not known how the catch is distributed, processed, how much is consumed or by whom. This research was conducted as a service learning opportunity for students in an undergraduate course on Marine Fisheries Economics and Policy offered at the University of Connecticut. This research was produced for fisheries managers at the Marine Fisheries Division of the Connecticut Department of Energy and Environmental Protection (CTDEEP) who desired more information on the nature of subsistence fisheries including a characterization of participating harvesters and the extent of harvesters' knowledge regarding a new enhanced shore-based fishing program that allows higher catch limits for certain species. Within the framework of the course, students developed and executed a survey of coastal fishermen to assess the extent of subsistence harvesting and consumption of coastal resources, including fish, shellfish, algae and plants, demographic information on harvesters and the extent and source of their knowledge regarding fishing regulations and health advisories.

Results: The majority of respondents consume their harvests and also share it with others. None sold their harvest. Of those who consumed their harvest, most consumed at least one meal a month. An overwhelming majority of respondents share their harvest with people living in their household. Despite high levels of consumption by Connecticut harvesters, less than half of respondents had heard of the term "subsistence fishing." The dissemination of information within the fishing community appears to be largely successful, as nearly all respondents were aware of the fishing regulations and health advisories and how to obtain the information if they needed to find it. The main sources of information are signs at fishing sites and the CTDEEP Web site. However, only half of respondents were aware of the new Enhanced Opportunity Shore Fishing Program, perhaps because of its newness. Most surveying was conducted in southeastern Connecticut and captured little demographic diversity.

Conclusions: Harvesters of Connecticut's marine and coastal resources appear to lack familiarity with the concept of subsistence, and few consider their activities to be subsistence-based. However, most consider the provision of food to be an important reason driving their participation, either directly consuming or sharing their harvests with others to consume. The harvest of marine resources for subsistence may not be labeled as such, counted rather as recreational catch. It may not be quantified or visible to regulators, but it certainly exists. It fills a provisioning role within the overarching cash-based economy and also affords individuals the opportunity to engage in a pleasurable and relaxing activity that reinforces familial relationships. Continued and expanded surveying will be conducted to better assess the nature and magnitude of subsistence harvesting of coastal resources in Connecticut.

Keywords: Subsistence, Fisheries, Shellfishing, Resource harvesting

*Correspondence: Syma.ebbin@uconn.edu
Department of Agricultural and Resource Economics, University of Connecticut, Avery Point Campus, 1080 Shennecossett Rd, Groton, CT 06340, USA

Background

Subsistence fishing in Connecticut is invisible. By this I mean that though there are people who fish and dig and rake and gather living resources along and within the coastal waters of Connecticut for food, subsistence harvesting of coastal resources is not recognized as a fishery by the state of Connecticut and is therefore not directly regulated. Little is known about its magnitude, diversity of species harvested or consumption rates [1].

Global reliance on coastal resources is high. The United Nations Food and Agriculture Organization (UN FAO) publishes statistics about the magnitude of consumption of these resources, their nutritional value and reliance on these resources as primary sources of protein. But it does not publish statistics on the fishery origins of this consumption, whether the seafood has been bought or caught by the consumer, and if the latter, in what kind of fishery; i.e., commercial, recreational or subsistence [2]. The UN FAO goes as far as to note that, "[s]tatistics from artisanal and subsistence fisheries are a particular concern and many key statistics are missing at the global level, e.g. economic and social data, discards, fishing capacity" [2]. Aside from a focus on subsistence activities taking place in the Arctic and mostly by indigenous residents, there is a paucity of attention and data. As Berkes notes, "To a large extent, the subsistence fishery is not being reported in fishery statistics, monitored, assessed or regulated" [3: 35].

Nationally, this trend is similar. Aside from the few tribal subsistence fisheries, and subsistence fisheries in Alaska—where an entire division of the Alaska Department of Fish and Game is devoted to subsistence activities—there has been little regulatory attention paid to this fishery. It has simply piggybacked atop commercial and recreational fisheries. The situation on land appears to be quite similar: Subsistence activities take place, the reliance on these products is important, but they are not recognized or acknowledged [4].

For managers, this situation poses a problem since it is difficult to manage that which is not known. There would seem to be an urgent need to learn more about the human dimensions of these activities: Who is fishing, where they are fishing, what they are catching and consuming and what they prefer to consume, in order to know whether it is necessary to tailor management efforts. Effective and equitable management needs to be flexible and responsive to context. The research effort described in this paper seeks to shed some light on the context of subsistence activities in the coastal zone of a largely urbanized region, to make visible a largely invisible activity.

The FAO defines subsistence fisheries as those "where the fish caught are shared and consumed directly by the families and kin of the fishers rather than being bought by intermediaries and sold at the next larger market" [2]. In Connecticut, however, there is no regulatory category for coastal subsistence harvesting activities; there are simply recreational and commercial fisheries. While fisheries managers in the Connecticut Department of Energy and Environmental Protection (CTDEEP) collect information on commercial and recreational harvests, no comparable information is collected on subsistence harvests or consumption, although data on the numbers of fish caught and released in recreational fisheries are estimated. It is unclear how many fish are eaten, how they are distributed, processed, how much is consumed and by whom. In addition, there are some coastal resources for which no harvest data are collected. Thus, assessing fishers' understandings and perceptions of the concept of subsistence is essential to characterizing subsistence and gauging whether harvesters' needs are being met and regulations are appropriately configured. For example, are harvesters able to satisfy their subsistence preferences and needs with the current configuration of bag limits, size restrictions, seasonal and area-specific openings and closures? The objective of the research outlined in this paper is to assess the extent to which subsistence harvest and consumption of coastal resources, defined to include fish, shellfish and coastal and marine plants and algae, exist in Connecticut and to characterize and define this practice.

The literature on subsistence fisheries in the USA or other activities aimed at harvesting coastal resources for consumption is thin, especially for non-indigenous populations. As Schumann and Macinko note, the little that does exist is focused primarily on Alaska and the Arctic [5]. The limited studies that have been conducted in other places did not find much identification or familiarity with the concept of subsistence [6, 7]. Most subjects in these studies seemed to identify as recreational fishermen, although some acknowledge fishing for food or income. Self-identified subsistence harvesters in the coastal zone appear to be invisible or at most a minority. Despite this, there is a growing literature that focuses on foraging wild foods on land, including in urban environments [4, 8, 9]. This literature highlights the widespread and persistent harvest and reliance on wild harvesting as important sources of foods, medicines and materials used in a variety of ways including arts and crafts.

Berkes defined subsistence as "local, non-commercial fisheries, oriented not primarily for recreation but for the procurement of fish for consumption of the fishers, their families and community" [3: 35]. Practically, however, the definition of subsistence is fraught with complexities. Schumann and Macinko [5: 707–710] teased out four distinct spheres of meaning from the academic literature. These included definitions encompassing subsistence as

(1) economic activities or systems sustaining a basic level of livelihood, not profit, but not precluding exchange; (2) economies of sharing, not selling; (3) social and cultural institutions creating an economy of non-market distribution; and (4) culturally significant food producing and distributing activities. Subsistence is a multidimensional concept that encompasses a set of diverse behaviors, motivations and values.

The delineation and naming of certain practices or behaviors can be problematic when the motivations driving them are multifaceted and differ among individuals and populations. The conflation of the complex and nuanced suite of motivations within a singular label such as subsistence, recreational or commercial fisheries elevates some motivations and may omit others. Managers label, define and manage in terms of one primary value. Yup'ik fishers that I interviewed eschewed recreational fishing which they associated with catch and release and found culturally repugnant [10]. However, it was clear to me in our communications that they very much enjoyed their subsistence fishing activities.

Steinhoff elaborated a framework of value in which he distinguished between the value of the "experience" and that of the "organism" [11]. Coastal harvesters, whether commercial, recreational or subsistence, value multiple and diverse attributes of the harvesting experience and the resources harvested to different degrees. The overall value of fishing can be thought of as the aggregate of the value of the activity of fishing and the value of the fish. Valued attributes of fishing might include relaxation, enjoyment, being outdoors, on or near the water, spending time with friends and/or family, being close to nature. Valued aspects of coastal resources might include the value of the harvest as food, money, fertilizer, medicine, etc. As adapted from Brown and Burch, this can be conceptualized as:

$$V_F = V_A + V_P$$

where V_F is the aggregate or "total" value of organism and experience, V_A is the value of fishing or harvesting as an activity, V_P is the value of the coastal resources which are harvested [12].

Managers label commercial fisheries by the primary value of the coastal resource V_P (economic exchange) and recreational fisheries by the value of the experience V_A (enjoyment). However, this Connecticut-focused study confirmed the existence of multiple motivations for harvesting—with food and enjoyment as the top two. Coastal subsistence activities in Connecticut are not conducted by individuals who supply the bulk of their nutrition through fishing and foraging in the coastal margin, they are, rather, comprised of individuals who engage in the activity because they enjoy it and are able to produce food for themselves and their families. In Connecticut, subsistence falls into a muddy middle ground with emphasis placed by practitioners on both the experience (enjoyment, etc.) and resource (food). In mixed subsistence economies, the engagement of individuals in a seasonal round of activities shifting among target resources and capture strategies has economic value. In market economies, the seasonal round of activities may include regular employment along with a range of non-market provisioning strategies, the latter of which may increase in importance as unemployment rates rise and wages stagnate or decline.

Thus, if the question of defining subsistence becomes too knotty, it might be prudent to leave that task alone, and examine instead how much of the recreational harvest is consumed, or the importance of consumption in motivating individuals to engage in harvesting activities. If consumption is large or a significant motivating force, then this has implications for management policies and regulatory approaches implemented to mitigate the human health risks associated with this consumption due to potential contamination by chemical pollutants and other toxins or pathogens [13–15]. These approaches include species-specific bag limits, size limits and the timing of open and closed seasons. In Connecticut, the state has issued statewide health advisories related to the consumption of striped bass (*Morone saxatilis*), large (in excess of 25 in. in length) and small (between 13 and 25 in. in length) bluefish (*Pomatomus saltatrix*) and weakfish (*Cynoscion regalis*). Pregnant or nursing women as well as children under the age of six are advised to consume no striped bass or large bluefish, and only one serving of small bluefish or weakfish per month; others are advised to eat no more than one serving of each of these per month. Information is posted in multiple languages at fishing sites, online and in print format at various locations around the state.

Methods

This study was conducted by undergraduate students as part of a service learning project. Many different models of service learning have been developed; one model is problem-based, identifying students as "consultants" who work for a community-based "client." In this project, the consultants were the students enrolled during the Fall 2013 semester in Marine Fisheries Economics and Policy (ARE 3437) at the University of Connecticut Avery Point Campus. The clients were fisheries managers with the Marine Fisheries Division of the CTDEEP.

The CTDEEP actively desired to have more information regarding these harvests because they had just instituted a new "Enhanced Opportunity Shore Fishing Program" to improve "the opportunity to catch the evening's meal" [16]. The program is aimed at increasing consumption of scup or porgy (*Stenotomus chrysops*) and summer flounder (*Paralichthys dentatus*) taken at 46 specific fishing sites within 18 coastal communities ranging east to west from Stonington to Westport, by decreasing size limits for shore-based anglers.[1] The CTDEEP has also initiated a "Bonus Striped Bass Fishing Program" for striped bass (*M. saxatilis*), which allows anglers the ability to obtain a voucher for two "bonus" stripers between the lengths of 22 and 28 in. [17]. Through discussions with CTDEEP fisheries personnel, we identified several specific data gaps and areas where they desired more information. With the CTDEEP's data needs delineated, the stage was set to develop a "problem-based" service learning project focused on subsistence use of coastal resources as a component of ARE 3437.

Students enrolled in the class were assigned background reading on subsistence activities in New England and provided with EPA reports on efforts to survey subsistence harvesting and consumption. The class developed and pre-tested a structured survey instrument comprised of closed- and open-ended questions. Students were trained in human subject research protocols, recruiting and interview techniques. The ten students in the class were asked to survey five individuals each with a goal of conducting 50 interviews in total and allowed to work in teams of two if desired. Students selected different enhanced shore-based fishing locations from the coastwide list of 45, for soliciting interviews to avoid overlaps. Most surveys collected were obtained along the eastern part of the coast, within 20 miles of the Avery Point Campus, primarily targeting the 16 enhanced shore-based fishing locations in New London County, although several of the 12 in New Haven County were sampled. Surveys were conducted anonymously. A few students chose to have respondents complete the survey online using a survey form designed specifically for this function. Completed surveys were entered by students into an online data entry portal created using SurveyMonkey. I assembled a poster showcasing the student's work products which was exhibited on campus at the end of the semester. The students presented only the results of close-ended questions compiled automatically by SurveyMonkey (i.e., percent distributions and frequency graphics) and did not assess any of the open-ended responses in the poster. I combined the two databases, comprised of self-reported and student-solicited responses, analyzed the responses to both close- and open-ended questions from the combined dataset and presented these data along with the poster to CTDEEP fisheries personnel. This article presents these analyses.

[1] Scup may be taken at 9" minimum length rather than the usual 10.5 in. limit, and summer flounder may be taken at a minimum of 16 inches in length rather than 18 in. [17].

Results

Students obtained 47 completed surveys: 40 in-person, 90% of which were from shore-based locations, and 7 self-administered online. Given the low number of respondents and lack of geographic representation for all of coastal Connecticut, this research effort should be considered a pilot study and the results treated as a qualitative assessment of coastal subsistence activities, focused primarily in New London County.

The majority of respondents were male (91%), white (80%), residents of Connecticut (91%) and characterized their hometown as "suburban" (60%). Roughly one-quarter of respondents were in each of the following age ranges: 18–24, 25–34, 35–44 and between 45 and 64; none over 64 were interviewed. Respondents came from 24 different Connecticut towns, 2 were from Rhode Island, and one each from New Hampshire and Texas.

The survey opened with the open-ended question *Why do you fish?* Answers were qualitatively analyzed, coded into thematic categories. The most common answer included expressions of enjoyment of the activity of fishing; 25 respondents indicated this reason first and 6 others mentioned this as a second or third reason. Twenty-five respondents answered with a composite answer composed of multiple reasons. These were coded into multiple relevant response categories, arranged to reflect the order in which the individual thematic categories were noted. For example, "For fun...it provides a relaxing, yet productive way to spend the day. I enjoy outdoorsy, recreational activities...this has a bonus of a meal. I can't complain" contains references to enjoyment, food, relaxation and being outdoors. The second most common answer involved the value of harvest activities to produce food to be eaten, or specifically referenced the intended consumption of the harvest; 11 individuals noted this as the primary reason for their harvest activities and an additional 9 indicated fishing for food as a secondary or tertiary reason. One individual elaborated on this, discussing his enjoyment of the many ways his wife cooked his catch, "I like to fish to see how many bluefish I can catch. My wife cooks them many different ways for my family to eat. We also eat porgy, they are fun to catch on small lines, light tackle." The third most common reason for fishing focused on relaxation associated with the activity (13 total). The fourth most frequently mentioned reason involved family (7 total), followed by reasons involving appreciation of nature or being outdoors. Only one individual mentioned (a lack of) money as a reason to fish.

Half-way through the interview, respondents were asked to rate the importance on a Likert scale ranging from 1 (not important at all) to 5 (very important) of a defined list of diverse reasons which might motivate someone to harvest coastal resources. The resulting index of importance reflected the responses received to the open-ended answers. Fun/recreation and relaxation were considered very important (resulting in indices of 4.66 and 4.43, respectively). Better health and nutrition and food were considered to be moderately important (2.91 and 2.74, respectively). Of some importance was money and providing for one's family (1.60 and 2.15, respectively), while spending time with family and friends was specifically listed by some as an "other" reason of importance.

Respondents were asked what specifically was done with their harvest: 76% consumed their catch, 68% shared it, no individuals sold it, three indicated they used it for bait. Of those who ate their catch, only 6% ate their catch at one meal or more a day, while 21% ate at least one meal of caught seafood each week, 30% ate at least one meal of caught seafood each month, 27% ate a meal of caught seafood every 6 months, and 15% ate at least one meal of caught seafood each year. A few respondents indicated that their consumption is seasonal and that they eat more seafood during the summer. Others noted they would eat more if they could catch more.

Of those that shared their catch, 91% indicated that they shared it within their household, and 60% distributed their catch to friends and 43% to family members who did not live with them. Within the household, harvests were primarily eaten by spouses (74%) and children (68%) and some by parents (29%). Siblings, roommates and pets were also noted to consume the catch. Of children within the household under 18 years of age who consumed seafood, 38% were 6–9 years old, 31% were 10–15, while the remaining 31% was split among youth younger than 6 or older than 15.

All those who consumed their catches ate the meat, 26% ate the whole fish or shellfish, 8% each consumed the skin or organs, 5% the head, and 3% the roe. One student noted of her interview subject, "He always uses the entire fish, due to his religion and the usefulness of all the parts. He only takes what is needed." This information is important when estimating potential exposure to risks via contaminated seafood.

A series of questions were fashioned to explore the economic aspects of coastal harvesting and consumption practices. A slight majority (55%) answered that they saved money on food through their coastal harvesting activities, while 45% did not believe there was a financial benefit. Additionally, 71% believed that they spent less money on their harvesting activities than they would have spent to buy seafood at a store. This likely reflects the relative high cost of seafood at retail markets along with the fact that many people do not purchase seafood. A minority of 29% thought that the cost of their harvesting activities was more expensive than the cost of purchasing seafood at the store would have been. However, 83% of respondents noted that they would still go fishing even if there was little to no chance of catching something, indicating that the experience of harvesting coastal resources was valued as much, if not more than the fish itself. Respondents were asked under what conditions would they choose not to retain their catch: 77% would not keep a fish that was too small, 55% would not keep a sick or injured fish, 43% would discard undesirable species, as one fisherman explained "some fish just aren't good tasting, or they may be good eating, but they may be too bony." The same percentage of respondents (43%) noted that they would not retain landed catch that exceeded their bag limit or fish in excess of need. Some harvesters (21%) noted they were exclusively catch and release.

Harvesters were asked whether they were familiar with the term subsistence fishing. Slightly fewer than half of respondents (47%) were familiar with the term but only two considered themselves to be subsistence fishermen. Respondents indicated that they had engaged in harvesting activities for a relatively long period of their lives in relation to the age distribution, with 30% each having harvested between 10 and 20 or 20 and 40 years.

Most respondents were taught to fish by their fathers (78%) or grandfather (11%), and 79% had taught others to fish, primarily their children and friends (60% each). Respondents indicated that they were roughly equally likely to fish with family (72%), friends (68%) or alone (66%). Generally, the individuals surveyed tend to fish fairly frequently. Roughly equal numbers of respondents noted that they fish once a week or more (28%), at least once a month (32%) or once every two months (28%).

Respondents were asked whether they have a preferred species of fish/shellfish which they harvest. Most respondents (62%) did have one or more favorite target species. The most frequently preferred species to catch was striped bass (*M. saxatilis*) (17 out of 28). Clams (*Mercenaria mercenaria*) were second (6). Tied for third were fluke (*P. dentatus*) (5) and bluefish (*P. saltatrix*) (also 5 including a few anglers who expressed a preference for (juvenile) snapper blues). Scup (*S. chrysops*) (4) and blackfish/tautog (*Tautoga onitis*) (4) tied for fourth. Other marine and coastal species identified included eel (*Anguilla rostrata*), sea bass (*Centropristis striata*), oysters (*Crassostrea virginica*) and cod (*Gadus morhua*). Several freshwater species were also identified as favorites: trout, blue crappy, small and large mouth bass and sunfish. By far the most important reason for favoring the harvest of a given species was a stated preference for that species as food (78%) or a stated preference expressed by others that they enjoy eating that species (44%). Almost 30% preferred harvesting "trophy" fish; a few specified a preference for fish that "put up a good fight."

Respondents were asked to estimate the specific number of each species, including fish, invertebrates, coastal and marine plants and seaweeds that they harvest in a given year (see Table 1). Overall, respondents acknowledged harvesting 18 different species of fish, shellfish and macroalgae. Several noted that they had recall problems in providing the numbers harvested, so this information should be considered as general estimates rather than as precise counts. Confirming its status as the preferred target species, striped bass (*M. saxatilis*) was the most frequently harvested species by the greatest number of fishers (36 fishers reported harvesting). However, in terms of the numbers harvested it ranked in fifth place, with 205 fish recorded. Scup (*S. chrysops*) were the second most frequently harvested species (26 fishers reported harvesting) and were also the second ranked species in terms of numbers harvested (610 fish reported). The species which recorded the largest harvests were clams (*M. mercenaria*) with 1685 clams reported harvested by 10 harvesters. This spike is most likely related to the relatively large daily limits allowed for this species and its relatively small size. Summer flounder (*P. dentatus*) were the fifth most frequently harvested species (22 fishers reported harvesting) and the third ranked in terms of numbers (285 landed). Tautog (*T. onitis*) were the third most frequently harvested (25 fishers) but ranked sixth in terms of numbers (139 fish landed). Bluefish (*P. saltatrix*) was ranked fourth in both frequency of fishers (23) and numbers landed (235). Black Sea Bass (*C. striata*) were ranked sixth in terms of landing frequency (14 fishers) and seventh in terms of numbers landed (100 fish landed). Other species that were noted to be caught with some frequency by some harvesters include Blue crabs (*Callinectes sapidus*), Winter flounder (*Pseudopleuronectes americanus*), Menhaden (*Brevoortia tyrannus*) and Macroalgae/seaweeds (sea lettuce (*Ulva Lactuca*) was specified by two respondents but may include other species), oysters (*C. virginica*), mussels (*Mytilus edulis*) and eel (*A. rostrata*). Other species specified with less frequency by harvesters included trout

Table 1 The number of each species that respondents harvest throughout a year

Species	Average catch (individuals caught)	Total catch (individuals caught)	Respondents (number of fishermen landing)
Scup/porgy (*Stenotomus chrysops*)	23.46	610	26
Striped bass (*Morone saxatilis*)	5.69	205	36
Bluefish (*Pomatomus saltatrix*)	10.22	235	23
Atlantic Salmon (*Salmo salar*)		0	
River herring (alewives/blueback herring) (Alosa spp.)		0	
Menhaden (*Brevoortia tyrannus*)	3.25	26	8
Black Sea Bass (*Centropristis striata*)	7.14	100	14
Blackfish/tautog (*Tautoga onitis*)	5.56	139	25
Eel (*Anguilla rostrata*)	2.00	14	7
Winter flounder (*Pseudopleuronectes americanus*)	2.70	27	10
Fluke/summer flounder (*Paralichthys dentatus*)	12.95	285	22
Other flounder spp. (specify below)		0	
Other fish spp. (specify below)	6.00	36	6
Lobster (*Homarus americanus*)		0	
Whelks (*Busycon* spp.)		0	
Oysters (*Crassostrea virginica*)	2.20	11	5
Mussels (*Mytilus edulis*)	2.20	11	5
Clams (*Mercenaria mercenaria*)	168.50	1685	10
Bay scallop (*Argopecten irradians*)		0	
Periwinkles/snails (*Littorina littorea*, etc.)		0	
Blue crabs (*Callinectes sapidus*)	7.14	50	7
other crabs (Asian shore/green) (specify below)		0	
Other invertebrate/shellfish spp. (specify below)		0	
Macroalgae/seaweeds (specify below)	4.29	30	7
Coastal plant species (specify below)		0	
Other species (specify below)	4.00	20	5
Answered question	41		
Skipped question	6		

(*Oncorhynchus* spp., *Salmo* spp. *and Salvelinus* spp.), silver shad (*Alosa* spp.) and sea robin (*Prionotus carolinus*).

Since our class was also studying the impacts of climate change on Long Island Sound marine species, students included a question on the incidence of uncommon species in the catch. A quarter of harvesters noted that they had indeed caught species not commonly found in local waters. Most commonly reported were Triggerfish (*Balistes* spp.), Jack (*Caranx* spp.), Marlin (species unclear), Bowfin (*Amia calva*; a freshwater fish species), Cobia (*Rachycentron canadum*), Snowy and Black groupers (*Epinephelus niveatus* and *Mycteroperca bonaci*), Fourspot flounder (*Paralichthys oblongus*) and Blowfish (*Sphoeroides maculatus*). One individual noted he had seen and videotaped an ocean sunfish (*Mola mola*). This list will be interesting to maintain into the future to track any increases in warm water or subtropical fish entering Long Island Sound.

In order to get at the CTDEEP's interest in learning how well their communications were being picked up by harvesters and to track awareness of fish health advisory information, we asked several questions related to harvesters' sources of information and level of awareness of coastal resource-related information. An overwhelming majority (93%) of respondents were aware that eating fish and shellfish is good for you. Ninety-two percent were aware of CTDEEP recreational fishing regulations, and 83% (39 respondents) were knowledgeable of Connecticut fish health advisories regarding consumption of some species in some areas; somewhat fewer respondents (33) actively followed these advisories, but most respondents (41) knew where to get health advisory information if they wanted it. Countering this relatively high level of awareness, some respondents noted a lack of concern regarding these advisories, "We eat tuna every day and it has mercury, I'm not going to worry about my catch if I won't worry about that"; another noted that, "I will eat more than one meal of stripers in a week to eat all of my catch."

However, only half of respondents indicated that they had any knowledge of the CTDEEP's new Enhanced Opportunity Shore Fishing Program with lower minimum size limits for summer flounder and scup. Of those who did know of the new program, information on it was obtained primarily through two sources: signage at fishing sites (52%) and the CTDEEP's Web site (48%). Additional sources of information included marinas and fishing tackle stores (28%) and CTDEEP offices and personnel (17%). A few noted that fishing magazines, local newspapers and other fishermen were sources of information. One respondent noted that, "the tackle shop hands out about 10,000 of their [CTDEEP] information pamphlets a year. Fishermen find them very useful."

The survey concluded with an open-ended question asking whether there was anything else regarding their fishing, shellfishing or other coastal/marine foraging activities that they would like to share. Among the diverse responses elaborating on their harvesting activities were several that focused on perceived declines in abundance of fish resources and a desire to maintain these resources into the future through conservation efforts. One individual noted after some discussion of the idea of subsistence fishing that "there should be special regulations for people that decide to live off of marine resources." Although this was only one individual's comment, it may be useful to include a targeted query along these lines if this project extends into the future.

Discussion

The student-conducted surveying effort appears to be a useful means of collecting information on the use and consumption of coastal resources in Connecticut. The information was produced at no cost to the state but with limitations on the ability to quantify the magnitude of the state's subsistence fishery or extrapolate the data to a larger scale. The effort revealed that there is a substantial amount and wide variety of coastal resources harvested and consumed by Connecticut residents. This confirms state and federal estimates of Connecticut marine recreational harvest numbers. Of importance, however, this study identified that consumption is an important driver behind these harvesting efforts.

There were limitations with the demographic diversity of respondents. Minority populations and those who likely rely heaviest on subsistence resource harvests and consumption were not readily interviewed. Language barriers, social and cultural divisions and insularity, were factors in this situation. Similarly, Avery Point University of Connecticut students are for the most part geographically stationed near campus and thus the vast majority of interviews took place in the southeastern part of the state. No data were collected from fishing locations in Fairfield or Middlesex Counties. This limited the ability to generate broad geographic insight into subsistence use of coastal resources. Coastal Connecticut counties vary demographically from each other with different percentages of White, Hispanic, Black, Asian and American Indian populations [18].[2] Because of these differences, the results reveal trends primarily in New London

[2] 2010 US census data (census.gov) identifies New London County as having 78.3% White, 8.5% Hispanic, 5.8% Black, 0.9% American Indian/Alaska Native, 4.2% Asian; Fairfield County as 66.2% White, 16.9% Hispanic, 10.8% Black, negligible American Indian/Alaska Native, 4.6% Asian; New Haven County as having 67.5% White, 15% Hispanic, 12.7%, negligible American Indian/Alaska Native, 3.5% Asian; and Middlesex County as having 86.4% White, 4.7% Hispanic, 4.7% Black, negligible American Indian/Alaska Native and 2.6% Asian [18].

County. Students targeted coastal shore-based recreational fishermen and thus did not survey subsistence use by other individuals (i.e., boat-based sport fishermen, charter and party boat fishermen, commercial fishermen). These limitations coupled with the small sample size, lack of a sampling frame as well as the fact that a number of individuals declined to be interviewed might have skewed the sampling, possibly impacting the results drawn from the data.

Subsistence does not appear to be a widely recognized concept in Connecticut, just under one half (47%) were familiar with the term, and only two individuals considered themselves to be subsistence fishermen. Relaxation and recreation associated with coastal harvesting activities were noted to be of primary importance by respondents. However, it appears from the results presented here that the use of coastal resources as food is an important driver of harvesting effort and of the quantity harvested for a majority of those interviewed. Since three-quarters of respondents eat their catches, with over a quarter eating at least a meal a week, and since a substantial number of harvesters share their catch with others, primarily family members, coastal resources appear to be important food items for Connecticut families.

The data suggest that the economic dimensions of harvesting for consumption are not strong components of the endeavor, although the monetary value of harvests was acknowledged by a few. The limitations of the sampling discussed previously may have obscured a better understanding of the economic imperatives that drive some to harvest, consume and perhaps sell or barter their catches.

The recreational aspects of the fishing experience are preeminent, and the harvest does not constitute an economic activity that sustains their basic livelihood. This is emphasized by the fact that most participants would fish even if they knew there was nothing to catch. However, the insights into harvesters' orientations and motivations for harvesting do indicate that there is substantial sharing, and that their practice is embedded within a social and cultural context, and food production and distribution are important; thus, the harvesters interviewed meet definitions two, three and four of subsistence harvesting.

To what extent did this problem-based service learning project address the informational needs of the CTDEEP Marine Fisheries Division outlined at the beginning of this article? The dissemination of regulatory fishing information seems to be quite effective. Fishermen typically get their information from the signs located by their fishing location or the internet, and many get information from tackle shops. CTDEEP fishing pamphlets and signs are a successful way to spread information to the fishermen. One tackle shop employee claimed to have distributed 10,000 pamphlets throughout the year. A large percentage of the people surveyed had these pamphlets with them during interviews, attesting to their effectiveness. Fishermen were generally well-informed about legal sizes and closures, although roughly half had not heard of the new fishing initiative for summer flounder and scup. This may be because the new program was initiated in 2012, announced in the 2012 Connecticut Angler's Guide published by the CTDEEP, online and at locations where fishing permits can be purchased. It may be that when the program has been in existence for a longer period of time, knowledge of it will increase accordingly. Although a greater percentage of respondents were found to be aware of health advisory information in this study (83%) than in Balcom et al. [15], it is concerning that the number of individuals informed and acting on this knowledge is not higher. One area of communication that could be improved is the number of individuals who follow the health advisories. Perhaps the development of an innovative social marketing campaign might deter consumption of fish and seafood subject to health advisories.

CTDEEP marine fisheries staff provided feedback on the project noting,

> *From what I have read, your students collected interesting information concerning fish consumption, fishing DEEP regulation knowledge, etc. As you know, collecting subsistence angling information can be challenging, especially in Connecticut. I know you have survey time constraints and students travelling to more urban cities to gather subsistence angling data problematic. Most subsistence anglers more than likely fish in obscure areas close to where they live in highly urbanized areas along the coast such as New Haven and Bridgeport. Accessing these areas are very difficult and unsafe in my opinion. I think I would look at the positive attributes your students collected in the report and expand on those.*

> *Student feedback was evaluated to examine the extent to which the incorporation of this problem-based service learning project focused on subsistence use and consumption of coastal resources enriched student's education. All students who completed the evaluations believed that the course had met their expectations and 80% considered the subsistence-focused service learning project to be a valuable learning experience and believed it had enhanced their understanding of marine fisheries economics and policy. One noted, "It allowed me to learn about a side of fisheries management not very publicized." Another commented that it, "showed that fisher-*

ies assessments are not perfect and revealed that there are still new things to discover about fisheries." Another explained that it, "helped me understand subsistence in my backyard."

According to students, the educational benefits associated with the project included the realization "that law enforcement may want to be a little more lenient when dealing with true subsistence fishing (family feeding, small scale)"; " learning about local fishing practices"; "I ended up getting a research job due to my involvement with the learning project"; and "[I] learned about local species and fishing communities." Other comments included: "I didn't realize that small scale subsistence fishing was a primary source of food for some families in CT" and "it felt cool being able to contribute to the scientific community." Finally, one student commented that the service learning project was valuable "*in theory, yes.* [but] *In execution, no.*"

Several drawbacks were noted and suggestions for improvements given. These are incorporated with feedback and reflections provided by students during the course of the semester. Time constraints were problematic due to the scheduling of the class in the fall and the need to first understand the topic and then jointly develop a survey. This meant that surveying did not begin until mid-November and posed some difficulties in locating fishermen due to bad (rainy and cold) weather, shorter days and less angling activity. One noted, "Due to the time of year and school class schedule, I think we missed a good chunk of time when people would actually be out fishing and willing to talk." The semester is also constrained to 14 weeks, and other academic and economic pursuits take the time of students, thus some students encountered time constraints, a very few were unable to complete portions of the project, and as with any group assignment, work allocation and completion can be problematic. One student noted that the "project felt too rushed, some classmates didn't do their fair share."

Several students noted that many harvesters chose not to be interviewed and that finding harvesters to interview was difficult. Individuals declined due to needing to catch a certain stage of the tide, not having time or simply not wanting to be involved in the survey. One older student, a retired prison guard, attempted to interview Hispanic fishermen in an urban fishing location in East Haven, and they indicated they did not want to be interviewed and that they did not speak English. It was not possible to calculate a response rate because of incomplete recording of failed interview attempts. It is evident that many fishers declined the interview but unclear whether this skewed the demographic makeup of respondents.

Conclusions

Overall, and despite some bumps in the execution, I believe that service learning approaches are valuable adjuncts to classroom teaching and educational benefits do accrue to students engaged in these projects. I think this subsistence-focused service learning project was able to produce some useful data, although not necessarily representative of Connecticut's subsistence harvesting effort. Perhaps more importantly, the effort provided some useful lessons for future implementation.

Although not labeled as subsistence, the practice of fishing for food appears to have substantial social and cultural significance to those who engage in the activity. This can be evidenced by the high percentage of individuals who share their harvest, distributing it through family and other social networks. Fishing appears to be integrally related to the maintenance of family relationships, evidenced in the large number of individuals who emphasized the importance of fishing as a time for families to be together. The transmission of fishing knowledge was also noted to occur through family members. The opportunity that fishing activities provide for family members to spend extended periods of time together, transmit skills and knowledge and strengthen familial and cultural ties, is clearly evidenced and significant.

Subsistence use is not a widely understood or embraced concept in Connecticut. Nonetheless, it appears to be common. Of those individuals interviewed, directly consuming and sharing their harvests with others, is quite important, filling a provisioning role within the overarching cash-based economy and also affording individuals the opportunity to engage in a pleasurable and relaxing activity that reinforces familial relationships. Fish and shellfish harvesting activities are strongly tied to cultural, social and especially familial imperatives. Economic imperatives associated with fishing for food appear less important from the surveys conducted, though the limitations of the sample make it impossible to draw definitive conclusions. Harvesting marine resources for subsistence may not be labeled as such, may not be visible to regulators, but understanding the multifaceted nature of subsistence practices will aid regulators in understanding regulatory shortfalls and provide insights to craft more appropriate policies which address the particular needs of those who fish for food.

Abbreviations

CTDEEP: Connecticut Department of Energy and Environmental Protection; UN FAO: United Nations Food and Agriculture Organization.

Authors' information

The author is affiliated with the University of Connecticut, Department of Agricultural and Resource Economics, as well as the Connecticut Sea Grant College Program, and can be reached at syma.ebbin@uconn.edu.

Acknowledgements
This research was conducted as part of a class held at the University of Connecticut, Avery Point Campus in Marine Fisheries Economics and Policy (ARE 3437), during the Fall 2013 semester. Students engaged in a service learning project to understand the nature of coastal resource subsistence harvest and consumption in Connecticut. The interviews conducted and data collected by my students: Ashley Hogan, William Kima, Heather Krassler, Corey Leamy, Sydney Marcks, Tiffany Rich, Nicole Stanley, Ronald Tardiff, Ashley Tougas, Edward Waido, were integral to this study and must be acknowledged. Much of the research and writing was conducted as part of my position at the University of Connecticut, Avery Point, and I acknowledge this institutional support.

Competing interests
The author declares that she has no competing interests.

References

1. Howell, P. CTDEEP personal communication, 2013.
2. UN FAO. http://www.fao.org.
3. Berkes F. Native subsistence fisheries: a synthesis of harvest studies in Canada. Arctic. 1990;43(1):35–42.
4. Emery M, Pierce A. Interrupting the telos: locating subsistence in contemporary US forests. Environ Plan A. 2005;37:981–93.
5. Schumann S, Macinko S. Subsistence in coastal fisheries policy: what's in a word? Mar Policy. 2007;31:706–18.
6. Macinko S, Schumann S. Searching for subsistence: in the field in pursuit of an elusive concept in small-scale fisheries. Fisheries. 2007;32(12):592–600.
7. Steinback A, Wallmo K, Clay P. Saltwater sport fishing for food or income in the Northeastern US: statistical estimates and policy implications. Mar Policy. 2009;33:49–57.
8. McLain R, Hurley P, Emery M, Poe M. Gathering "wild" food in the city: rethinking the role of foraging in urban ecosystem planning and management. Int J Justice Sustain. 2014;19(2):220–40.
9. Poe M, McLain R, Emery M, Hurley P. Urban forest justice and the rights to wild foods, medicines, and materials in the city. Hum Ecol. 2013;41:409–22.
10. Ebbin SA. Emerging cooperative institutions for fisheries management: equity and empowerment of indigenous peoples of Washington and Alaska. Yale University, Graduate School, PhD. Dissertation; 1998. 710 pp.
11. Steinhoff H. Analysis of major conceptual systems for understanding and measuring wildlife values. In: Shaw W, Zube E, editors. Wildlife values. Center for Assessment of Noncommodity, Natural Resource Values, Inst. Series Report No. 1. Proceedings of Workshop. Tucson, AZ Oct 1979; 1980. p. 11–21.
12. Brown T, Burch E Jr. Estimating the economic value of subsistence harvest of wildlife in Alaska. In: Peterson G, Swanson C, McCollum D, Thomas M, editors. Valuing wildlife resources in Alaska. Boulder, CO: Westview Press; 1992. p. 203–54.
13. Corburn J. Combining community-based research and local knowledge to confront asthma and subsistence fishing hazards in Greenpoint/Williamsburg Brooklyn, New York. Environ Health Perspect. 2002;110(2):241–8.
14. Toy K, Polissar N, Liao S, Mittelstaedt G. A fish consumption survey of the Tulalip and Squaxin Island Tribes of the Puget Sound Region. Marysville, WA: Tulalip Tribes, Department of Environment; 1996.
15. Balcom N, Capacchione C, Wright Hirsch D. Quantification of fish and seafood consumption rates for Connecticut. Connecticut Sea Grant Program Report to the CT DEP OLISP; 1999.
16. CTDEEP Enhanced Opportunity Shore Fishing Program website. http://www.ct.gov/deep/cwp/view.asp?a=2696&q=514534&depNav_GID=1647.
17. CTDEEP Marine Recreational Fisheries webpage. http://www.ct.gov/deep/cwp/view.asp?a=2696&q=322744&deepNav_GID=1647.
18. 2010 US census data website. www.census.gov.

Nutrient intake, digestibility and growth performance of Washera lambs fed natural pasture hay supplemented with graded levels of *Ficus thonningii*(*Chibha*) leaves as replacement for concentrate mixture

Yeshambel Mekuriaw and Bimrew Asmare*

Abstract

Background: The huge livestock resources of the country could not be used to its potential due to lack of good-quality feed throughout the year. Consequently, finding alternative feed supplements for livestock is an important step to sustain livestock production in the country. *Ficus thonningii* leaf is one of the potential feed resources for ruminants in the area. The experiment was conducted with the objective of evaluating the nutrient intake, digestibility and growth performance of local lambs fed natural pasture hay (NPH)-based diet supplemented with graded levels of *Ficus thonningii* dried leaves (FTL) and determining of the optimum level for replacing of concentrate mixture (CM). Twenty intact male yearling Washera lambs with initial body weight of 17.06 + 1.43 (mean + SE) kg were used in experiment. The experiment was conducted using randomized complete block design having four replications with five dietary treatments. The treatments were T1 = NPH + 0 g FTL +215 g CM; T2 = NPH + 112.5 g FTL +161.25 gCM; T3 = NPH + 225 g FTL +107.5 g CM; T4 = NPH + 337.5 g FTL + 53.75 g CM; and T5 = NPH + 450 g FTL +0 g CM. Data were analyzed using the general linear model procedures of SAS (version 9.3; 2002).

Results: The results showed that FTL are a good source of crude protein (CP) (126.9 g/KgDM) and fibers (377–426 g/Kg DM). Results also showed that supplementation of higher-level FTL significantly ($P < 0.001$) increased total DM, OM and fibers intake ($P < 0.001$), but no significant difference ($P > 0.05$) observed in CP and ME intakes among treatments. Increasing level of FTL supplementation, however, reduced significantly ($P < 0.001$) the digestibility coefficient of all nutrients. In terms of ADG, nonsignificant ($P > 0.05$) differences were observed in lambs assigned up to 75% replacement of CM. The lowest performance was observed in sole FTL supplementation (T5), despite increased nutrients intake.

Conclusion: It was concluded that indigenous fodder tree *Ficus thonningii* leaves could serve as alternative CP supplement in NPH-based feeding of Washera sheep and can be used to replace CM up to 75% to improve performance.

Keywords: Concentrate mixture, *Ficus thonningii* leaves, Grass hay, Washera lambs

Background

Ethiopia has huge and diversified livestock population than any African countries [1]. Among these livestock species existing in the country, sheep is the dominant one next to cattle [1].Recently, about 29 million sheep population has been estimated and reported in the country and the population is still increasing. Sheep are an important sub-sector of livestock farming, which is an integral part of the livelihood of the household level and contributes significantly to the national economy of Ethiopia [2].

*Correspondence: limasm2009@gmail.com
Department of Animal Production and Technology, Bahir Dar University, P.O. Box 79, Bahir Dar, Ethiopia

Food, manure, skins, risk aversion are some of the important resources obtained from sheep for smallholder farmers. However, the contribution of the sub-sector both in smallholder farmer's and the country's economy remains below its potential due to low productivity of animals than the regional and continental average. The average reported carcass yield for Ethiopian sheep is 10 kg which is lower than the neighboring African countries sheep carcass weight such as Sudan (12 kg); Kenya (13 kg) and Djibouti (14 kg) [3]. This low productivity of animals could be reflected by many factors, but shortage of feed in terms of quality and quantity is the critical one in the country.

Nowadays, the most important livestock feed resources in Ethiopia are natural pasture, crop residues and grass hay [1, 3]. The reliability of natural pasture as a source of feed is, however, restricted to the wet season [4] and most of it is degraded because of overgrazing [5]. Moreover, the resource is continuously converted to crop land to satisfy the food demand of an alarmingly increasing human population. Consequently, the natural pasture existing in the country has been characterized as poor in botanical composition, low nutritional quality and biomass yield. On the other hand, grass hay harvested from such areas could not satisfy the nitrogen requirements of sheep. Similarly, crop residues are inherently low in quality [6]. Hence, multipurpose fodder trees were introduced in the country just four decades ago aimed to supplement these major feed resources. However, these improved forage species production and utilization in the country is very minimal, with less than one percent [1]. Therefore, the basal diets like grass hay should be supplemented with commercial protein concentrates to satisfy maintenance and production requirements for critical nutrient like protein to enhance the rumen microbes which facilitate the degradation of fibrous feeds. In reality, protein concentrates are not easily available and affordable for the poor farmers in Ethiopia. Hence, looking other alternative protein sources from multipurpose fodder trees which are available especially during the dry season is vital and highly commendable.

The fodder value of *Ficus thonningii* tree leaves to different livestock species has been studied and reported [7–9]. Moreover, Balehegn et al. [10] reported that *Ficus thonningii* is a potential replacement for commercial concentrate for goat production. The proximate, in vitro and *in sacco* analysis studies made by scholars [11–13] indicated *Ficus thonningii* leaves to be a potential protein source (11–14%) in ruminant nutrition. However, very limited information is available on the *Ficus thonningii* leaves for sheep production. Hence, the study was designed to evaluate the effect of graded level replacement of concentrate mixture with dried leaves of FT on nutrient intake, digestibility and performance of Washera lambs fed natural pasture hay basal feed.

Methods

Study area

The experiment was conducted at Zenzelima campus of Bahir Dar University, which is located between latitude and longitude of 11° 37′ N and 37° 28′ E coordinates and an elevation of 1912 m above sea level. The average daily minimum and maximum temperatures were 7 and 29 °C, respectively. The average annual rainfall is 1445 mm. The main rainy season is from June to September. The soils are dominantly nitosols (fertile agricultural soil).

Experimental animals management and treatments

Twenty intact male yearling Washera sheep with initial body weight of 17.06 + 1.43 (mean + SE) were purchased from Achefer and Dangila local markets. The age of the sheep was determined by dentition and information given from the owners. The animals were quarantined for 15 days at the experimental site in order to let the animals get accustomed to the specific environment and to observe their health situations. Meanwhile, the animals were vaccinated against common prevalent diseases of the area with the consultation of veterinarians from the college of agriculture and environmental sciences, Bahir Dar University. During the quarantine period, each animal was treated against internal parasites using albendazol and external parasite using ivermectin. Animals were continuously observed for incidence of any ill health and disorders during the experimental period.

The experimental animals were housed in individual pens in a ventilated shade that was built for this purpose. The pens were equipped with feeding trough for hay and plastic buckets for concentrates supplements and watering separately. The experimental sheep was identified with neck collars. Experimental animals were allowed to adapt the experimental diets for 15 days before the commencement of the actual data collection.

The design of the experiment was randomized complete block design (RCBD). Experimental animals were grouped into four blocks of five animals each based on their initial body weight, which was determined by the average of two consecutive weightings after overnight fasting. The experimental animals from each block were randomly assigned to each treatment feed. Natural pasture hay was provided as a basal diet and offered ad libitum with 20% refusal. The five experimental diets used both for feeding and digestibility trials are presented in Table 1.

Table 1 Experimental treatment

Treatments	Basal diet	Supplements		
		CM (g/d)	FTL (g/d)	CP (g)
T1	NPH	215 (100%)	0	57.62
T2	NPH	161.25 (75%)	112.50 (25%)	57.49
T3	NPH	107.50 (50%)	225.00 (50%)	57.36
T4	NPH	53.75 (25%)	337.50 (75%)	57.23
T5	NPH	0	450.00	57.11

NPH natural pasture hay, *FTL Ficus thonningii* leaves, *CM* concentrate mixture; T1 = NPH + 0 g FTL + 215 g CM; T2 = NPH + 112.5 g FTL + 161.25 g CM; T3 = NPH + 225 g FTL + 107.5 g CM; T4 = NPH + 337.5 g FTL + 53.75 g CM; T5 = NPH + 450 g FTL + 0 g CM

Feeds and feeding management

The grass hay used in this study was natural pasture harvested in November 2014. *Ficus thonningii* leaves were collected in January 2015 (dry season) from farmers' arable land and backyard from different ages of the plant. Information was not obtained about the age of the trees. The leaves collected were transported to the experimental site and dried under shed. After drying, the leaves were thoroughly mixed to minimize variation. The natural pasture hay was chopped manually to the size of 3–10 cm so as to improve intake by experimental sheep. Noug seed cake and wheat bran were purchased from oil extraction and floor meal factories in Bahir Dar town. The grass hay was offered ad libitum along with free access of water and salt licks. The concentrate supplements, i.e., noug seed cake and wheat bran were mixed in 50:50 ratio, and the FTL was offered based on the proportions indicated in Table 1.

Chemical analysis

The dried FTL and NPH samples were ground to pass through a 1-mm sieve (Willy mill) for chemical analysis. These ground samples were stored individually in an airtight plastic bag pending chemical analysis. Dry matter, OM and CP contents of FTL, NPH and CM were determined following the procedures described by AOAC [14]. The NDF, ADF and ADL contents were analyzed following the recommendations of VanSoest et al. [15].

Feed intake

Daily feed offered and refusals were collected from each treatment throughout the experimental period. The feeding or growth trial was lasted for 90 days. The daily feed intake of each experimental animal was calculated as the differences between feed offered and refused.

Body weight change and feed efficiency

Animals were weighted weekly in the morning after overnight fasting of the animal using Salter Scale with a sensitivity of 100 g. The average daily weight gains (ADG) were calculated on a weekly basis. It was calculated as the difference between final BW and initial BW divided by a number of feeding days. The feed efficiency of experimental animals was determined by dividing the ADG into the amount of feed consumed.

Digestibility

The digestibility trial was conducted at the end of the feeding or growth trial. In the digestion trial, each sheep was fitted with fecal collection bags for 4 days of acclimatization period with fecal collection bags prior to the actual collection of feces for 7 consecutive days. Feces voided was thoroughly mixed, weighed and recorded every morning, and 20% of the representative samples was taken and frozen at − 10 °C, and pooled over the collection period for each animal. At the end of the collection period, each sample was mixed and dried at 60 °C for 72 h at the college of Agriculture and environmental sciences laboratory. Then the samples of feces were taken to Debre birhan research center Animal Nutrition Laboratory for chemical analysis. The apparent digestibility of DM, OM, CP, NDF and ADF was determined using the following equations:

Apparent DM digestibility coefficient
= DMI − Faecal DM output
DMIApparent nutrient digestibility coefficient
= Nutrient intake − Faecal nutrient output
Nutrient intake
where: DM = dry matter; DMI = dry matter intake

Data analysis

Feed intake, nutrients digestibility and performance of lambs' data were analyzed using the General Linear Model (GLM) procedures of SAS [16]. Means were separated using the least significant difference for variables when *F* values declared a significant difference. Differences were considered statistically significant at 0.05% significance level.

The analysis model was:

$$Y_{ij} = \mu + t_i + b_j + e_{ij}$$

where Y_{ij} = dependent variables, μ = overall mean, t_i = the effect of the *i*th treatments, b_j = effect of the *j*th block and e_{ij} = random error of *i*th treatment in the *j*th block.

Results

Chemical composition of the treatment feeds

The chemical composition of feed components consisting treatment diets is presented in Table 2. Almost similar amount of dry matter (DM) is obtained in the feed components of the treatment diet. Higher crude protein (CP) content was recorded in FTL than natural pasture hay (NPH). Similarly, FTL showed lower fiber (NDF, ADF and ADL) contents than NPH. The concentrate mixture of wheat bran and noug seed cake (50:50) proportion had better nutritive value when evaluated in terms of CP and fiber contents as compared to FTL.

Feed and nutrient intake

The daily DM and nutrient intake of Washera lambs fed natural pasture hay and supplemented with different levels of ficus leaves and concentrate mixture is presented in Table 3. There was a significant difference ($P<0.001$) between T1 and T5 in terms of total DM (TDM), DM (% BW) and OM intake. In general, intake parameters increased as the level of FTL increased. On the other hand, lowest TDM, (% BW) and OM intake were observed in sole CM-supplemented group (T1) while the highest for sole FTL-supplemented group (T5). Similarly, fibers (NDF, ADF, ADL) intake also significantly ($P<0.001$) varied among treatments and their intake increased as the FTL inclusion increased. However, a nonsignificant difference ($P>0.05$) was observed in CP and ME intake among treatments.

Dry matter and nutrient digestibility coefficient

Apparent digestibility coefficients of DM and nutrients in Washera lambs fed NPH and supplemented with noug seed cake and wheat bran mixture and dried FTL is presented in Table 4. A significant difference ($P<0.001$) was observed between sole CM supplement (T1) and sole.

FTL supplement (T5) in terms of DM, OM and fibers digestibility coefficient. However, significant difference ($P>0.05$) was not observed among treatments (T1, T2, T3 and T4) in terms of digestibility coefficients.

Table 2 Chemical composition of feed used in the experiments

Feeds	Chemical composition (g/Kg for DM and g/Kg DM for others)					
	DM	OM	CP	NDF	ADF	ADL
NPH	890.0	830.0	38.8	768.6	448.6	150.0
FT L	900.0	789.0	126.9	426.5	377.8	126.8
NSC	900.0	922.0	356.4	406.5	282.6	108.8
WB	910.0	952.5	194.4	286.9	173.9	90.6
CM	910.0	913.0	226.8	366.8	260.8	106.8

NPH natural pasture hay, *FTL Ficus thonningii* leaves, *CM* noug seed cake: Wheat bran (50:50)

Table 3 Feed intake of Washera lambs fed natural pasture hay supplemented with different levels of Ficus leaves and concentrate mixture

Parameters	Treatments					SEM	SL
	T1	T2	T3	T4	T5		
Basal DMI (g/d)	472.62^{a}	452.79^{ab}	445.93^{ab}	421.61^{b}	424.82^{ab}	11.073	***
Supplement DM intake (g/d)	195.7^{e}	248.0^{d}	300.3^{c}	352.7^{b}	405.0^{a}	0.000	***
Total DMI (g/d)	668.27^{d}	700.48^{cd}	746.78^{bc}	774.27^{b}	829.82^{a}	11.073	***
TDMI (%BW)	2.72^{c}	2.99^{bc}	3.27^{b}	3.44^{ab}	3.90^{a}	0.114	***
OMI (g/d)	570.90^{d}	589.68^{cd}	619.21^{bc}	634.25^{ab}	672.15^{a}	9.190	***
CPI (g/d)	55.59^{a}	54.99^{a}	54.82^{a}	54.23^{a}	54.09^{a}	0.357	ns
NDFI (g/d)	367.02^{b}	372.09^{b}	385.38^{ab}	387.56^{a}	407.30^{a}	7.063	***
ADFI (g/d)	222.56^{d}	233.72^{cd}	249.70^{bc}	259.18^{b}	278.89^{a}	4.123	***
ADLI (g/d)	77.918^{d}	80.81^{cd}	85.32^{bc}	87.650^{ab}	93.41^{a}	1.378	***
[1]MEI (MJ/Kg DM)	7.12^{a}	7.16^{a}	7.29^{a}	7.31^{a}	7.48^{a}	0.174	ns

a, b, c, d, e = means within rows having different superscript are significantly different at *** = ($p<0.001$); *ADF* acid detergent fiber, *BW* body weight, *CM* concentrate mixture, *CP* crude protein, *DMI* dry matter intake, *NDF* neutral detergent fiber, *OM* organic matter, *SL* significant level, *NPH* natural pasture hay, *FTL Ficus thonningii* leaves, *CM* concentrate mixture; T1 = NPH + 0 g FTL + 215 g CM; T2 = NPH + 112.5 g FTL + 161.25 g CM; T3 = NPH + 225 g FTL + 107.5 g CM; T4 = NPH + 337.5 g FTL + 53.75 g CM; T5 = NPH + 450 g FTL + 0 g CM

[1] Estimated Metabolizable Energy (ME) = 0.0157 × Digestible Organic Matter Intake (DOMI) [46]

Table 4 Apparent digestibility coefficients of nutrients in sheep fed natural pasture hay mixture as a basal diet and supplemented with noug seed cake and wheat bran mixture and ficus leaves

Parameters	Treatments (%)						
	T1	T2	T3	T4	T5	SEM	SL
DMDC	0.77^{a}	0.75^{a}	0.73^{ab}	0.71^{ab}	0.69^{b}	0.012	***
OMDC	0.79^{a}	0.78^{ab}	0.75^{abc}	0.74^{bc}	0.71^{c}	0.011	***
CPDC	0.79^{a}	0.77^{a}	0.69^{b}	0.62^{c}	0.60^{c}	0.013	***
NDFDC	0.66^{a}	0.58^{ab}	0.54^{ab}	0.51^{ab}	0.44^{b}	0.040	***
ADFDC	0.63^{a}	0.56^{ab}	0.53^{abc}	0.49^{bc}	0.42^{c}	0.042	***

$^{a\,b\,c}$ Letters in the same row having different superscript are significantly different; *DMDC* dry matter digestibility coefficient, *OMDC* organic matter digestibility coefficient, *CPDC* crude protein digestibility coefficient, *NDFDC* neutral detergent fiber digestibility coefficient, *ADFDC* acid detergent fiber digestibility coefficient; *** ($P<0.001$); *SEM* standard error of mean; *SL* significance level, *Ns* nonsignificant, *NPH* natural pasture hay, *FTL Ficus thonningii* leaves, *CM* concentrate mixture, T1 = NPH + 0 g FTL + 215 g CM; T2 = NPH + 112.5 g FTL + 161.25 g CM; T3 = NPH + 225 g FTL + 107.5 g CM; T4 = NPH + 337.5 g FTL + 53.75 g CM; T5 = NPH + 450 g FTL + 0 g CM

Significant difference ($P<0.001$) was observed in CP digestibility coefficient among treatments in the following ranking (T1 = T2 > T3 > T4 = T5).

Body weight gains and feed efficiency

The average daily gain and feed efficiency of Washera lamb are presented in Table 5. A significant difference ($P<0.001$) was observed between sole CM-supplemented (T1) and sole FTL-supplemented (T5) groups in terms of ADG and FE. However, significant difference ($P>0.05$) was not observed among treatments (T1, T2, T3 and T4) and in FTL-supplemented groups (T2–T5) in terms of ADG. Among the FTL-supplemented groups, T5 had the lowest FE value.

Discussion

Chemical composition of treatment feeds

The natural pasture hay (NPH) used in the current study had very low CP content (38.8 g/kg DM) which is below the CP requirement for ruminant animals for proper rumen function [17, 18]. CP content (38.8 g/kg DM) of NPH obtained in current study is similar with the value reported by [19–21] and lower than the values (70.2–92 g/kg DM) reported by [21–24].The obtained fiber content of NPH in this study is within the expected ranges of most tropical roughage feedstuffs, but the NDF values are comparable with the values reported by [19, 23] but lower than the values reported by [20, 22] and higher than the value reported by Asmare et al. [21].

Similar values were obtained with regard to ADF, by many authors [20, 23, 24] and lower than the values reported by Mekuriaw et al. [22] Ephrem et al. [19] and higher than the value reported by Asmare et al. [23]. Higher level of ADL recorded in this study was comparable with the values reported by [20–22, 24] and Asmare et al. [23], but lower than the values reported by Ephrem et al. [19],but similar with value reported by Asmare et al. [23]. This fiber contents difference among studies could be explained by altitude, rainfall, harvesting stage, fertility, soil and cropping intensity variation [25, 26]. The fiber fraction of hay in the current study is within the range of most tropical forages (<45% for high quality, 45–65% for medium quality and >65% for poor quality roughages) [27] indicating that the hay is poor in quality for optimum rumen condition and supplementation of this natural pasture hay with alternative protein source was commendable.

Table 5 Body weight measurements and feed efficiency of sheep fed on natural pasture hay and supplemented with noug seed cake and wheat bran in 50:50 ratio and ficus leaves

Parameters	Treatments						
	T1	T2	T3	T4	T5	SEM	SL
IBW	16.28^{a}	17.53^{a}	17.53^{a}	17.95^{a}	17.33^{a}	1.331	ns
FBW	22.80^{a}	22.30^{a}	21.98^{a}	21.55^{a}	21.13^{a}	1.175	ns
ADG	67.50^{a}	63.89^{ab}	56.39^{ab}	55.28^{ab}	53.62^{b}	4.212	***
FE	0.10^{a}	0.083^{ab}	0.073^{bc}	0.060^{cd}	0.050^{d}	0.014	***

a, b, c = means within rows having different superscript are significantly different at $P<0.001$; *ADG* average daily gain, *BWC* body weight change, *FBW* final body weight, *IBW* initial body weight, *FCE* feed conversion efficiency, *NPH* natural pasture hay, *FTL Ficus thonningii* leaves; *CM* concentrate mixture; *** ($P<0.001$); T1 = NPH + 0 g FTL + 215 g CM; T2 = NPH + 112.5 g FTL + 161.25 g CM; T3 = NPH + 225 g FTL + 107.5 g CM; T4 = NPH + 337.5 g FTL + 53.75 g CM; T5 = NPH + 450 g FTL + 0 g CM

The CP content of 126.9 g/kgDM noted in the current study for *Ficus thonningii* was quite similar with data ranging from 102.4 to 154.1 g/kg DM [12, 13, 28–31], but lower than the values 164.7 to 248.8 g/kgDM as reported by [7, 10, 32–35]. This difference could be explained by the season and age of the plants where samples were collected. The same author Funmilayo et al. [28] reported the effect of season on the nutritional value of FTL. Balehegn et al. [10] also reported the nutritional value of leaf of Ficus plant could be affected by the age at which the sample is collected. In general, FTL had better CP content being beyond the maintenance requirement of ruminant animals. The CP content 8% (80 g/kg DM) obtained in FTL can provide the minimum ammonia levels required for microbial activity in ruminants [36].

Fiber contents of FTL (426.5, 377.8 and 126.8 g/Kg DM) for NDF, ADF and ADL, respectively, obtained in the current study were lower than the ficus leaves used by Kassahun et al. [12]. (NDF, ADF, ADL = 47.61, 39.49, 20.28 g/kg DM), Dupe and Olaniyi [29] (NDF, ADF, ADL = 640, 480, 120 g/kg DM), Njidda and Ikhimioya [32] (NDF, ADF, ADL = 512; 412; 100 g/kg DM), Bruh and Destalem [13] (NDF, ADF, ADL = 640, 392, 74.4 g/kgDM), Yousuf and Ogundun [30] (NDF and ADL = 419.7; 291.1 g/kg DM) in shed ficus leaf, Balehegn et al. [13] (NDF, ADF, ADL = 427, 365, 162.3 g/kg DM) and NDF = 557.9 [34]. The possible explanation for these differences could be the environment where plants are grown, the season or age of the plant [28]. Higher CP (226.6 g/kg DM) and lower fiber contents (260–366 g/Kg DM) were observed in CM as compared to the FTL in terms of these nutrients. The overall variability in fiber nutrient composition could be attributed to various factors such as plant age, harvesting regime, season and location. These factors should be considered when carrying out the chemical evaluation of fodder trees.

Dry matter and nutrient intake

Natural pasture grass hay intake by all treatment groups obtained in the current study (424–472 g/day) was higher than the value (415 g/day) reported by Ephrem et al. [19] for natural pasture hay and by Tefera et al. [37] for Rhodes grass consumed by the same sheep breed. The total DM intake was more than sufficient to meet maintenance requirement for DM (2–2.5% BW). In line with the current study (put the value of the fiber content of CM here), Balehegn et al. [10] also observed improved DM and nutrient intake when FTL was replaced by concentrate supplement in the highland goats. The total DM intake of sheep in the current study was similar with the results obtained by Kassa et al. [38] for the same breed of sheep fed hay as a basal diet and supplemented with *F. sycomorous* leaf, fruit and their mixtures. However, the total DM intake was lower than the reports of Tadelle et al. [39] for the same breed of sheep fed natural pasture hay supplemented with processed lupin grain (*Lupinus albus*).

Total DM intake of the current study is lower than the values reported (666–788 g/day) for Farta sheep supplemented with wheat bran, noug seed cake and their mixtures [20]. The mean total DM intake of sheep in the current study was 58.08 g DM kg-1 W0.75 which was similar with the results of other works [40, 41] who reported total DM intake of sheep being within the range of 58.6–82.2 g DM kg^{-1} W0.75. The total DM intake expressed as percent of body weight obtained from the current study is in agreement with that of Kitaw et al. [42] with the value being in the range of the recommended dry matter intake of ruminants (2–6%) by ARC [43].

Apparent nutrient digestibility

The lower apparent DM and nutrient digestibility in the sole FTL indicated that sole FTL might not be used solely as a supplement of sheep in such type of basal diet (Table 4). However, FTL can be used as supplement up to 75% in total supplements as there were relatively better digestibility coefficients in such treatment diets except for 100% FTL supplement group.

The better digestibility coefficient up to 75% FTL supplement might be due to the fact that the higher protein content might have increased the digestibility of the crude fiber of the feeds. If the protein-rich feeds are added to balance low protein roughages, there will be an increased population of the microorganism which increased the rate of fermentation of the crude fiber component [44]. The lower digestibility of nutrients observed as FTL supplementation increased from 0 to 100% might be associated with the higher fiber fraction both in the basal diet and FTL supplement. This observed lower digestibility of nutrients among sheep assigned under higher FTL might be due to the higher fiber intake. This idea was supported by Gemeda and Hassen [45] who reported negative effects on voluntary feed intake and nutrient digestibility.

Body weight change

The positive weight gain observed in the control group in the current study was due to the inclusion of CM (Table 2) beyond their microbial nutrient requirement (VanSoest, 1994). ADG of Washera lambs assigned in FTL of the current study showed similar gains by Ephrem et al. [19] using similar sheep breed feeding sweet lupin as a protein supplement. Similarly, increasing the level of FTL supplementation improved ADG ($P < 0.001$) of Washera lambs (47.8–74.7 g/day). Balehegn et al. [10] also reported similar results on goats fed ficus leaf in

Tigray, Ethiopia. The same authors reported that in goat up to the level of 50% concentrate replacement with FTL improved intake and performance of goat. In the current study even though the level of FTL inclusion increased the values are almost the same with sole CM but the lowest level of digestibility and growth performance observed in the sole FTL. Contrary to the current study, Yousuf and Ogundun [30] reported fallen FTL fed goat showed negative weight gain. However, practical limitation was not observed due to utilization of FTL solely in ruminant feeding [34].

Conclusion

From the current study, it is possible to conclude that supplementation of Washera lambs with *Ficus thonningii* leaves by replacing concentrate mix up to the 75% inclusion for optimum performance of improved nutrient intake, digestibility and growth of sheep and this multipurpose indigenous fodder tree (*Ficus thonningii*) leaves could serve as an alternative CP supplement in the natural pasture hay-based feeding system of Washera sheep in the dry season when most forages could not supply enough nitrogen for the function of rumen in ruminant animals.

Abbreviation

ADF: acid detergent fiber; ADG: average daily gain; ADL: acid detergent lignin; AOAC: association of official analytical chemists; BW: body weight; CP: crude protein; CSA: central statistical agency; DM: dry matter; DMI: dry matter intake; DMD: dry matter digestibility; DOM: digestible organic matter; DOMI: digestible organic matter intake; FT: *Ficus thonningii*; FTL: *Ficus thonningii* leaves; FCE: feed conversion efficiency; GLM: general linear model; ME: metabolizable energy; MJ: mega joule; NDF: neutral detergent fiber; NPH: natural pasture hay; NSC: noug seed cake; OM: organic matter; OMDC: organic matter digestibility coefficient; OMI: organic matter intake; RCBD: randomized complete block design; SAS: statistical analysis system.

Authors' contributions

BA contributed in research proposal writing, data collection, data analysis, data interpretation and article writing. YM contributed in research data collection, data analysis, data interpretation and article writing. Both authors read and approved the final manuscript.

Authors' information

Asmare is an Assistant Professor of Animal Feeds and Nutrition, currently teaching and conducting research in the area of animal feeds and feeding, animal production and productivity improvement in Ethiopia. Mekuriaw is an Associate Professor of Animal Feeds and Nutrition, currently teaching and conducting research in the area of animal feeds and feeding, animal production and productivity improvement in Ethiopia.

Acknowledgements

The authors acknowledge Bahir Dar University, College of Agriculture and Environmental Sciences for the financial support to do the feeding and digestibility trials and Debre Berhan Agricultural Research Center Animal Nutrition Laboratory for assisting in laboratory chemical analysis of feeds and feces. The authors also would like to extend our thanks to reviewers who gave very crucial points to improve the paper.

Competing interests

Both authors declare that they have no competing interests.

Funding

Bahir Dar University, College of Agriculture and Environmental Sciences.

References

1. CSA. Agricultural sample survey, vol. II report on livestock and livestock characteristics. CSA (Central Statistical Agency), 2013, Addis Ababa, Ethiopia.
2. Gizaw S, van Arendonk JAM, Komen H, Windig JJ, Hanotte O. Population structure, genetic variation and morphological diversity in indigenous sheep of Ethiopia. AnimalGenetics. 2007;38:621–8.
3. Tegegne F, Assefa G. Feed resources assessment in Amhara National Regional State.Ethiopia Sanitary and Phytosanitary Standards and Livestock and Meat Marketing Program(SPS-LMM, Texas Agricultural Experiment Station (TAES). Final Report, Bahir Dar, Ethiopia. 2010: 81–94.
4. Zinash S, Seyoum B, Lulseged G, Tadesse T. Effect of harvesting stage on yield and quality of natural pasture in the central highlands of Ethiopia. In: Proceedings of third National Conference of the Ethiopian Society of Animal Production (ESAP), pp. 316–322,1995, Addis Ababa, Ethiopia.
5. Alemayehu M. Country Pasture/Forage Resource Profiles: Ethiopia. 2006 Available at http://www.fao.org/AGp/agpc/doc/counprof Ethiopia.htm. Accessed on 10 Aug 2016.
6. Wondatir Z, Mekasha Y. Feed resources availability and livestock production in the central rift valley of Ethiopia. Int J Livest Prod. 2014;5:30–5.
7. Jokthan GE, Afikwu EV, Olugbemi TS. The utilization of fig (*Ficus thonningii*) and mango (*Mangifera indica*) leaves by rabbits. Pak J Nutr. 2003;2:264–6.
8. Tegbe TSB, Adeyinka IA, Baye KD, Alawa JP. Evaluation of feeding graded levels of dried and milled *Ficus thonningii* leaves on growth performance, carcass characteristics and organs of Weaner Rabbits. Pak J Nutr. 2006;5:548–50.
9. Bamikole MA, Ikhatua UJ, Arigbede OM, Babayemi OJ, Etela I. An evaluation of acceptability as forage of some nutritive and antinutritive components and of the dry matter degradation profiles of five species of ficus. Trop Anim Health Prod. 2004;36:157–67.
10. Balehegn M, Eik LO, Tesfay Y. Replacing commercial concentrate by *Ficus thonningii* improved productivity of goats in Ethiopia. Trop Anim Health Prod. 2014;46:889–94.
11. Berhe DH, Tanga AA. A. nutritional evaluation of *Ficus thonningii* Blume leaves as ruminant livestock feed in the Ahferom district of Tigray, Ethiopia. Afr J Range Forage Sci. 2013;30:149–54.
12. Desalegn K, Meksasha Y, Assefa G. Identification and nutritional value assessment of the major browse species in Chilega District, North Gondar. Glob Vet. 2016;16:06–17.
13. Bruh W, Destalem G. Identification and determination of chemical composition of some potential browses as livestock fodder in the Central and North Western Zones of Tigray, Ethiopia. Livest Res Rural Dev. 2015: 27.
14. AOAC 1990 Official method of analysis. AOAC (Association of Official Analytical Chemists) 15th ed. AOAC Inc., Arlington, Virginia, USA. 1990:12–98.
15. Van Soest PJ, Robertson JB, Lewis BA. Methods for dietary fiber, neutral detergent fiber and non-starch polysaccharides in relation to animal nutrition. J Dairy Sci. 1991;74:3583–97.
16. SAS. Statistical Analysis System, Version 9.3. SAS (Statistical Analysis System), Institute Inc., 2002. Cary, NC, USA.
17. Van Soest PJ. Nutritional ecology of the ruminant. 2nd ed. London: Cornell University; 1994. p. 244–52.
18. McDonald AA, Edwards R A, Greenhalgh JFD, Morgan CA. Animal nutrition, 6th ed. pp. 2002:187–198, 544–549.
19. Ephrem N, Tegegne F, Mekuriaw Y, Yiheyis L. Nutrient intake, digestibility and growth performance of Washera lambs supplemented with graded levels of sweet blue lupin (*Lupinus angustifolius* L.) seed. Small Rumin Res. 2015;130:101–7.

20. Bishaw F, Melaku S. Feed utilization and live weight change of Farta sheep supplemented with noug seed meal and wheat bran on feed intake, digestibility and body weight change. Trop Anim Health Prod. 2008;40:597–606.
21. Asmare A, Solomon D, Taye T, Firew T, Jane W, Barbara R. Determinants of the utilization of desho grass (Pennisetum pedicellatum) by farmers in Ethiopia. Trop Grassl.-Forrajes Trop. 2016;4:112–21.
22. Mekuriaw Y, Urge M, Animut G. Intake, digestibility, live weight changes and rumen parameters of Washera sheep fed mixtures of lowland bamboo (*Oxytenanthera abyssinica*) leaves and natural pasture grass hay at different ratios. Pak J Nutr. 2012;11:322–31.
23. Asmare B, Melaku S, Peters Kurt J. Supplementation of Farta sheep fed hay with graded levels of concentrate mix consisting of noug seed meal and rice bran. Trop Anim Health Prod. 2010;42:1345–52.
24. Nega A, Melaku S. Feed intake, digestibility and body weight change in Farta sheep fed hay supplemented with rice bran and/or noug seed (Guizotia abyssinica) meal. Trop Anim Health Prod. 2008;41:507–15.
25. Denekew Y, Tamir B, Melaku S. Effect of harvesting date on composition and yield of natural pasture in north western Ethiopia. Trop Sci. 2005;45:19–22.
26. Kitaba A, Tamir B. Effect of harvesting stage and nutrient levels on nutritive values of natural pasture in central highlands of Ethiopia. Agric Tropica et Subtrop. 2007;40:1–10.
27. Lemus R Hay storage: Dry matter losses and quality changes. Mississippi State University Extension Service, USA. 2009. Accessed 25 Sept 2016.
28. Funmilayo B, Olaniyi B and Adegboyega A. Feed resources and seasonal nutrient composition of predominant forages for small ruminant production in iwo local Government Area of Osun state, Nigeria. J Biol Agric Healthc. 2013:17.
29. Dupe OO, Olaniyi JB. The effect of forage based diets on milk composition, lactation stages and growth rate kids from West African dwarf (WAD) goat in South West Nigeria.
30. Conference on International Research on Food Security, Natural Resource Management and Rural Development. ETH Zurich, Sept 14–16, 2010.
31. Yousuf MB, Ogundun NJ. Feeding value of shed or fresh Ficus (*Ficus thonningii*) leaves for the West African dwarf. J Centrepoint. 2005;13:51–9.
32. Yusuf AO, Muritala RO. Nutritional evaluation and phytochemical screening of common plants used in smallholder farming system. Pac J Sci Technol. 2013;14:12–8.
33. Njidda AA, Ikhimioya I. In vitro gas production and dry matter digestibility of semi-arid browses of North Eastern Nigeria. Slovak J Anim Sci. 2010;43:154–9.
34. Otitoju GTO, Nwamarah JU, Otitoju O, Iyeghe LU. Nutrient composition of some lesser known green leafy vegetables in Nsukka Lga of Enugu State. J Biodivers Environ Sci. 2014;4:233–9.
35. Gemeda BS, Hassen A. Effect of Tannin and species variation on In vitro digestibility, gas, and methane production of tropical browse plants. Asian Australas J Anim Sci. 2015;28(2):188–99.
36. Musibau A, Bamikole U, Julius I. Nutritional evaluation of *Ficus thonningii*-Panicum maximum mixtures in West African dwarf goats. Nutr Food Sci. 2010;40:280–8.
37. Norton BW. Anti-nutritive and toxic factors in forage tree legumes. In Forage Tree Legumes in Tropical Agriculture. Edited by: Gutteridge RC, Shelton HM. CAB International, Wallingford, Oxford; 1994: 202–215.
38. Tefera G, Tegegne F, Mekuriaw Y, Melaku S and Tsunekawa A. Effects of different forms of white lupin (*Lupinus albus*) grain supplementation on feed intake, digestibility, growth performance and carcass characteristics of Washera sheep fed Rhodes grass (*Chloris gayana*) hay-based diets. Trop. Anim. Health Prod. 2015:47.
39. Kassa A, Tadele Y, Mekasha Y. Ficus sycomorus (Sycamore Fig or Shola) leaf, a potential source of protein for ruminants: a review. J Fisheries Livest Prod. 2015;3:4–11.
40. Tadele Y, Mekasha Y, Tegegne F. Supplementation with different forms of processed Lupin (*Lupinus albus*) grain in hay based feeding of Washera sheep: effect on feed intake, digestibility body weight and carcass parameters. J Biol Agric Healthc. 2014;4:27–35.
41. FAO. Successes and failures with animal nutrition practices and technologies in developing countries. Proceedings of the FAO (Food and Agricultural Organization of the United Nations) Electronic Conference, 1–30 September 2010. Editor H P S Makkar. FAO Animal Production and Health Proceedings. No. 11. Rome, Italy.
42. Alem M, Tamir B, Kurtu MY. Feed utilization of Ethiopian highland lambs on a basal diet of eleucine coracana straw and supplemented with variously sourced protein mixed with wheat bran. Trop Anim Health Prod. 2011;43:115–20.
43. Kitaw G, Dejene M, Kahilew A, Assefa G. Comparative evaluation of tree lucerne (Chamaecytisus palmensis) over conventional protein supplements in supporting growth of yearling horro lambs. Livest Res Rural Dev. 2012:24.
44. ARC. The nutrient requirements of ruminant livestock, technical review Agricultural Research Council working party. ARC (Agricultural Research Council), 1980.
45. Ranjihan S K Animal nutrition in the tropics. 4th eds. Vikas publishing House P. Ltd. New Delhi, India. 1997, p. 557.
46. AFRC. Energy and protein requirements of ruminants. AFRC (Agricultural Food and Research Council), An advisory manual prepared by Agricultural Food and Research Council technical committee on responses to nutrients. CAB int. Wallinaford, UK. 1993, p. 342.

15

Fisheries, food, and health in the USA: the importance of aligning fisheries and health policies

David Clifford Love[1,2*], Patricia Pinto da Silva[3], Julia Olson[3], Jillian Parry Fry[1,2,4] and Patricia Mary Clay[3]

Abstract

Background: Food availability, access, and utilization are the three pillars of food security and need to be aligned in order to support a healthy population. United States (US) fisheries policy plays an important role in seafood *availability*. US health policy impacts *access* and *utilization* of seafood in various ways; however, health policies are often disconnected from fisheries policy. Aligning fisheries and health policies is imperative to improve food security. We address two questions with our work: (1) how would US federal fisheries policy be different if our fisheries were managed with beneficial health outcomes for Americans as clear objectives; and (2) how would US health policy be different if one of its goals was to support sustainable domestic fisheries and aquaculture?

Results: We report how fisheries policies and health policies are additive, synergistic, or antagonistic with regard to seafood, and provide illustrative examples of collaboration between health and fisheries communities at different levels of the food system (federal and state policies, corporate partnerships, and civil society). We also develop a list of topics for future research, and opportunities to align and integrate fisheries and health policies.

Conclusions: Managing fisheries to promote optimal nutrition and efficient food production likely requires a different approach to fisheries management—new outcomes will need to be monitored, new approaches found, and fisheries, aquaculture, and health policies better integrated. Health policies rarely consider the source of fish, their connections with US fisheries systems, and global distribution of seafood. Change can begin where the most promising opportunities exist, such as institutional food procurement, Farm to School programs, social marketing campaigns, and private sector start-ups. Continued development in fisheries and health policies, however, will need to occur at multiple levels of federal policy, and across the different domains and dimensions of the food system (e.g., social, political, biophysical, economic).

Keywords: Aquaculture, Fisheries, Food, Health, NOAA, Policy, USDA

Background

Fish and other types of aquatic animal and plant species are an integral part of human economies, cultures, and health. Here, we refer to those species using the term "seafood" to include freshwater and marine species, principally those caught in commercial fisheries or grown through aquaculture.[1] Seafood supports the livelihoods of 10–12% of the world population [1]. Seafood is considered part of a healthy diet, providing nearly 3 billion people with 20% of their dietary protein needs and

*Correspondence: dlove8@jhu.edu
[2] Department of Environmental Health and Engineering, Bloomberg School of Public Health, Johns Hopkins University, 615 N. Wolfe Street, Baltimore, MD, USA
Full list of author information is available at the end of the article

[1] We have limited our focus to federal fisheries because we are connecting federal fisheries policies to federal food policies. Furthermore, we concentrate on commercial fisheries because much recreational and subsistence fishing is regulated at the state level. To the extent federal recreational fishing regulations exist, comparable recreational catch data are not yet available for all states. Meanwhile, subsistence fishing that is governed by federal regulations is often governed under special community-oriented programs that are not easily comparable with standard fisheries governance structures.

micronutrients such as iodine, potassium, selenium, B vitamins, and vitamin D [1, 2]. The omega-3 polyunsaturated fatty acids (PUFAs) in seafood support childhood development and reduce risks from heart disease [3, 4]. Underscoring the importance of seafood, the United States (US) and other national dietary recommendations call for increasing seafood consumption, particularly fish high in omega-3 PUFAs and low in mercury [5, 6]. Yet at the same time, fisheries experts and health experts are raising questions about whether national dietary recommendations are in fact achievable, equitable, and sustainable, given what some see as a diminished seafood supply from the world's oceans and general movement of seafood products from low- and middle-income countries to high-income countries [1, 7–11]. The underexplored intersection of these policy spheres, with different concerns and goals, has significant implications for future scenarios in both health and sustainability.

For millennia, wild capture fisheries were often viewed as inexhaustible resources. Today, scientists believe that the global fishing fleet reached its "peak catch" in 1996 at 130 million metric tons (MMT), with subsequent harvests declining by about 1.2 MMT each year [12]. Others have challenged the idea that global fisheries collapse is as imminent as predicted, and authors of both competing views found a patchwork of recovery and decline in different fisheries and regions around the world [13]. A combination of factors is driving many (but not all) fish stocks to decline, including: increased human demand due to population growth and rising incomes; overfishing; use of better technology to catch seafood by large fishing fleets; illegal, unregulated, unreported (IUU) fishing; poor management decisions in some fisheries; and the impacts of climate change (e.g., increasing ocean acidification and ocean temperatures) and biodiversity loss [13–17]. Humans have entered a new epoch, the Anthropocene, where human influence permeates the planet and decreasing wild harvests affect global food security [12, 18].

Aquaculture, or aquatic animal and plant farming, has helped fill the gap between demand for seafood and declining wild fisheries [19]. Aquaculture has buoyed the global edible seafood supply and now contributes equal amounts to wild capture, in terms of seafood used for human consumption [20]. However, despite the different trajectories for aquaculture and wild capture fisheries, fed aquaculture (i.e., species requiring feed inputs) cannot be considered independently from wild capture fisheries or terrestrial crop production. Carnivorous farmed species like Atlantic salmon, trout, and marine shrimp still rely on feeds containing ingredients from wild-caught forage fish, such as anchovies, herring, and sardines, which tightly couples fisheries and aquaculture and introduces natural resource trade-offs [21, 22] that may impact food system resilience. The use of fishmeal and oil has greatly decreased in carnivorous farmed species, and the use of alternative feeds from terrestrial sources (e.g., soybean meal and oil, corn, other vegetable oils, animal byproducts) has increased in herbivorous, omnivorous, and carnivorous species. This links aquaculture with terrestrial agricultural systems and creates trade-offs that potentially impact the resilience of global food systems [23, 24]. Given these multiple connections, national policies related to fisheries and agriculture will have ripple effects on aquaculture production costs, and the human food supply and nutrition [23].

A recognition of the complex and interrelated set of challenges in the commercial seafood sector provides the contextual foundation for this paper. We follow these challenges as a means of exploring how current fisheries and health policies are additive, synergistic, or antagonistic with regard to seafood. Admittedly, we are necessarily juxtaposing different institutions' ideas and practices about best uses for seafood; as we will discuss later, ideas and practices in one US-based institution can be inconsistent with another, or even internally inconsistent, due in part to differing priorities reflecting divergent values and social beliefs. Throughout this paper, we take the perspective that fish (including finfish and shellfish) is food and, more specifically, that seafood is a key ingredient in a food system [25]. Our aim is to highlight the policy intersections of fisheries and health and raise important questions while exploring this juncture.

In this study, we ask two questions: (1) how would US federal fisheries policy be different if our fisheries were also managed with health outcomes for Americans as clear management objectives; and (2) how would US health policy be different if one of its goals was to support sustainable domestic fisheries and aquaculture? As we begin to explore these questions, looking at past and current activities within US public health and fisheries management, we provide illustrative examples of collaboration and disconnects between health and fisheries management communities at different levels of the food system (federal and state policies, corporate partnerships, and civil society). Finally, we discuss opportunities to align fisheries and health policy and suggest priority areas for further research.

Methods

The study began with an exploratory phase consisting of six months of regular dialogue among an interdisciplinary research team, composed of two public health experts (Love and Fry) and three fisheries experts (Pinto da Silva, Olson, and Clay). This dialogue gave study team members an opportunity to better understand each others' disciplines and identify gaps in knowledge across the disciplines. Following the exploratory phase, the study team formulated two overarching questions related to the interactions between fisheries and public health. To answer the questions, the research team identified relevant background information about health policy and fisheries policy through federal agency Web sites, communicating with experts/stakeholders, reading literature available on PubMed and Google Scholar, and sharing our expert knowledge. This information was compiled in the "Background," "Health policy," and "Fisheries policy" sections.

In the health policy section, data in Table 1 on US nutrition assistance programs were compiled using the US Department of Agriculture Web site and in consultation with Janice Fitzgerald, Household Programs Operations Branch Chief at the US Department of Agriculture, Food, and Nutrition Service. To identify American seafood consumption trends for Table 2, a search was performed in PubMed for the literature on seafood consumption using the NHANES dataset from the 1980s to present, and data from five sources were extracted and compiled. In "Fisheries policy" section, the annual commercial catch of forage fish in the USA was found in the National Marine Fisheries Service (NMFS) "Commercial Fisheries Statistics" database (in metric tons and value) and we report averages of yearly data from 2003 to 2013. Data on US edible and industrial seafood supply were procured from the 2013 edition of the NMFS annual report "Fisheries of the United States, 2013."

We then developed five case studies to provide readers with tangible examples of synergy between fisheries and public health. These case studies were carefully selected to represent different levels of the food system and different entry points into the supply chain. To write the case studies, the team conducted background research and informal interviews with an environmental advocate, a fishmonger, a seafood wholesaler, a university food service provider, a canned fish business owner, and a fishing industry representative. Notes from these interviews were compiled and drafted into case reports. We followed up with some interviewees to check facts and ask additional questions.

Finally, the study team jointly developed a set of recommendations and conclusions based on the research conducted. The team shared these recommendations with external reviewers, a fisheries economist and an agriculture/health policy expert, for comments and revisions.

Results and discussion

Health policy

Nutritional benefits and seafood safety

To understand health policy related to seafood, it is useful to understand its scientific underpinnings in two of the most important disciplines: nutrition and food safety. Among the most significant health benefits of seafood are

Table 1 US nutrition assistance programs that support seafood purchasing, 2013

Federal program	Approved seafood	People participating annually (millions)	Total cost (million US dollars)	Seafood cost (million)	Seafood quantity purchased (million pounds)
Supplemental Nutrition Assistance Program	Any seafood[a]	47.4	79,936	n/a	n/a
Child Nutrition Programs					
National School Lunch Program	Numerous products[b]	30.6	12,220	7.5	1.6
School Breakfast Program	n/a	13.2	3514		
Summer Food Service Program	n/a	2.4 (daily)	428		
Child/Adult Care Program	n/a	3.6 (daily)	2994		
Women, Infants, and Children Program	Canned chunk light tuna, pink salmon, sardines	8.7	6488	n/a	n/a
The Emergency Food Assistance Program		n/a	693	13.7	3.8
	Canned pink salmon			3.8	1.7
	Unbreaded catfish fillets			9.9	2.1
Commodity Supplemental Food Program	Canned pink salmon	0.58	203	3.0	1.3
Food Distribution on Indian Reservations Program	Canned pink salmon	0.09	119	0.43	0.2
National Services Incentive Program (elderly feeding)	Catfish, frozen fillet strips	0.076	100	0.19	0.04
Total		100.6	106,695	24.8	6.9

Data source: [102]

[a] Except hot food or food that can be eaten in the store

[b] Fresh/frozen clams, crabs, crawfish, fish fillets and steaks, fish portions, fish sticks, mackerel, oysters, scallops, squid, canned mackerel salmon, chunk light tuna, sardines, shrimp

Table 2 Consumption of seafood by Americans from 1988 to 2010, NHANES

Study	Per capita mean intake (g/week)		Among seafood consumers (past 30 days) mean intake (g/week)		Year
	Men (+19 years)	Women (+19 years)	Men (+19 years)	Women (+19 years)	
NCI [104]	138.9	99.2	–	–	2007–2010
Jahns et al. [62]	–	–	179.2	138.6	2005–2010
Papanikolaou et al. [103]	140.1	101.2	–	–	2003–2008
Tran et al. [105]	128.7	86.3	186	127.8	1999–2006
Wang et al. [106]	115.5	84.7	–	–	1999–2004
Wang et al. [106]	141.4	100.8	–	–	1988–1994

Data sources: [62, 103–106]

those derived from the omega-3 polyunsaturated fatty acids (PUFAs) present in many types of seafood. Fish high in omega-3 fatty acids include herring, mackerel, sardines, and salmon. In infants, there is strong evidence of gains in childhood development when pregnant and nursing mothers consume seafood, and limited evidence for a protective effect for asthma and eczema [3]. For adults, some of the best evidence that seafood consumption provides health benefits is the prevention of cardiac deaths in individuals with preexisting coronary heart disease [4]. There is limited evidence that seafood consumption later in life improves bone health and is protective against dementia; however, more studies are needed [2]. Seafood is a lean protein and can serve as a replacement for meat high in saturated fat. Importantly, fish oil supplements are not always associated with the same health benefits as eating a meal of a whole seafood product [3, 26]. In addition to nutrition, there are psychological, psychosocial, cultural, health, and food security benefits of consuming locally produced seafood [27–29].

Some of the health benefits of seafood are offset by food safety risks, such as naturally occurring biotoxins, metals, microbiological agents, other pollutants, and allergens present in some seafood. The most well-known contaminant associated with seafood is methylmercury (MeHg), a neurotoxin that enters the ocean primarily by deposition from burning fossil fuels and accumulates in the marine food web. Individuals eating large amounts of long-lived marine fish (e.g., shark, swordfish, king mackerel) and bottom-feeding fish (e.g., tilefish, catfish) are most at risk of negative health effects, particularly fetuses and infants exposed via their mother's intake of contaminated seafood [3]. Persistent organic pollutants (POPs), metals, and pesticides can be present in both saltwater and freshwater fish near urban centers, and people who practice subsistence fishing can be exposed to unsafe levels of pollutants [30, 31]. Biotoxins can include toxins produced by algae and incorporated into shellfish (i.e., paralytic shellfish poisoning) or fish (i.e., ciguatera), and naturally occurring marine bacteria (i.e., *Vibrio* spp.) that produce toxins in shellfish. Foodborne diseases from seafood are not uncommon, occurring most frequently through consumption of filter-feeding shellfish, raw fish products, or improperly handled or stored processed or imported seafood [32, 33].

On balance, the benefits of seafood outweigh the risks [34], which undergirds many health policies related to seafood. In fact, many policies are intended to increase seafood consumption to meet national dietary guidelines.

Policies related to seafood and health

There are many US health policies regarding seafood, primarily related to access and use of seafood but also food safety issues related to production and processing. To start this section, we familiarize readers with the federal agencies involved in seafood as it relates to human health, and describe each type of policy—be it dietary advice, labeling, or nutrition assistance. Most of these policies are inclusive of fisheries and aquaculture, and we note where differences exist. Additional detail, specifically on fisheries policy that affects seafood availability, is provided in the "Fisheries policy" section.

Overview of federal agencies involved in seafood as a food source Federal oversight of fisheries and aquaculture production is split among three agencies. The National Marine Fisheries Service (NMFS), which is a line office within the National Oceanic and Atmospheric Administration (NOAA), has regulatory purview over marine fisheries and aquaculture in federal waters and the Great Lakes [35, 36]. Freshwater and inland aquaculture is overseen by the US Department of Agriculture (USDA) [37]. Freshwater fishing outside of the Great Lakes is overseen by the Fish and Wildlife Service; however, their focus is not on food but rather on conservation.

Specific to aquaculture, the Food and Drug Administration (FDA) regulates veterinary drug use in farmed aquatic animals [38]. Aquaculture operations receive assistance from the USDA Animal and Plant Health Inspection Service (APHIS) for disease control, and live organisms originating from fish hatcheries are inspected by APHIS for diseases before sale across state or national borders [39]. The USDA is also in the process of developing organic standards and a USDA Organic label for aquaculture producers, which would limit the use of synthetic chemicals, certain feed ingredients, and some production methods; however, the draft regulation has not been released as of time of publication for this article [40]. Organic standards for aquaculture have come much later than organic standards for meat, dairy, fruits, and vegetables, in part because of the complexity and variety within aquaculture production methods and the relatively small size of the aquaculture industry in the US.

More generally, for all fisheries and aquaculture products, the FDA—along with support from NOAA, and USDA (for catfish)—inspects a small proportion of domestic and imported seafood and seafood processing plants [41]. NOAA will implement a new Seafood Import Monitoring Program focusing on traceability for 13 commercial species beginning in 2018 [42]. The USDA specifies grades and standards for various processed seafood products (such as canned or breaded processing standards) [43]. In some cases, physical or chemical inspection occurs, and in other cases, inspection of paperwork such as trade documents, permits, and hazard analysis and critical control points (HACCP) plans are reviewed by one of these agencies [44, 45]. Because shellfish carry a higher risk for causing disease, the FDA, in collaboration with states and the National Shellfish Sanitation Program, has special regulations concerning where shellfish can be grown or harvested in the US, and specific permits and regulations for shellfish processing and interstate trade [46]. Finally, at the end of the supply chain, state and local health departments inspect food preparation, restaurant, and retail locations to make sure they are meeting regulations for safe food handling. The Centers for Disease Control (CDC) also plays a role in food safety by assisting with outbreak investigations for foodborne disease [47].

Consumption-focused policies The Departments of Health and Human Services (HHS) and Agriculture (USDA) produce the US Dietary Guidelines every five years with dietary advice related to foods including seafood [5, 6]. The Dietary Guidelines are developed based on input from an advisory committee that reviews the current health and nutrition literature and makes recommendations for dietary patterns [48]. The Dietary Guidelines indicate that seafood intake is below recommended levels for all age groups and should be increased. In 2015, the Dietary Guidelines Advisory Committee made recommendations about sustainability, which mainly focused on reducing meat consumption but also included some discussion of fisheries and aquaculture. Sustainability language and recommendations were not included in the final report, despite broad support from the US public [49]. The Dietary Guidelines have an impact on Americans' diets via nutrition education, food purchasing done using federal funds, and nutrition assistance programs (described below).

Product labeling is an example of a consumer-facing policy. The only government-required form of product labeling for seafood (outside of nutrition and ingredient labeling) is country of origin labeling (COOL). COOL labels report whether the product is wild-caught or farm-raised and its country of origin. COOL is required for all minimally processed seafood sold in retail locations, and some consumers infer quality attributes from these labels [50]. Third-party labels, such as eco-labels, complement or fill gaps left by federal labeling laws. Food labels are important because they inform consumers about products in ways that are more direct and tangible than other approaches, such as methylmercury warnings, which are not present on seafood products.

Beginning in the early 2000s, the FDA, Environmental Protection Agency (EPA), and states began working together to advise consumers about methylmercury (MeHg) in both commercial and self-caught seafood. To help recreational anglers, the EPA worked with all 50 states to develop recreational seafood consumption advisories for the general public and women of childbearing age to reduce human exposure to MeHg and certain pesticides [51, 52]. The FDA and EPA also produced seafood MeHg advisories for retail seafood that target women of childbearing age and young children to reduce MeHg exposures [53]. Unfortunately, the original MeHg advisories had the unintended consequence of decreasing maternal seafood consumption, thereby reducing the potential benefits for these populations of eating seafood [54]. A more recent study indicates that the MeHg advisories' message may now be better received, as women of childbearing age are eating seafood but switching to products with lower levels of mercury [55]. The FDA and EPA are in the process of revising seafood advisories for MeHg in light of the expert consensus that the benefits of seafood consumption outweigh the risks [34, 56].

Nutrition assistance programs The USDA Food and Nutrition Service administers 15 nutrition assistance and nutrition education programs. Table 1 presents the amount and cost of seafood purchased under nutrition assistance programs by individuals (i.e., cash purchases) and by the USDA (i.e., entitlement commodity purchases). The largest program is the Supplemental Nutrition Assistance Program (SNAP, commonly known as food stamps), which in 2014 supported 46 million Americans with low incomes [57]. Individuals can spend SNAP funds to purchase any type of fresh, frozen, or processed seafood, as long as it is not a hot or prepared food that can be eaten in the store. The Nutrition Program for Women, Infants, and Children (WIC) distributes funds to eight million pregnant, low-income women and their children. WIC participants can use electronic benefit transfer (EBT) cards (similar to debit cards) to purchase canned light tuna, pink salmon, and sardines, among other food items [58].

There are also several entitlement programs in which state agencies select what types of commodity foods they want from a list of allowable products that the USDA then purchases and distributes to the states. Examples of these entitlement programs are the Emergency Food Assistance Program, which distributed over 600 million pounds of products to food banks in 2015 [59], the Food Distribution Program on Indian Reserves, which supports 276 tribes [60], and the Commodity Supplemental Food Program, which supports approximately 500,000 low-income elderly Americans annually. Rules require that the USDA purchase only domestically produced and processed foods under these entitlement programs. The only seafood product purchased by these entitlement programs currently is canned pink salmon from Alaska, with $7.2 million (3.2 million pounds) purchased in 2013 (Table 1). Canned tuna will be approved for purchase beginning in 2016. While catfish is an allowable item, it is not typically selected by states due to its expense; instead, items like catfish tend to be purchased using the USDA bonus program, which is used to correct oversupply in the market and does not count against state entitlement spending. In 2013, the USDA purchased $17.6 million (3.7 million pounds) of catfish (Table 1).

The Child Nutrition Program (CNP) is a hybrid program that provides both cash reimbursements for students to purchase school food and entitlement funds for schools to purchase that food. The National School Lunch Program (NSLP) is one example within the CNP; it provides five billion free and reduced-price lunches to 30.5 million school-age children through the age of 18 [61]. The list of allowable seafood products purchased through the NSLP is larger than any other nutrition assistance program (Table 1), and while schools are asked to purchase domestic food with their funds, they may choose imported products that are cheaper or more widely available. Cost is a major barrier for seafood purchasing by nutrition assistance programs, which may cause commodity purchasing programs and participants of SNAP or WIC to purchase more economical animal and plant proteins. It would be useful to identify the barriers, in addition to cost, preventing nutrition assistance programs from incorporating more domestic seafood into food banks and feeding programs. In addition, given the low consumption of seafood among women and low income Americans, consideration should be given to what role the federal government can play in encouraging seafood consumption by SNAP and WIC participants [62].

Consumption trends in the USA

Americans make food choices based on a variety of reasons. Seafood purchasing is primarily driven by consumer taste preference, cost, income, gender, age, and for some an interest in meeting national dietary recommendations [63, 64]. American seafood consumption is as diverse as our population; however, in aggregate, the most commonly eaten seafood products are whitefish (pangasius, domestic catfish, cod, pollock, tilapia), shrimp, farmed salmon, and canned tuna. The US imports 88% of the seafood it consumes, which is equally split among wild-caught and farmed sources [65]. Most popular seafood products (with the exception of tuna, tilapia, and pollock) have farmed and wild-caught versions available in stores. Yet among the top seafood products, only salmon is high in omega-3 PUFAs, the nutritional element that drives seafood consumption guidelines. Forage fish such as anchovies, herring, and sardines are high in omega-3 PUFAs, low in contaminants like mercury, and low on the food chain; however, these products rarely appear on nutrition assistance program lists and have decreased in popularity among American consumers over the last 50 years.

Indeed, the sheer number of nutrition and health policies described above appears to encourage access to seafood that is safe, affordable, low in mercury, and shelf-stable for key groups such as pregnant and breastfeeding women, children, the elderly, and individuals with low incomes. Nonetheless, from 1988 to 2010, Americans have consistently eaten much less seafood than government-recommended levels of 2–3 meals per week (227 g per week, or 8–12 oz per week) (Table 2) [5]. One study found that while four out of five Americans eat seafood on a monthly basis, most people (80–90%) were not meeting dietary guidelines [62]. A very low proportion of Americans meet dietary recommendations in other areas, too, such as fruit and vegetable intake [66]. Women, young people (19–30 years old), and individuals with lower levels of income and education all consume less seafood than the average American [62]. These findings highlight potential inequalities in access to seafood (and/or related factors such as preferences and familiarity) among certain groups, which could contribute to health disparities.

Moreover, federal health policy does not mention the goal to support sustainable domestic fisheries and aquaculture, or include language that indicates how food systems are linked to healthy fisheries. In this sense, fisheries and health policies are intrinsically linked while being effectively "siloed." Below, we explore US marine fisheries policies to understand how US wild seafood resources are currently managed, and consider what changes to federal fisheries policies/management might lead to optimal food production, human nutrition, and health outcomes for Americans.

Fisheries policy

Fisheries policy and management

Early efforts to develop US fisheries policy began over a century ago. In 1871, Congress passed the first legislation recognizing a Federal role in conservation of natural resources: the Joint Resolution for the Protection and Preservation of the Food Fishes of the Coast of the United States [Act of February 9, 1871, Sess. III, Res. 22, 16 Stat. 593–94 (1871)]. The resolution recognized that "the most valuable food fishes of the coast and the lakes of the United States are rapidly diminishing in number, to the public injury, and so as materially to affect the interests of trade and commerce." The US Fish Commission was created to protect, preserve, and study these "food fishes." After many decades of organizational changes and realignments, duties were divided between the U.S. Fish and Wildlife Service (FWS), which governs freshwater species, and the National Marine Fisheries Service (NMFS), which governs saltwater species. In 1970, NOAA was established and "directed to improve our understanding of the nation's living marine resources, the environment in which they are found, and the interaction between the two," and NMFS (thereafter also known as NOAA Fisheries) was placed within NOAA [67].

The primary law governing marine fisheries management in federal waters is the Magnuson–Stevens Fishery Conservation and Management Act (MSA; 16 U.S.C. ß 1801 et seq.). While the MSA refers to the importance of fish as food in its definition of optimum yield (OY),[2] the first of ten National Standards included in the document, fisheries regulations generally focus on rebuilding overfished stocks and ensuring that overfishing is not occurring through controls on inputs (gear restrictions/area closures) and/or outputs (limits on catch/landings) [25], but do not consider what happens to fish after it is landed or how regulatory actions might impact nutrition or food systems [68]. The US Fish and Wildlife Service regulates freshwater fishing outside of the Great Lakes and operates fish hatcheries, fishery resource offices, and ecological services field stations [69], where the primary concern is conservation of resources and not fish as food. All federal agencies must comply with the National Environmental Policy Act (NEPA), which requires agencies to assess the environmental impacts of and alternatives to proposed actions. These assessments are conducted and shared with stakeholders and the public as Environmental Assessments (EA) or, when it is determined that a Federal action will have a significant effect on the human environment, Environmental Impact Statements (EIS). For example, EPA reviewed a large gold mine planned upstream of Bristol Bay, Alaska, and found reason for concern due to potential impacts to fisheries and the livelihood or cultural identity of fishermen. Human health could be impacted by stress from loss of income or identity; however, health issues related to how the fish are marketed after landing was not considered. In this case, the EPA conducted an EA and determined that a more detailed EIS was not required.

Fisheries policy extends to aquaculture as well, especially regarding use of wild-caught forage fish as aquaculture feed (discussed below); however, our discussion of federal aquaculture-specific policy (i.e., developed by the NOAA Office of Aquaculture or in the NMFS Aquaculture Strategic Plan) is limited in this document. NMFS directs attention to aquaculture in federal waters, where it is primarily concerned with "developing and implementing policies that enable marine aquaculture and works to ensure that aquaculture complies with existing federal laws and regulations that NOAA enforces under its marine stewardship mission" [70]. The USDA gathers data on economics and animal health of US aquaculture products, conducts a census of aquaculture operations every five years [71], and tracks US imports/exports of aquaculture products [72]. It is important to note that US government agencies are working to expand the domestic aquaculture industry, which currently contributes less than one percent of global aquaculture production [20, 73, 74]. Therefore, an opportunity exists to incorporate sustainability and health outcomes, as well as alternative supply chains, as the industry grows. In addition, integration with food systems, public health, and alternative supply chains may also serve to expand the number of stakeholder groups supporting aquaculture development in the US.

Outcomes and markets

Seafood is the most valuable traded food commodity in the global marketplace, more valuable than coffee, tea, or spices [1]. The US is a major player in global seafood trade, routinely buying and selling to/from over

[2] The amount of fish which will provide the greatest overall benefit to the nation, particularly with respect to food production and recreational opportunities, and taking into account the protection of marine ecosystems; is prescribed as such on the basis of the maximum sustainable yield from the fishery, as reduced by any relevant economic, social, or ecological factor; and in the case of an overfished fishery, provides for rebuilding to a level consistent with producing the maximum sustainable yield in such fishery [16 United States Code (USC) §1802(33)].

Table 3 Annual catch of forage fish in the US, mean of 2003–2013

Forage fish species (region)	Metric tons	Value (USD)
Anchovy (Northern)	6796	883,554
Mackerel (Scomber)	4	4750
Menhaden (Atlantic) (Brevoortia)	659,579	93,930,294
Sardine (Pacific)	80,682	12,286,746
Sardine (Spanish)	591	228,673
Herrings (Atlantic)	72,694	32,398,361
Herring (Atlantic Thread)	645	236,890
Herring (Blueback)	9	10,550
Herring (Lake or Cisco)	397	554,303
Herring (Pacific)	38,761	17,262,876
Herrings	487	217,018
Total forage fish	860,645	158,014,015
Total US landings	4,174,308	4,399,172,575

Data source: [107]

100 countries in multiple regions of the world [75]. The major trading partners for the US are Asia and Canada [76]. This trade includes two types of products that are important to differentiate: edible products and industrial products. Edible products are what we commonly think of as the seafood that is sold in grocery stores and restaurants. Industrial fisheries products, made primarily from forage fish and fish scraps, are used as animal feed, fish feed, or fertilizer. Forage fish are small oily fish that live in schools within the water column along coasts, oceans, and lakes and are caught with large nets. In the US, over 800,000 metric tons of forage fish are caught each year, which is a fifth of US landings by weight, but just 4% by value (Table 3). Forage fish caught in the US are often turned into fishmeal and fish oil. It is difficult to convert quantities of forage fish into amounts of fishmeal and oil produced because rendering removes water weight, but NMFS reports that the US produces over 200,000 metric tons of fish meal and 70,000 metric tons of fish oil each year [1].

Once aquatic animals are removed from a body of water, they enter seafood supply chains consisting of hundreds of processors, wholesalers, distributors, and retailers that employ 1.3 million full- and part-time workers in the US [77]. Edible and industrial fisheries products are both traded, but in different ways. For example, the US imports more edible products than it exports (i.e., a negative trade balance), while the opposite occurs with industrial products—the US exports three times more industrial products than it imports (Table 4). Overall, these trading patterns have led to a food system where nine-tenths of the US edible seafood supply is imported, sometimes as US fish that has been processed abroad, and a feed system where large amounts of forage fish are extracted from US territorial waters and exported. As we will explore later in this paper, these complex interactions between supply and demand in the use of natural resources have important outcomes for human health, as well as implications for society, the economy, and ocean ecosystems.

The Atlantic herring fishery, formerly a food fishery and now a bait fishery for the New England lobster fishery, illustrates the evolution of social and economic outcomes influenced by fisheries policy. Small herring used to be canned in Maine in the nineteenth and twentieth centuries, peaking at 400 canneries that employed thousands. The last cannery in Maine closed in 2010 [78]. Changing markets and consumer preferences played some role, for example, in the dietary replacement of canned herring and sardines with canned tuna, but how Atlantic herring is caught also determines what it is used for. Fishing methods have changed since herring was used as a human food and changes in fishing methods and technology have accompanied changes in the management and use of herring. Early methods (traps called weirs and relatively small nets called seines) caught herring without damaging the fish, which was ideal for using herring as human food. Over time, regional fisheries management council regulations have allowed large trawlers and pair trawlers (trawlers working together with a net between them), to become the dominant gear types; these vessels pump large catches out of nets into onboard refrigerated seawater tanks. Fish caught this way are not marketable as human food, as they are often damaged during the process of extraction. Concurrent with these technological shifts, Atlantic herring shifted from being a food fish to a baitfish, used almost exclusively as bait for today's American lobster fishery (Pinto da Silva, Olson and Benjaman, in review). While herring was once a low-cost source of high quality protein and other nutrients, the fishery it now supports, lobster, is a high-cost (i.e., less affordable) food source. While these changes from food to bait have been exacerbated by a confluence of economic, social, technological, and cultural shifts, regulatory measures have likely contributed to this situation as well by not considering the impacts of regulatory changes on the regional and national food system.

Examining management and the use of herring and other forage fish raises the following questions: Should forage fish be managed in a way that recognizes the benefit of leaving a certain amount in the ocean to provide food for other fish higher on the food chain? Should people eat these fish instead of using it for industrial purposes, including as aquaculture feed or bait, where there is a conversion loss of protein and biomass? Herring, as well as menhaden and other forage fish, expose some of the trade-offs that are inherent in fisheries management,

Table 4 US edible and industrial seafood supply in 2013 (million metric tons)

Fisheries and aquaculture product	Domestic landings	Domestic aquaculture	Imports	Exports	Total supply
Edible products	3.7	0.25	4.8	3.2	5.5
Industrial products	0.83	n/a	0.27	0.84	0.26
Total	4.5	0.25	5.0	4.0	5.7

Data source: [65]

trade-offs that will only be increasingly highlighted as regulatory agencies begin to grapple with ecosystem-based approaches to management. The important point in this context, however, is that how we choose to manage our fisheries will also influence the food supply in the US and beyond, making fisheries policy part of US food policy and requiring more explicit recognition of the intersection of, and especially the trade-offs between, fish, food, and health policy. As the following examples help demonstrate, expanding the universe of those involved with fisheries management can shift the objectives of concern and change the questions stakeholders are asking.

Synergy between fisheries and health policy, illustrative examples

Here, we present four illustrative examples of policies, initiatives, and business models where fisheries and health policies are purposefully connected. Each example targets a different level (federal and state government, civil society, economy) as a way to acknowledge that hybrid, multi-level interventions are needed to solve complex problems. The examples we selected are: anchovy consumption in Peru; an Alaskan government program to purchase local seafood for public school lunches; a public–private partnership between a New England university, seafood distributor, and nonprofit working to support local fishers; and a California-based canned seafood business.

Social marketing: Peruvian anchovies, changing culture takes time

Along the West Coast of South America in the Pacific Ocean lies the Humboldt Current, where a rich upwelling of nutrients supports an abundance of marine life [79]. As a result, Peru has some of the best fishing grounds for small pelagic fish, including anchovies and sardines. At the peak of anchovy fishing in 1970, 12 million metric tons of anchoveta, a type of anchovy, were harvested off the coast of Peru, constituting an astounding 20% of all global fish harvests [80]. Peruvian anchoveta stocks have fluctuated due to El Niño conditions and overharvesting, but most recent statistics indicate that they provide a third to half of the world supply of both fishmeal and oil [81]. Peruvian anchoveta are almost exclusively harvested for export as a feed ingredient (as fishmeal and oil). Only 2% of Peruvian anchoveta are used for human consumption [79] despite food insecurity and micronutrient deficiencies in Peru and surrounding countries [11, 79].

In 2006, the Oceana Center for Environmental Sustainability (CES) at the Cayetano Heredia University in Lima set out to reshape the way anchoveta were used in Peru by promoting them as human food via the campaign "Semana de la Anchoveta" (Anchovy Week) (personal communication, Patricia Majluf, Oceana). Working with Peru's top celebrity chef on a nationally recognized TV cooking show, CES was able to broadcast their message about the sustainability, health benefits, and flavor of Peru's national fish, the anchoveta. The story caught the attention of the president of Peru, and after a meal with him at the Presidential Palace, the campaign had the backing of the government. A subsequent law allocated 8% of Peru's government spending on food toward anchoveta. Anchovy Week was repeated in 2007, 2009, and 2012, though the founder of the campaign admits it never reached her expectations in terms of public participation (personal communication, Patricia Majluf, Oceana). A lesson we can draw from Anchovy Week is that enacting food system changes requires adjustments not only in policy, but also throughout the supply chain, and among consumers. For example, the anchoveta fishing quota is solely based on reduction fishmeal uses, not producing food on food-grade anchovy. If vessels in the industrial fleet could use their catch for human consumption, the ensuing supply could be at a scale that would make production for human consumption economically viable. Fresh and frozen product forms, and new more convenient presentation formats (i.e., dried, minced) are needed to attract more consumers beyond those currently eating canned anchovies. These issues may include environmental, social, and economic trade-offs [81] and require unraveling and replacing current incentives in the fisheries policy that maintain the status quo [82].

State-level policy: Farm to School programs

In the section on health policies, we describe several nutrition assistance programs that include seafood. One innovative method for introducing seafood into nutrition assistance programs is through the National School Lunch Program (NSLP) using a concept called "Fish to School" (related to "Farm to School" programs). The State of Alaska has been proactive in navigating the bureaucratic requirements that must be fulfilled in order to source, purchase, and serve local fish like pollock, salmon, cod, halibut, and rockfish in school cafeterias [83]. Like other states, the Alaska state government requests that USDA purchases bulk quantities of commodity seafood (mainly canned pink salmon and catfish) for entitlement programs and through schools with CNPs. Alaska also provides grants to reimburse schools that purchase Alaskan-caught seafood through the Nutritional Alaskan Foods in Schools (NAFS) program. The NAFS helps schools better align meals with the national dietary guidelines and has its underpinnings in a 2010 state law to promote local agriculture in public schools [84]. The Alaska Fish to School effort aims to improve student health via seafood consumption, help students understand where their food comes from, provide traditional foods for schools, and acknowledge the cultural and economic significance of purchasing local seafood [83]. A study of the Alaska Fish to School program found that school fish meals were well aligned with the local foodways of the community, and purchasing local seafood engendered goodwill with parents and the community [84]. Challenges for schools implementing Fish to School in Alaska were finding seafood suppliers, ordering appropriate quantities of fish, and finding suitable recipes [84]. Implementation of the program also requires that there is demand for new foods, that they are available domestically from a pool of vendors, and that there is capacity within industry to provide necessary quantities. Oregon, Massachusetts, and New Hampshire are also working on similar Fish to School programs [85–87], which highlights an opportunity for repeating these innovative programs in other coastal states.

Food procurement: connections among universities, hospitals, and regional fisheries

Universities, hospitals, and other groups with institutional food procurement can play a key role in connecting local fisheries to improved diets [88]. Universities and hospitals are increasingly interested in the sustainability of their physical campuses as well as of the services they offer. Sustainability has become a "core value" of these institutions that requires mainstreaming throughout their operations. As part of this process, most universities have developed sustainability plans that cover a range of issues from procurement and building practices to improving energy and water use efficiency and reducing their overall carbon footprint. Food service procurement is often seen as a component of these plans. Some universities are rethinking their relationships with local economies as well as their ability to provide fresh local health food options for students. While the focus of these efforts has been on the local and regional procurement of agricultural products, some institutions are connecting with local fisheries.

In Boston, Massachusetts, improved technology around fish marketing is enabling regional universities (such as Boston University, Harvard University, and the University of Connecticut) to buy thousands of pounds of locally caught fish at competitive prices. Red's Best, an innovative seafood dealer, provides fish caught on day boats from Rhode Island to Maine directly to universities through negotiated partnerships [89]. By using QR codes that track where, when, and who caught the fish, universities are able to confirm the chain of custody and ensure that they are purchasing seafood caught locally or regionally. While universities continue to purchase much of their seafood from national distributors, purchases from Red's Best are growing as a share of total university seafood purchases in participating institutions.

Hospitals are also reflecting on the social and environmental impacts of their purchases. The quality as well as the source of hospital food can influence regional agricultural systems and markets for regionally caught seafood, not to mention health outcomes for patients. Like universities, hospitals are also developing sustainability plans, which often include increasing the amount of locally sourced food as a component of sustainability. Healthcare Without Harm is an international coalition of hospitals, community groups, environmental health organizations, and others that are working together to improve the overall contribution to health that a hospital can make beyond the medical services offered to patients [90]. The food procurement components of these plans have most often focused on agricultural products, leaving seafood sourcing to traditional mechanisms. However, through partnerships with local NGOs, some hospitals in New England are engaging with their suppliers on where their seafood comes from, as well as purchasing seafood directly through local community-supported fisheries (CSFs) [91], which are currently being encouraged in the Northeast USA through support from the Northwest Atlantic Marine Alliance (NAMA). These partnerships are helping to achieve "optimum yield" in fisheries by encouraging the purchase of US caught seafood in regional institutions.

Private sector: business opportunity

Bill Carvalho is the founder of Wild Planet, a canned seafood company. In 2001, a tour of the Monterey Bay Aquarium led to a realization of the economic opportunities available in exclusively selling sustainably harvested seafood, including forage fish, that are low in mercury and high in omega-3 PUFAs (personal communication, Bill Carvalho) In 2004, Wild Planet began selling canned tuna that met these standards, and later added canned sardines, anchovies, and mackerel. In 2008, Wild Planet received backing from a private equity firm involved with the Packard Foundation, enabling it to grow into one of the only American companies selling domestic forage fish, among other products, to the domestic mass market. Wild Planet products are now featured in many major grocery store chains in the US alongside traditional canned seafood.

The company's emphasis on sustainability has meant that product sourcing and supply chain management require constant attention. Tuna is purchased directly from US tuna fishermen or from the Japanese pole and troll fleet. Mackerel is also purchased from Japan. Originally, sardines were purchased from West Coast bait vendors; however, unreliable sizes of fish and a recent moratorium on sardine harvests meant the company had to source sardines primarily from Japan, and the USA as available. Most of Wild Planet's fish are canned in Vietnam, in part because the labor prices are more competitive, but also because Carvalho has had difficulty trying to revive the US West Coast seafood canning industry. Wild Planet also sources and cans anchovies in Peru, in an effort to support the Peruvian food fish canning industry. The story of Wild Planet highlights the international nature of the seafood supply, and the challenges for one company that attempts to source and process environmentally sustainable domestic seafood that is of high nutritional value.

Recommendations

US national health policy can have a significant impact on seafood access, utilization, and consumer seafood purchasing and consumption patterns. However, many of these policies, regulations, and guidance documents have been developed without any cross-integration to fisheries policies and regulation. As a result, government efforts to improve population health miss many opportunities for sourcing domestic, sustainably caught or farmed seafood. Others have raised similar concerns about misalignment of nutrition policy and agriculture policy [92].

Fisheries policy is also detached from, but integral to, public health. Fisheries and aquaculture policies refer frequently to the importance of fish as food for Americans; however, the emphasis of fisheries regulations and the science produced to support it have been on sustaining the resource and reducing impacts on fishing businesses without considering the broader outcomes, such as the production of food for the nation, that are defined in the MSA as necessary to achieve not just maximum sustainable yield but the overarching goal of optimum yield, or OY [MSA Sect. (3)(33)].[3] There is also a general disconnect between fisheries and consumers, such that the average citizen is often unaware of where their seafood comes from [93].

Aligning fisheries and health policies is imperative to achieving food security. Change will need to occur at multiple levels of policy and across various domains and dimensions of the food system (e.g., social, political, biophysical, economic) [94]. Institutions including seafood-related businesses, hospitals, state government agencies, NGOs, and academia can also play a role in better weaving together fisheries and health priorities (as we describe in "Synergy between fisheries and health policy" section). Drawing from these examples and our own professional experiences, we have developed a set of recommendations for integrating fisheries, aquaculture, and health policies.

The following bullets describe a number of ways to connect US health policy to fisheries and aquaculture policy, within the context of food production.

- *Strengthen federal agency linkages* Government agencies often work on similar topics but approach them from different disciplines or have divergent outcome measures. We see the following opportunities for stronger agency linkages: (1) integrate seafood consumption and nutrition guidance from other federal agencies with NOAA Fish Watch, a national database for sustainable seafood [95]; (2) include sustainable sourcing in the US Dietary Guidelines, with input from NMFS and others on seafood sustainability; (3) coordinate activities between FDA's seafood import inspection program and NOAA's seafood traceability initiative; (4) integrate state advisories for recreational fishing with state guidelines for self-caught fish consumption with support from EPA, FDA, NMFS, and FWS [52]; (5) connect seafood to the Obama White House backed initiative "Local Foods, Local Places" that connects multiple agencies around agricultural products [96]; and (6) move toward a national food policy that recognizes how fisheries and agriculture policy can support human health and the environment [97].
- *Use federal agency purchasing power* The federal government has substantial purchasing power, which could be leveraged to purchase sustainable seafood through: (1) increasing the diversity of bulk domestic

[3] Some economists have also argued for the inclusion of maximum economic yield (MEY) in definitions of OY. However, given that MEY leads to lower fishery yields than MSY, this risks contradicting the requirement to consider food production (see [108]).

seafood products included in USDA Nutrition Assistance Programs; (2) supporting states in developing "Fish to School" programs to purchase domestic sustainably harvested and farmed seafood with National School Lunch Program funds; and (3) changing institutional food procurement polices [88] (e.g., the Federal Food Service Purchasing Guidelines) for food sold at federal buildings and in the military so they support sustainably harvested and farmed seafood.

- *Foster federal agency investments in people* Investments are commonly made by NMFS on the harvesting side of fisheries—vessel buybacks, and payments for involvement in cooperative research. But sustainable fisheries and sustainable seafood require new skills and relationships for fishermen and others involved in the supply chain in order to build new connections that will enable selling locally, regionally, to institutions, or using other supply chains that serve as an alternative to the current high import/export model. Innovations in fisheries, such as traceability technology, community-supported fisheries, and direct marketing, could be expanded rapidly with targeted government funds.
- *Shift assumptions about fisheries management* (1) Expand the scope of fisheries management's general consideration of fish as being a food to more specifically include food systems and supply chains that form part of the broader socio-ecological connections integral to the viability and resilience of fishing businesses [98] and to think of fishing communities as part of this system rather than simply the home of fishermen or a landing site for vessels. This is a radical shift from the knowledge and training that most individuals who work in fisheries possess, and would require new collaborations. (2) Recognize that markets are not always the most effective way to distribute seafood to achieve MSA objectives beyond revenue and profit. In some fisheries, markets may serve objectives of food production and regional distribution better than others. Fisheries assessments will need to recognize that achieving optimum yield requires consideration of factors beyond fish stock sustainability and social and economic impacts to the harvest sector and fishing communities. Given that food is one of the elements of optimum yield, NMFS must consider the broader food system and seek partners who can operate in realms beyond its limited regulatory purview.
- *Change how success is measured* How we measure success in fisheries management matters. Focusing on ecological sustainability (the Fisheries Stock Sustainability Index or FSSI, for example) is an incomplete picture of whether the greatest benefits to the nation are being achieved. A stock might be in good health but might be under performing in terms of its contribution to the food system. Fishery performance measures could overlap with other government agency performance measures to reinforce and clarify connections between food policy and fisheries policy, expanding beyond the harvest sector to include seafood distribution, food access, supply chain, and consumption patterns.
- *Foster new social science research* Fisheries social science research has focused primarily on the harvesting sector and fishing communities. New areas of research that combine traditional social science fisheries research with food systems, health policy, and social justice research are necessary. Additional studies are needed on, for example: consumer preferences and consumption patterns with respect to lesser-known seafood products harvested from domestic sustainable fisheries and domestic aquaculture operations; interventions for reducing consumer seafood waste, such as clear package labeling or education [99]; and policy studies of how the Affordable Care Act can be leveraged to purchase local seafood for private hospitals and healthcare facilities [100]. We need to better integrate data on seafood access, consumption, and local markets with fisheries EAs and EISs. These research projects will necessarily involve collaborations across disciplines and research specialties.
- *Create new relationships* The perspectives used to inform the fisheries management process will need to be broadened to reflect the range of stakeholders associated with the larger fisheries system and food system. Currently, Fishery Management Council meetings are attended mainly by harvester organizations, environmental NGOs and, to differing degrees by region, processing representatives. Consumer groups, public health professionals, civil society organizations, and local seafood marketers are largely absent from the collaborative management process, in part at least because they are not targeted for announcements of these meetings and may therefore be unaware of them. Increased outreach by Regional Fishery Management Councils might help to alleviate this. One promising evolution in the Northeast Region is the creation of a Seafood Marketing Group within the NMFS Greater Atlantic Regional Fisheries Office [101] that brings together processors, harvesters, chefs, scientists, educators, state food system representatives, and others to consider how to add value to the fish harvested in the region. Another new relationship worth exploring and supporting in a broader context is a joint policy statement from

health professional associations (American Public Health Associations, American Medical Association, and American Society of Nutrition) and fisheries and aquaculture associations (American Fisheries Society, US Aquaculture Association) about the importance of synergy between the fisheries and health fields.

Conclusions

Managing fisheries to promote optimal nutrition and efficient food production likely requires a different approach to fisheries management—new outcomes will need to be monitored, new approaches found and fisheries, aquaculture, and health policies better integrated. The Magnuson–Stevens Act as well as NMFS strategic guiding documents all highlight the importance of fish as food, as do the USDA/HHS nutritional guidelines. However, the implementation of the MSA has not focused on food production outcomes. What happens to fish beyond the harvesting and processing level is currently beyond the scope of factors considered by the councils. Council decisions are informed by impact assessments that focus on harvesters, fishing businesses, and to some degree on impacts to fishing communities.

Simultaneously, while health policies such as the US Dietary Guidelines indicate the need to consume more fish, little consideration is given to the sustainability or source of fish, their connections with US fisheries systems, and global distribution of seafood. Nonetheless, some states have implemented "Fish to School" programs in school districts, and more opportunities for bulk purchasing of sustainably caught or raised seafood exist in nutrition assistance programs and institutional food-buying programs. Seafood businesses, universities, hospitals, NGOs, and other actors are also beginning to make connections between public health, fisheries, and aquaculture. These efforts are a positive sign because change will need to occur at multiple levels and across various domains (federal, state, and local) and dimensions of the food system (e.g., social, political, biophysical, and economic). Aligning fisheries, aquaculture, and health policies should begin first where the most promising opportunities exist. There are numerous small steps to begin this process with shared responsibility among government, industry, and civil society. We also call on researchers to continue addressing questions and issues raised here, and develop and test theories of change that move us in a direction toward a more just, equitable, and sustainable food system.

Abbreviations

APHIS: Animal and Plant Health Inspection Service; CDC: Centers for Disease Control; CNP: Child Nutrition Program; DOC: Department of Commerce; HHS: Department of Health and Human Services; EPA: Environmental Protection Agency; EEZ: Exclusive Economic Zone; FWS: Fish and Wildlife Service; FDA: Food and Drug Administration; HACCP: hazard analysis and critical control point; IUU: illegal, unregulated, unreported; MSA: Magnuson–Stevens Fishery Conservation and Management Act; MeHg: methylmercury; MMT: million metric tons; NMFS: National Marine Fisheries Service; NOAA: National Oceanic and Atmospheric Administration; NSLP: National School Lunch Program; NAMA: Northwest Atlantic Marine Alliance; WIC: Nutrition Program for Women, Infants, and Children; PUFAs: omega-3 polyunsaturated fatty acids; POPs: persistent organic pollutants; SNAP: Supplemental Nutrition Assistance Program; USDA: Department of Agriculture; US: United States.

Authors' contributions

All authors contributed to early brainstorming, developing the study questions, designing the study, and outlining the manuscript; DCL and PPdS wrote a large portion of the manuscript with help from PMC and JO; DCL and JPF developed tables and figures; and all authors reviewed and edited the final manuscript. All authors read and approved the final manuscript.

Author details

[1] Johns Hopkins Center for a Livable Future, Johns Hopkins University, 615 N. Wolfe Street, Baltimore, MD, USA. [2] Department of Environmental Health and Engineering, Bloomberg School of Public Health, Johns Hopkins University, 615 N. Wolfe Street, Baltimore, MD, USA. [3] Social Sciences Branch, Northeast Fisheries Science Center, National Marine Fisheries Service, 166 Water Street, Woods Hole, MA, USA. [4] Department of Health, Behavior, and Society, Bloomberg School of Public Health, Johns Hopkins University, 624 N. Broadway, Baltimore, MD, USA.

Acknowledgements

The authors would like to thank the following individuals for their helpful discussion: Dr. Patricia Majluf, Vice President of Oceana Peru; Paul Johnson, owner of Monterey Fish Market; Bill Carvalho, owner of Wild Planet; and Janice Fitzgeral, Household Programs Operations Branch Chief at the US Department of Agriculture, Food, and Nutrition Service. The authors also thank Eric Thunberg and Michael Simpson of NOAA Fisheries, Shawn McKenzie and Robert Martin at the Johns Hopkins Center for a Livable Future, and Anthony So at the Johns Hopkins Bloomberg School of Public Health for reviewing a draft of the manuscript.

Competing interests

The authors declare that they have no competing interests.

Declarations

The authors received permission from participants in the case studies to publish their data.

Funding

Support for DCL and JPF was provided by the Johns Hopkins Center for a Livable Future with a gift from the GRACE Communications Foundation. PMC, JO, and PPdS were supported by NOAA Fisheries, or NMFS. The funders had no role in study design, data collection and analysis, decision to publish, or preparation of the manuscript.

References

1. Food and Agriculture Organization of the United Nations. The state of world fisheries and aquaculture 2014. 2014: Rome.
2. Weichselbaum E, et al. Fish in the diet: a review. Nutr Bull. 2013;38(2):128–77.

3. Mahaffey KR, et al. Balancing the benefits of n-3 polyunsaturated fatty acids and the risks of methylmercury exposure from fish consumption. Nutr Rev. 2011;69(9):493–508.
4. Sanders TA. Protective effects of dietary PUFA against chronic disease: evidence from epidemiological studies and intervention trials. Proc Nutr Soc. 2014;73(01):73–9.
5. U.S. Department of Health and Human Services and U.S. Department of Agriculture. 2015–2020 Dietary guidelines for Americans. 2015.
6. U.S. Department of Health and Human Services and U.S. Department of Agriculture. Dietary guidelines for Americans, 2010. 2010.
7. Hilborn R, et al. State of the world's fisheries. Annu Rev Environ Resour. 2003;28(1):359.
8. Jenkins DJ, et al. Are dietary recommendations for the use of fish oils sustainable? Can Med Assoc J. 2009;180(6):633–7.
9. Brunner EJ, et al. Fish, human health and marine ecosystem health: policies in collision. Int J Epidemiol. 2009;38(1):93–100.
10. Greene J, et al. Fish oils, coronary heart disease, and the environment. Am J Public Health. 2013;103(9):1568–76.
11. Golden C, et al. Nutrition: fall in fish catch threatens human health. Nature. 2016;534(7607):317.
12. Pauly D, Zeller D. Catch reconstructions reveal that global marine fisheries catches are higher than reported and declining. Nat Commun. 2016;7:10244. doi:10.1038/ncomms10244.
13. Worm B, et al. Rebuilding global fisheries. Science. 2009;325(5940):578–85.
14. Worm B, et al. Impacts of biodiversity loss on ocean ecosystem services. Science. 2006;314(5800):787–90.
15. Rockström J, et al. A safe operating space for humanity. Nature. 2009;461(7263):472–5.
16. Pauly D, et al. Fishing down marine food webs. Science. 1998;279(5352):860–3.
17. Costello C et al. Global fishery prospects under contrasting management regimes. In: Proceedings of the national academy of sciences, 2016. p. 201520420.
18. Waters CN, et al. The Anthropocene is functionally and stratigraphically distinct from the Holocene. Science. 2016;351(6269):aad2622.
19. Tidwell JH, Allan GL. Fish as food: aquaculture's contribution. EMBO Rep. 2001;2(11):958–63.
20. FAO. State of the world fisheries and aquaculture in 2014. Rome: United Nations FAO Fisheries and Aquaculture Department; 2014.
21. Naylor R, Burke M. Aquaculture and ocean resources: raising tigers of the sea. 2005;30:185–218.
22. Metian AGTM. Fishing for feed or fishing for food: increasing global competition for small pelagic forage fish. AMBIO J Hum Environ. 2009;38(6):294–302.
23. Fry JP, et al. Environmental health impacts of feeding crops to farmed fish. Environ Int. 2016;91:201–14.
24. Troell M, et al. Does aquaculture add resilience to the global food system? Proc Natl Acad Sci. 2014;111(37):13257–63.
25. Olson J, Clay PM, Pinto da Silva P. Putting the seafood in sustainable food systems. Mar Policy. 2014;43:104–11.
26. He K, et al. Accumulated evidence on fish consumption and coronary heart disease mortality a meta-analysis of cohort studies. Circulation. 2004;109(22):2705–11.
27. Loring PA, Gerlach SC. Searching for progress on food security in the North American North: a research synthesis and meta-analysis of the peer-reviewed literature. Arctic. 2015;68(3):380–92.
28. Kittinger JN, et al. From reef to table: social and ecological factors affecting coral reef fisheries, artisanal seafood supply chains, and seafood security. PLoS ONE. 2015;10(8):e0123856.
29. Islam D, Berkes F. Indigenous peoples' fisheries and food security: a case from northern Canada. Food Secur. 2016;8(4):815–26.
30. Tilden J, et al. Health advisories for consumers of Great Lakes sport fish: is the message being received? Environ Health Perspect. 1997;105(12):1360.
31. Steinback S, Wallmo K, Clay P. Saltwater sport fishing for food or income in the Northeastern US: statistical estimates and policy implications. Mar Policy. 2009;33(1):49–57.
32. Gould LH, et al. Surveillance for foodborne disease outbreaks—United States, 1998–2008. MMWR Surveill Summ. 2013;62(2):1–34.
33. Lipp E, Rose J. The role of seafood in foodborne diseases in the United States of America. Revue Scientifique et Technique—Office International des Epizooties. 1997;16(2):620–40.
34. Nesheim MC, Yaktine AL. Seafood choices: balancing benefits and risks, ed. I.o. Medicine. 2007: National Academies Press, Washington, DC.
35. NOAA Fisheries. 2016 April 12, 2016. http://www.nmfs.noaa.gov/.
36. NOAA Office of Aquaculture. 2016. http://www.nmfs.noaa.gov/aquaculture/.
37. USDA. Aquaculture. 2016 April 12. 2016. http://www.usda.gov/wps/portal/usda/usdahome?navid=AQUACULTURE.
38. FDA. Animal and veterinary. 2016. http://www.fda.gov/AnimalVeterinary/.
39. USDA-APHIS. Focus on fish health. 2016 April 12, 2016. https://www.aphis.usda.gov/focusonfish/aphis-aquaculture-program.php.
40. USDA. Organic aquaculture. 2016 April 12, 2016. https://afsic.nal.usda.gov/aquaculture-and-soilless-farming/aquaculture/organic-aquaculture.
41. Love DC, et al. Veterinary drug residues in seafood inspected by the European Union, United States, Canada, and Japan from 2000 to 2009. Environ Sci Technol. 2011;45(17):7232–40.
42. NOAA, Magnuson–Stevens Fishery Conservation and Management Act Seafood Import Monitoring Program. 2016. pp. 88975–88998.
43. USDA. Meat, poultry, fish, and shellfish CIDs. n.d. April 12, 2016. https://www.ams.usda.gov/grades-standards/cid/meat.
44. FDA. The imported seafood safety program. 2016 April 12, 2016. http://www.fda.gov/Food/GuidanceRegulation/ImportsExports/Importing/ucm248706.htm.
45. FDA. Guidance for industry: HACCP regulation for fish and fishery products; questions and answers for guidance to facilitate the implementation of a HACCP system in seafood processing. 2016 April 12, 2016. http://www.fda.gov/Food/GuidanceRegulation/GuidanceDocumentsRegulatoryInformation/Seafood/ucm176892.htm.
46. National Shellfish Sanitation Program. Guide for the control of molluscan shellfish 2013 revision. 2013.
47. FoodSafety.gov. Selected federal agencies with a role in food safety. n.d. April 12, 2016. http://www.foodsafety.gov/about/federal/index.html.
48. Millen B, Lichtenstein AH, Abrams S, Adams-Cambell L, Anderson C, Brenna JT, Campbell W, Clinton S, Foster G, Hu F, Nelson M, Neuhouser M, Perez-Escamilla R, Siega-Riz AM, Story M. Scientific report of the 2015 dietary guidelines advisory committee. 2015.
49. Wood-Wright N. New U.S. dietary guidelines ignore broad support for food sustainability. 2016, Johns Hopkins HUB.
50. Lagasse LP, Love DC, Smith KC. Country-of-origin labeling prior to and at the point of purchase: an exploration of the information environment in Baltimore City grocery stores. Ecol Food Nutr. 2014;53(1):58–80.
51. EPA. National listing of fish advisories. 2016 April 12, 2016. http://fishadvisoryonline.epa.gov/General.aspx.
52. Love DC, Hawes M, Harding J. State-level recreational fishing regulations and fish consumption advisories in the United States: identifying opportunities for improved interagency collaboration. J Public Health Manag Pract. 2013;19(4):E11–9.
53. FDA. What you need to know about mercury in fish and shellfish. 2004 April 12, 2016. http://www.fda.gov/food/resourcesforyou/consumers/ucm110591.htm.
54. Oken E, et al. Decline in fish consumption among pregnant women after a national mercury advisory. Obstet Gynecol. 2003;102(2):346.
55. Birch RJ, et al. Trends in blood mercury concentrations and fish consumption among US women of reproductive age, NHANES, 1999–2010. Environ Res. 2014;133:431–8.
56. FDA. Fish: What pregnant women and parents should know. 2014.
57. USDA. Supplemental nutrition assistance program (SNAP). 2016 April 12, 2016. http://www.fns.usda.gov/pd/supplemental-nutrition-assistance-program-snap.
58. USDA. Women, infants and children (WIC): links to state agency WIC approved food lists. 2016 April 12, 2016. http://www.fns.usda.gov/wic/links-state-agency-wic-approved-food-lists.
59. USDA. The emergency food assistance program (TEFAP). 2016 April 12, 2016. http://www.fns.usda.gov/tefap/about-tefap.
60. USDA. Food distribution program on Indian reservations (FDPIR). 2016 April 12, 2016. http://www.fns.usda.gov/fdpir/about-fdpir.

61. USDA. National school lunch program factsheet. 2016. http://www.fns.usda.gov/sites/default/files/NSLPFactSheet.pdf.
62. Jahns L, et al. Intake of seafood in the US varies by age, income, and education level but not by race-ethnicity. Nutrients. 2014;6(12):6060–75.
63. Oken E, et al. Which fish should I eat? Perspectives influencing fish consumption choices. Environ Health Perspect. 2012;120(6):790.
64. Verbeke W, Vackier I. Individual determinants of fish consumption: application of the theory of planned behaviour. Appetite. 2005;44(1):67–82.
65. NOAA. Fisheries of the United States 2013. Silver Spring: National Marine Fisheries Service Office of Science and Technology; 2014.
66. Moore LV, Thompson FE. Adults meeting fruit and vegetable intake recommendations—United States, 2013. Morb Mortal Wkly Rep. 2015;64(26):709–13.
67. Abbott-Jamieson S, Clay PM. The long voyage to including sociocultural analysis in NOAA's national marine fisheries service. Mar Fish Rev. 2010;72(2):14–33.
68. Loring PA, Gerlach SC, Harrison H. Seafood as local food: food security and locally caught seafood on Alaska's Kenai Peninsula. J Agric Food Syst Commun Dev. 2013;3(3):13–30.
69. FWS. About the U.S. fish and wildlife service. 2016. http://www.fws.gov/help/about_us.html.
70. NOAA. Aquaculture policy and regulation: NOAA's role. n.d. April 12, 2016. http://www.nmfs.noaa.gov/aquaculture/policy/13_policy_and_reg_homepage.html.
71. USDA. Aquaculture. n.d. April 12, 2016. http://www.usda.gov/wps/portal/usda/usdahome?navid=AQUACULTURE.
72. USDA-ERS. Aquaculture Data. n.d. April 12, 2016. http://www.ers.usda.gov/data-products/aquaculture-data.aspx.
73. NOAA. Marine aquaculture strategic plan FY 2016–2020. 2015.
74. Interagency Working Group on Aquaculture. National strategic plan for federal aquaculture research (2014–2019). 2014.
75. Kirkley JE et al. International trade in seafood and related products: An assessment of U.S. trade patterns. 2006, National Oceanographic and Atmospheric Administration, National Marine Fisheries Service. pp. 1–147.
76. McClenachan L, et al. Do community supported fisheries (CSFs) improve sustainability? Fish Res. 2014;157:62–9.
77. National Marine Fisheries Service. Fisheries economics of the United States, 2012, E.a.S.A. Division, Editor. 2014.
78. Seelye KQ. In Maine, Last Sardine Cannery in the U.S. is Clattering Out. In: New York Times. 2010.
79. Pikitch E et al. Little fish, big impact: managing a crucial link in ocean food webs., Lenfest Ocean Program, Editor. 2012: Washington, DC. p. 108.
80. Chavez FP, et al. The northern Humboldt Current System: brief history, present status and a view towards the future. Prog Oceanogr. 2008;79(2):95–105.
81. Avadí A, Fréon P, Tam J. Coupled ecosystem/supply chain modelling of fish products from sea to shelf: the Peruvian anchoveta case. PLoS ONE. 2014;9(7):e102057.
82. Christensen V, et al. Valuing seafood: the Peruvian fisheries sector. Mar Policy. 2014;44:302–11.
83. Gagnon T. A guide to serving local fish in School Cafeterias, S.C. Society, Editor. n.d.
84. Izumi BT, et al. Serving fish in school meals: perceptions of school nutrition professionals in Alaska. J Child Nutr Manag. 2015;39(1):n1.
85. Profita C. Oregon delivers local seafood from boat to school. In: Oregon Public Broadcasting. 2013.
86. Horwitz S. Sea to school: models of local, sustainable seafood for schools. 2015. http://www.farmtoschool.org/news-and-articles/sea-to-school-models-of-local-sustainable-seafood-for-schools.
87. Zeiber R. Program seeks to bring local seafood back to the school lunch menu. In: New Hampshire Sea Grant. n.d.
88. Fitch C, Santo R. Instituting change: an overview of institutional food procurement and recommendations for improvement. The Johns Hopkins Center for a Livable Future, Editor. 2016.
89. nd. Red's Best. 2016 June 23, 2016. https://www.redsbest.com.
90. Goad M. Maine hospitals look to local fish and vegetables to improve health and patient satisfaction, in Portland Press Herald. ME: Portland; 2015.
91. Brinson A, Lee M-Y, Roundtree B. Direct marketing strategies: the rise of community supported fishery programs. Mar Policy. 2011;35(4):542–8.
92. Siegel KR et al. Association of higher consumption of foods derived from subsidized commodities with adverse cardiometabolic risk among us adults. JAMA Intern Med. 2016.
93. Da Silva PP, Fulcher C. Human dimensions of marine fisheries: using GIS to illustrate land-sea connections in the northeast US herring, Clupea harengus, fishery. Mar Fish Rev. 2005;67(4):19–25.
94. IOM (Institute of Medicine) and NRC (National Research Council). A framework for assessing effects of the food system, N.A. Press, Editor. 2015.
95. NOAA. Seafood watch. 2016 April 21. http://www.fishwatch.gov/.
96. na. Local Food, Local Places: Bringing Expertise and Creative Thinking to Community Economic Development. 2014 June 23, 2016. https://www.whitehouse.gov/blog/2014/12/03/local-food-local-places-bringing-expertise-and-creative-thinking-community-economic-.
97. Bittman M et al. How a national food policy could save millions of American lives. In: Washington Post. 2015.
98. Stoll JS, et al. Expanding the 'geography' of resilience in fisheries by bringing focus to seafood distribution systems. Ocean Coast Manag. 2015;116:185–92.
99. Love DC, et al. Wasted seafood in the United States from production to consumption. Glob Environ Change. 2015;35:116–24.
100. Smith T, Palmer A, Martin R, Fitch C. Achieving the triple aim in health care reform: the importance of the food system. The Johns Hopkins Center for a Livable Future, Editor. 2016.
101. GARFO. Member links. 2016 May 17, 2016. http://www.greateratlantic.fisheries.noaa.gov/educational_resources/seafood/links/index.html.
102. USDA-FNS. 2015 explanatory notes. 2015.
103. Papanikolaou Y, et al. US adults are not meeting recommended levels for fish and omega-3 fatty acid intake: results of an analysis using observational data from NHANES 2003–2008. Nutr J. 2014;13(1):1.
104. National Cancer Institute (NCI). Usual intake of total seafood 2007–2010. Epidemiology and Genomics Research Program. Online. https://epi.grants.cancer.gov/diet/usualintakes/pop/2007-10/table_a27.html.
105. Tran NL, et al. Estimated long-term fish and shellfish intake—national health and nutrition examination survey. J Eposure Sci Environ Epidemiol. 2013;23(2):128–36.
106. Wang Y, et al. Trends and correlates in meat consumption patterns in the US adult population. Public Health Nutr. 2010;13(09):1333–45.
107. NMFS. Commercial fisheries statistics. nd June 13, 2016. http://www.st.nmfs.noaa.gov/commercial-fisheries/commercial-landings/annual-landings/index.
108. Dueri S, Guillotreau P, Jiménez-Toribio R, Oliveros-Ramos R, Bopp L, Maury O. Food security or economic profitability? Projecting the effects of climate and socioeconomic changes on global skipjack tuna fisheries under three management strategies. Glob Environ Change. 2016;41:1–12.

16

GIS modeling of potentially suitable sites for aquaculture development in the Lake Tana basin, Northwest Ethiopia

Workiyie Worie Assefa[1*] and Wubneh Belete Abebe[2]

Abstract

Background: Small-scale pond aquaculture has been practiced in various parts of the Lake Tana basin. However, its establishment and expansion have been carried out regardless of the suitability of the locations in terms of land, water, socioeconomic and other inputs available to the aquaculture development. In recent years, there is also an increased demand for aquaculture product, and the government and other private institutions have given much attention to the sector. In response to this need, it is necessary to evaluate the potential sites for aquaculture suitability based on biophysical and socioeconomic indicators. Consequently, GIS-based site suitability analysis was conducted at the basin level to identify and delineate areas, which are suitable for inland pond aquaculture, mainly for Nile Tilapia (*Oreochromis niloticus*). About 21 criteria, which are sub-grouped into socioeconomic, farm inputs and land use types, water availability, water and soil quality, and urban markets, were considered for aquaculture suitable site-selection analyses. They are converted into various thematic layers. The layers, then combined to develop their respective sub-model suitability maps, which in turn grouped to produce the final overall suitability maps using the weighted linear combination. The relative weights were generated through pairwise comparison by the analytical hierarchy process.

Results: The analysis indicates that the prevailing situation in the basin supports promising opportunity for the establishment and development of aquaculture. The GIS evaluation predicts that out of the total land identified for aquaculture site, 32,678.9 ha (2.7%) is highly suitable, whereas 1,166,594.5 ha (97.2%) is moderately suitable for small-scale pond aquaculture system. Similarly, the considerable part of the basin was moderately suitable with a total area of 1,183,253.0 ha (98.6%) and more than 1% (16,768.2 ha) was highly suitable for commercial aquaculture farming. The land area delineated for the marginally suitable category was negligible in both cases.

Conclusion: Therefore, the result confirms that the Lake Tana basin has sizeable potential room for aquaculture development at small-scale and commercial production levels.

Keywords: APH, Aquaculture, Ethiopia, GIS, Lake Tana basin, Land evaluation

Background

The Lake Tana catchment is the second largest sub-basin in the Blue Nile Basin in the country with an abundance of aquatic resources [1]. The fish from the lake and its tributaries support the livelihoods of many rural people and an important source of animal protein for the poor. However, the fish resources are highly threatened by fishing pressure. The total catches of commercial interest have been seriously declining during the last two decades, particularly the endemic *Labeobarbus* fish species [2]. Currently, the annual fish catch in the lake is about 10,000t from which Nile tilapia accounted for more than 50% by volume [3]. Nile Tilapia (*Oreochromis niloticus*) is the most preferred fish by the consumers and become an important commercial fish species in the region with a better market price. This species is widely cultured in

*Correspondence: workiye2008@gmail.com
[1] Department of Biology and Blue Nile Water Institute, College of Science, Bahir Dar University, P.O.Box 79, Bahir Dar, Ethiopia
Full list of author information is available at the end of the article

Africa and around the world because of its fast growth, and tolerance to a wide range of environmental conditions [4].

The overfishing problem of Lake Tana can be alleviated by introducing alternative livelihoods such as pond aquaculture. Therefore, a great effort is needed to advocate fish biodiversity conservation through developing the aquaculture. Aquaculture is the primary means of achieving an increase in fish supply, which could overcome the steady decline in capture fisheries production. Small-scale pond aquaculture has been practiced in various parts of Amhara Regional State since 2009 [5]. These ponds have an area ranging from 100 to 400 m^2. The stocking density practiced by most fish farmers was two Nile tilapia fingerlings/m^2. Usually, fish ponds are fertilized with poultry and cattle manure and fishes are also fed with leftover food from the livestock and agriculture [6]. Nevertheless, the establishment and expansion of the small-scale fish farming have been carried out regardless of the suitability of the locations in terms of land, water, socioeconomic and other inputs available to the aquaculture development. For instance, it was observed that some ponds in Amhara Regional State remained unproductive. These are mainly because of the construction of the ponds without adequate consideration of biophysical and socioeconomic assessment. Careful planning is, therefore, necessary to aquaculture development. Thus, the first step for the establishment of pond aquaculture is selecting suitable sites. Prioritizing locations for aquaculture ensures sustainable development, avoids conflict among resource users, minimizes pollution of the environment and ultimately maximizes the profitability of the sector [7]. Geographical information systems (GIS) and remote sensing (RS) are essential tools for the assessment of multiple potential sites in a rapid and systematic way since it combines the spatial data with their attributes and overlays them [8]. Applications of GIS and RS in choosing suitable locations for aquaculture planning and development were demonstrated for various cultures of finfish and shellfish species [7, 9–21].

Strategic fish farming potential assessments using GIS were accomplished in Africa including Ethiopia [9, 22]. The analyses revealed that 29%, 31%, 28% and 12% of the surface area of Ethiopia were classified as very suitable, suitable, moderately suitable and unsuitable for small-scale subsistence fish farming, respectively [22]. Although water was identified as the most limiting factor for inland fish farming [9], the major sources of water for fishponds in Africa were identified [22] to be rainfall runoff, perennial rivers, and streams. In Ethiopia, the preliminary site suitability analysis for pond production of Nile Tilapia (*Oreochromis niloticus*) using GIS was also reported [23]. Water availability, temperature, topography, soil texture, land use and cover, and economic factors were used in this analysis and predicted that highly suitable land for inland pond-based aquaculture to be 15,158 km^2. However, GIS and RS for selection of suitable sites for the development of Nile tilapia land-based pond aquaculture have not been analyzed for the Lake Tana basin, which is identified as one of the potential aquaculture areas of the region. The present study has, therefore, made an effort to identify the need for biophysical, socioeconomic and environmental information on aquaculture potential in the Lake Tana basin. It aims to identify potential sites for small-scale and commercial aquaculture development based on the use of GIS models in the Lake Tana basin, using data with higher resolution. The outcomes of the study will facilitate the decision-making process in the allocation of land for commercial fish farming and expanding small-scale household level aquaculture in the Lake Tana region.

Methods

Study area description

The biophysical and socioeconomic data for mapping aquaculture suitable locations were gathered from the Lake Tana basin in Amhara Regional State, Northwest Ethiopia. The Lake Tana basin is one of the largest watersheds in the Blue Nile basin comprising diverse topography, which occupies an area of 16,500 km^2. It lies between 10.95°N and 12.78°N latitude and from 36.89°E to 38.25°E longitude and its altitude ranges from 1660 to 2364 m above sea level (Fig. 1). The basin is rich in water resources encompassing the Blue Heart Lake Tana, many streams and rivers, wetlands and reservoirs. The lake alone accounts for 50% of the country's freshwater resources and makes it the largest lake in Ethiopia [24]. Seven major rivers and several seasonal streams drain from the upper watershed of the basin and heading into the Lake Tana. The major rivers such as Gilgel Abay, Gumara, Rib, and Megech contribute about 95% of inflow water to the lake [24], while only the Blue Nile River outflows from Lake Tana. These rivers and streams provide sanctuaries and habitats for various organisms, especially for *Labeobarbus* fish spawning grounds and very critical for the development of agriculture, hydropower electricity, aquaculture and fisheries. Because of these huge potentials, the basin has been chosen as a growth corridor at the national level since 2012 in order to achieve socioeconomic development within the region [1]. This is a good opportunity for the development of fish farming too. The basin also encompasses extensive wetlands that have been playing an important role in the livelihoods of many rural households. Swamps, floodplains and marshes are the dominant wetland resources within the Lake Tana basin. They provide a natural filter against

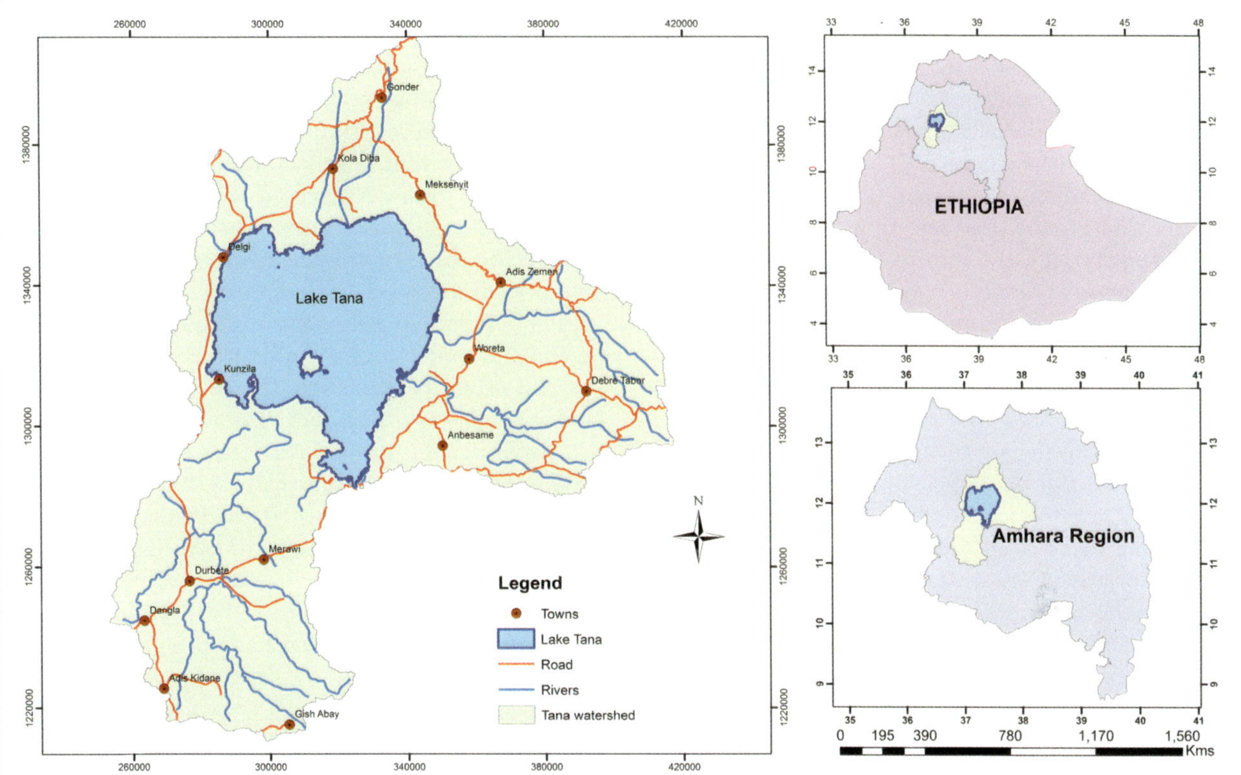

Fig. 1 Location map of the Lake Tana basin showing watershed area, inflowing rivers and *Woredas* (inset map shows location in Ethiopia)

sediments and pollutants for lake and rivers and serving as habitats for fish spawning, birds, wildlife and plants. The major types of soil in the basin are Luvisols, Vertisols, Nitisols, Leptosols, Gelysolsl and Fluvisols. Vertisols and Fluvisols are available in the land of flat plains and along the rivers [25]. In terms of texture, these soils are heavy clay and sandy loam [25], which are important for the interest of pond aquaculture establishment.

The Lake Tana basin receives an average annual rainfall of 1329 mm. The annual rainfall in the study area ranges from 900 to 2346 mm in a single rainy season, which mostly happens from June to September and also receives minor rain from April to May. The distribution of rainfall is higher in the South and Southeast mountainous side and relatively low in the northern and western parts of the basin. The average air temperature also ranges from 18 to 21 °C, which is high in the northern and western parts of the basin, whereas low temperature in the south and southeast of the study area. The mean monthly sunshine hours of the study area range from 5 to 9.5 h/day, which decreases from May to August and rise from September to January. Wind velocity also ranges from 0.6 to 1.7 m/s, with the strongest winds generally occurring in June, at about 2.0 m/s [26].

The major crops that are grown in the basin covered half a million hectares of land in 2014 [27]. The Lake Tana basin has approximately 10 million livestock population and supports more than 3.5 million people, of which 90% the population lives in the rural area the remaining 10% lives in urban areas in 2007 [27, 28]. The population density is about 228 people/km^2, which implies that resources are being severely exploited that in turn leads to eutrophication. This could ultimately cause off-flavor of Nile tilapia, declining of the endemic *Labeobarbus* fish species, and watershed degradation in the basin [29, 30]. The basin is also coined as a biodiversity hotspot because the lake harbors 15 unique species flock of *Labeobarbus,* the only remained cyprinid species flock in the world [31]; Nile Tilapia and African catfish, which are important aquaculture fish species and are widely cultivated in Africa, are inhabiting with good breeding capability. UNESCO has registered Lake Tana as a world biosphere reserve taking into account its fishery potential, harboring unique biodiversity, transportation, and tourism values [32]. This is another opportunity to develop aquaculture as an alternative fishery to improve the livelihoods of fishers and farmers.

Data sources and database generation

Primary and secondary data were (biophysical and socioeconomic) collected using social survey, document analyses and field measurements. For laboratory analysis, a total of 34 georeferenced triplicate soil samples were collected from 16 districts of the Lake Tana basin consisting of forest land, wetland, cultivated land, grazing land or farmland. According to the Food and Agriculture Organization (FAO) category [33], the major soil types identified are Vertisols, Luvisols, Alisols, Fluvisols, Gleysols, Cambisols, Regosols, and leptosols. The soil samples were air-dried and grounded (< 2 mm) for the determination of soil particle size distribution, pH, and electrical conductivity. The particle size distribution was determined by the hydrometer method following pre-treatment with H_2O_2 to remove organic matter, and the dispersion was aided by sodium hexametaphosphate [34]. Soil pH was measured potentiometrically using Maji multi-parameter (WAGTECH, UK) in supernatant suspension of the soil sample, 1 M KCl and deionized water with a ratio of 1:2.5 [35]. Water quality parameters such as pH, temperature, and dissolved oxygen were measured in situ at the head, middle and mouth of 6 major rivers and 10 streams during the dry season and post rainy seasons using a portable all-in-one meter (Maji multi-parameter, WAGTECH, UK). The population of Lake Tana basin for each district and city administration was obtained from Ethiopian Central Statistical Authority [28], while the crop cover and livestock populations (cattle, sheep, goats, poultry) were obtained from district administration agriculture offices. The population densities (individual/km^2) indicate potential markets of fish and other aquatic products, whereas district capital towns were considered as potential centers for the fish market. The livestock numbers were used as a surrogate measure of manure availability for fish feed and pond fertilization. The total potential available manure was calculated as described by Aguilar-Manjarrez and Nath [10]. The water quality (pH, dissolved oxygen, and temperature) of the Koga reservoir and Lake Tana were obtained from previous reports [36]. The number of fishers obtained from district agriculture offices was considered as a surrogate for factors such as production experience. Annual per capita fish consumption was used to measure the consumption behavior of the districts in the basin. Moreover, water-holding capacity of the soils, sufficient water availability with desirable quality, socioeconomic conditions, infrastructure facilities, and market availability were taken as influential factors for the development of Nile Tilapia (*Oreochromis niloticus*) farming in the study area.

Data analysis and software used

All the data were analyzed using ArcGIS10 for windows, and images were processed by ERDAS Imagine version 9.1 (Environmental Systems Research Institute Inc., USA). The spatial extension module was used for surface interpolation in the ArcGIS. Point data (population, livestock population, temperature, pH, dissolved oxygen, soil texture, crop coverage) were first changed into point shapefiles and then interpolated into a surface using inverse distance weighted tool (IDW) by the ArcGIS. River density was processed using ArcHydro tool in the ArcGIS where the river network and their lengths are determined; afterward, the length was divided by the area within the district possessing the rivers. Proximity (to perennial rivers, to road and access to market center) was mapped using the Euclidian distance tool. For the area land cover, the Jan 2013 image from Landsat 5 was used that had 30 meters ground resolution, and ground truth was brought from the field to classify the image using the ground truth as a signature. Hence, the supervised type of classification of the image in ERDAS IMAGINE software was applied. Slope and elevation data were derived from a digital elevation model (DEM). DEM was taken from open source USGS that had 30 m × 30 m resolution. The DEM directly holds elevation data and then; the slope was processed using a slope tool under the spatial analysis of ArcGIS toolset. All maps and images that were generated in the GIS models were registered on a WGS_1984_UTM_Zone_37N coordinate system in unit kilometers.

Classification criteria and suitability rating

The overall modeling of suitability was developed using the procedures shown in Fig. 2. Land suitability assessment for aquaculture development is usually based on a number of criteria. In this study, 21 criteria were identified based on data accessibility and relevance within the biophysical and socioeconomic conditions of the study area (Table 1). These criteria were scored for optimal site selection of Nile Tilapia farming by reviewing literature and judgment of aquaculture experts. Consumption behavior and production experience, which were not proposed and used by scholars, were included as important indicators for assessing aquaculture suitability in the study area. These factors were logically classified into six intermediate level sub-models, which includes socioeconomic and infrastructure, farm inputs, water availability, water quality, soil quality, and urban markets. Twenty-one base layers (thematic maps) were prepared for Nile Tilapia farming. Produced maps were awarded a suitability rating according to the land requirements of specified land use functions [37]. The suitability ratings for developing fish farming were classified into four classes; highly suitable (HS), moderately suitable (S), marginally suitable (MS), and unsuitable (US) [22, 37]. The HS level implies that the fish farming

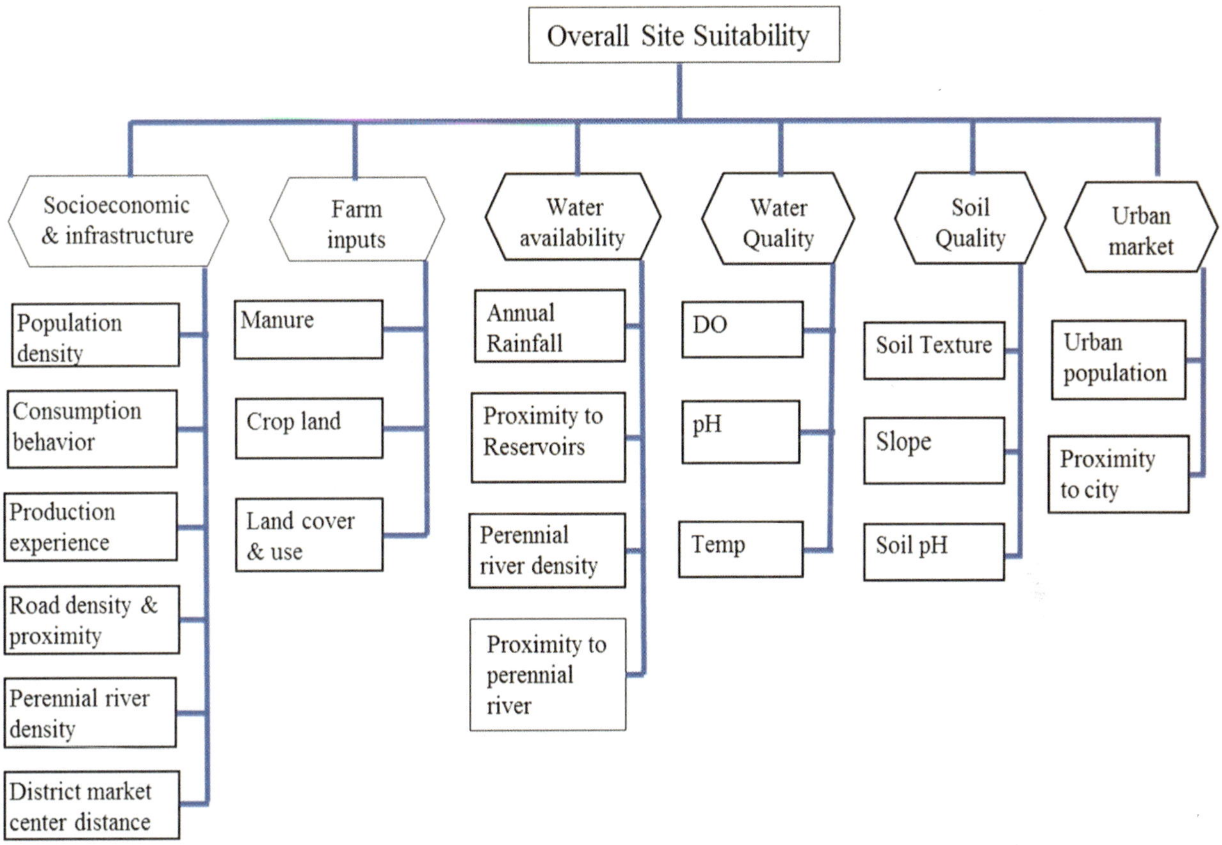

Fig. 2 Flowchart for modeling of fish farming site selection in the Lake Tana basin

is developed at minimum effort or cost, whereas S classification requires modest time and investment. The MS level indicates that substantial intervention on the land is a prerequisite. For the US level, the efforts in terms of the time and cost are too great to be worthwhile for developing fish farming [10]. Each criterion was assigned to a suitability rating by reviewing literatures and consulting expert's opinions to suit the local environmental condition (Table 1). The criteria were standardized using pairwise comparison technique to make the scores of the criteria compatible [38].

Weighting procedure

The model that was developed for decision-making (Fig. 2) comprised a number of criteria which are classified and developed into a series of sub-models. For instance, the criteria annual rainfall, perennial river density, proximity to perennial river and reservoir were grouped into a sub-model called water availability. Some other criteria were grouped into sub-models according to their shared characteristics [44]. Then, a relative weight for each factor was assigned using a pairwise comparison procedure employing analytical hierarchy process (AHP) method based on their influence to the success of Nile tilapia aquaculture development. Each category was also given a score according to suitability; most suitable scored 4, suitable scored 3, moderately suitable scored 2, and unsuitable scored 1 [44]. The degree of logical consistency among pairwise comparison can be determined by calculating a consistency index (CI) using the following formula:

$$\mathrm{CI} = \frac{\lambda_{\max} - n}{n - 1}$$

where—$\lambda_{\max}$ is the biggest eigenvalue that can be obtained associated with known eigenvector and n is the number of columns of the matrix A. Allowed inconsistency (10%) can be measured by calculating the consistency ratio (CR), which can be defined as CI/RI, where RI is the random index that depends on the number of elements being compared. If the CI/RI is smaller than 10%, the degree of consistency is satisfactory; if, however, the CI/RI is greater than 10%, there are inconsistencies and the AHP method may not yield meaningful results, which implies that the pairwise comparison matrix procedure should be revised [44, 45].

Table 1 The classification criteria for the selection of aquaculture potential suitable site in the Lake Tana basin

Land characteristics	Classification criteria and scale				References
	Highly suitable (4)	Moderately suitable (3)	Marginally suitable (2)	Unsuitable (1)	
Socio-economics sub-model suitability					
Population density (per km^2)	150–300	25–250	1–25	< 1 and > 300	[10, 39]
Consumption behavior (per capita fish consumption-g/year)	> 1000	500–1000	200–500	< 200	Based on local practice
Production experience (# of fishers/district)	> 600	300–600	100–300	< 100	Based on local practice
Market center distance (km)	< 5	5–10	10–15	> 15	Expert judgment
Road density (km/km^2)	≥ 0.36	0.20–0.35	0.04–0.19	< 0.04	[22]
Road proximity (km)	< 5	5–10	10–15	> 15	Expert judgment
Input suitability sub-model					
Total amount of livestock manure (in 1000 tons)	> 100	30–100	2–30	< 2	[10]
Crop land tonnage/km^2	80–100	50–79	20–49	< 20	[40]
Land use-land cover	Woodlands, shrub lands and grass-lands	–	Dense agricultural high lands	Large natural forest, and rocky barren	[23]
Water availability sub-model					
Annual rainfall (mm)	> 1199	1000–1199	700–999	< 700	[22]
Perennial river density (per district) (km/km^2)	≥ 0.36	0.2–0.35	0.04–0.20	< 0.04	[22]
Proximity to perennial river (km)	< 1	1–2.5	2.5–3.5	> 3.5	Expert opinion
Proximity lake and reservoirs (km)	< 1	1–2.5	2.5–3.5	> 3.5	Expert opinion
Water quality sub-model					
pH	6.0–8.0	5.0–6.0	4.0–5.0	< 4.0	[41]
DO (mg/l)	5.0–10.0	4.0–5.0	3.0–4.0	< 3.0	[41]
Temperature (°C)	25–32	22–25	18–22	0–18	[41]
Soil suitability sub-model					
Soil texture (%)	> 80	50–80	20–50	0–20	[42]
Slope (%)	0–2	2–5	5–8	> 8	[42]
Soil pH	6.5–8.5	5.5–6.5	4.0–5.5	1.0–4.0	[42]
Urban markets and proximity sub-model (commercial farming)					
Urban market population (in 1000)	> 1000	250–1000	50–250	< 50	[10]
Proximity to district capitals (km)	0–165	165–330	330–499	> 500	[43]

Combination of models

The individual thematic layer was produced based on the suitability score and saved with a common coordinate system. Then, the relative weights generated through pairwise comparison (Tables 2 and 3) using AHP method combined with the suitability score were used to generate sub-model suitability maps using the mathematical technique of weighted linear combination [46]. This step is principally known as multi-criteria evaluation (MCE), which involves the integration of multiple factors or criteria affecting the suitability of fish farming in one or another. Again, the final overall small-scale and commercial aquaculture suitability maps were mapped using a weighted linear combination. Each sub-model was multiplied by their corresponding weights derived from the pairwise comparison matrix (Tables 2 and 3) and then summed using map algebra algorithms:

$$Ai = \sum_{j}^{i} W_j X_{ij}$$

where X_{ij} is the score of the ith alternative with respect to the jth attribute, and the W_j is the normalized weight, so that $\sum W_j = 1$.

Model scenarios

The potential land areas for both small-scale and commercial fish aquaculture production levels were

Table 2 A pairwise comparison matrix for generating the relative weight of different factors for aquaculture potential in the Lake Tana basin

	PD (ind/km²)	Con.Bhr	Pod.Exp	D.MC (km)	RD (Km/km²)	RP (km)	DW (%)
Socioeconomic and infrastructure							
PD (ind/km²)	1	6	3	6/4	2	6/5	28.6
Con.Bhr	1/6	1	1/2	1/4	1/3	1/5	4.8
Prod.Exp	1/3	2	1	1/2	2/3	2/5	9.5
D.MC (km)	4/6	4	2	1	4/3	4/5	19.0
RD (km/km²)	1/2	3	3/2	3/4	1	3/5	14.3
RP (km)	5/6	5	5/2	5/4	5/3	1	23.8
Consistency ratio (CR) = 0.0000							
	ARA (mm)	**PRD (km/km²)**	**PPR (Km)**	**PWB (Km)**			**DW (%)**
Water availability							
ARA (mm)	1	2/5	2/3	2			18.2
PRD (km/km²)	5/2	1	5/3	5			45.5
PPR (km)	3/2	3/5	1	3			27.3
PWB (km)	1/2	1/5	1/3	1			9.0
Consistency ratio (CR) = 0.0000							
	TAM (tones)	**CrL**	**LU**				**DW (%)**
Farm inputs and land use types							
TAM (tones)	1	7/3	7				63.6
CrL	3/7	1	3				27.3
LU	1/7	1/3	1				9.1
Consistency ratio (CR) = 0.0000							
	pH	**DO (mg/l)**	**Temp (°C)**				**DW (%)**
Water quality							
pH	1	2	1/2				28.6
DO (mg/l)	1/2	1	1/4				14.3
Temp (°C)	2	4	1				57.1
Consistency ratio (CR) = 0.0000							
	Soil texture (%)	**Slope (%)**	**Soil pH**				**DW (%)**
Soil quality							
Soil texture (%)	1	2	6				65.3
Slope (%)	1/2	1	3				25.1
Soil pH	1/6	1/3	1				9.6
Consistency ratio (CR) = 0.02%							

PD population density (individual/km²), *Con.bhr* consumption behavior, *Prod.exp* production experience, *D.MC* distance to district market center, *RD* road density, *RP* road proximity, *ARA* annual rainfall amount, *PRD* perennial river density, *PPR* proximity to perennial river, *PWB* proximity to water bodies: lake and reservoirs, *DO* dissolved oxygen, *Temp* temperature, *TAM* total amount of manure form livestock (tones), *CrL* crop lands (suitability index), *LU* land use types, *DW* derived weight

analyzed based on the schematic design of the model scenario shown in Fig. 3. The purpose of small-scale fish farming is assumed to supply food for household consumption to improve food security, and the contribution to the local market is very limited. In this case, only socioeconomic and biophysical factors were analyzed assigning highest relative importance for water availability, water quality and soil quality for pond construction. In commercial fish farming, however, some factors, such as the size of the urban population and city proximity were added assuming that the production needed high market potential for substantial financial return.

Results and discussion

Criteria weight generated through pairwise comparison matrix are indicated in Tables 2 and 3. They were standardized and scored based on their influence over

Table 3 A pairwise comparison matrix for generating the relative weight of land requirements for aquaculture potential site suitability selection in the Lake Tana basin

	SEI	FI_LU	WA	WQ	SSP	UM	DW (%)
Small-scale aquaculture potential site suitability selection							
SEI	1	2	2/9	2/5	1/3		8.7
FI_LU	1/2	1	1/9	1/5	1/6		4.3
WA	9/2	9	1	1/8	9/6		39.1
WQ	5/2	5	8	1	5/6		21.7
SSP	3	6	6/9	6/5	1		26.2
Consistency ratio (CR) = 0.001							
	SEI	**FI_LU**	**WA**	**WQ**	**SSP**	**UM**	**DW (%)**
Commercial aquaculture potential site suitability selection							
SEI	1	2	2/7	1/3	2/5	1/4	6.9
FI_LU	1/2	1	1/7	1/6	1/5	1/8	3.4
WA	7/2	7	1	7/6	7/5	7/8	27.6
WQ	3	6	6/7	1	6/5	3/4	20.7
SSP	5/2	5	5/7	5/6	1	5/8	17.2
UM	4	8	8/7	4/3	8/5	1	24.2
Consistency ratio (CR) = 0.02							

SEI socioeconomic and infrastructure, *FI_LU* farm input and land use types, *WA* water availability, *WQ* water quality, *SSP* soil suitability for pond construction, *UM* urban market, *DW* derived weight

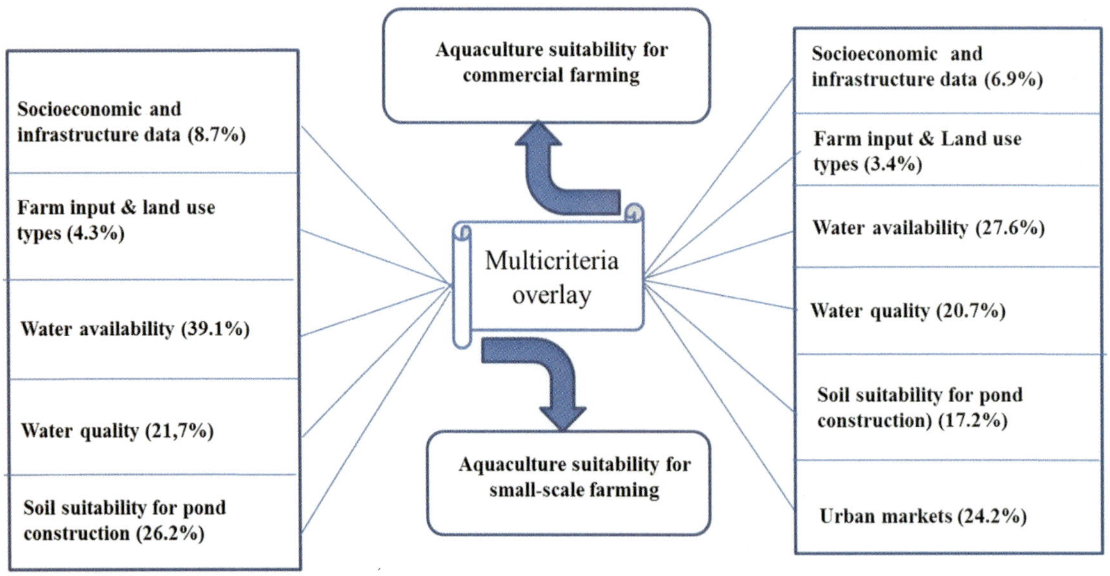

Fig. 3 Schematic design of the model scenario showing sub-models with derived weights obtained from APH method

the success of small-scale and commercial fish farming, mainly for Nile Tilapia in the Lake Tana basin. The consistency ratios (CR) ranged from 0 to 0.02 were within the acceptable inconsistency ratio less than 10% [46]. The outcomes of the sub-models for socioeconomic and infrastructure, farm inputs, water availability, water quality, soil quality, and urban market potential areas (ha) and percentage contribution in four suitable classes, including results of integrated models for both small-scale and commercial farming are summarized in Table 4. Their corresponding suitability maps are illustrated in figures from Figs. 4, 5, 6 and 7.

Table 4 Summary table, the areas (ha) and the percent suitability of lands for the factors, sub-models and integrated models for small-scale and commercial fish farming in the Lake Tana basin

Land characteristics	Highly suitable (4)		Moderately suitable		Marginally suitable		Unsuitable	
	ha	%	ha²	%	ha	%	ha	%
Socioeconomic s sub-model suitability								
Population density (per km²)	774,188.2	64.5	335,653.9	28.0	0	0	90,179.2	7.5
Consumption behavior	217,267.2	18.1	388,253.2	32.4	287,898.5	24.0	306,602.4	25.5
Production experience	184,883.6	15.4	396,381.9	32.6	145,035.8	12.1	478,720	39.9
Distance to district market center (km)	216,512.9	18.0	456,938.1	38.1	313,220.9	26.1	213,349.3	17.8
Road density (km/km²)	358,947.7	29.9	610,010.0	50.8	230,005.7	19.2	1057.9	0.1
Road Proximity (km)	1,185,997.9	98.8	4023.4	1.2	0	0	0	0
Sub-overall	327,705.3	27.3	803,184.1	66.9	69,031.9	5.8	0	0
Input suitability sub-model								
Total amount of manure from livestock (tones)	1,200,021.0	100	0	0	0	0	0	0
Crop land tonnage/km²	652,448.0	54.4	8433.7	0.7	539,139.6	44.9	0	0
Land cover and use	239,751.0	20.0	762,289.3	63.5	=	=	197,981.0	16.5
Sub-overall	657,983.4	54.8	542,037.5	45.2	0	0	0	0
Water availability sub-model								
Annual rainfall amount (mm)	826,360.8	68.9	352,028.8	29.3	21,631.7	1.8	0	0
Perennial river density (per district) (km/km²)	0	0	0	0	1,067,649.74	89.0	132,371.54	11.0
Proximity to perennial river (km)	194,365.8	16.2	268,294.8	22.4	150,233.4	12.5	587,127.3	48.9
Proximity to water bodies: lake and reservoirs (km)	54,104.2	4.5	58,120.9	4.8	38,721.8	3.2	1,049,074.3	87.4
Sub-overall	402,595.9	33.5	702,548.8	58.6	94,876.6	7.9	0	0
Water quality sub-model								
pH	1,503,300.87	100	0	0	0	0	0	0
DO (mg/l)	1,149,109.1	76.4	316,317.8	21.0	37,874.0	2.5	0	0
Temperature (°C)	40,857.2	2.7	1,434,139.9	95.4	28,303.8	1.9	0	0
Sub-overall	40,853.84	2.7	1,462,441.0	97.3	0	0	0	0
Soil suitability sub-model								
Soil texture (%)	63,364.8	63.6	310,541.1	25.9	99,842.2	8.3	26,273	2.2
Slope (%)	258,173.2	21.5	284,411.5	23.7	172,799.2	14.4	484,637.5	40.4
Soil pH	264,728.2	22.1	869,479.0	72.5	65,814.1	5.5	0	0
Sub-overall	326,417.6	27.2	661,935.4	55.2	204,973.7	17.1	6694.6	0.6
Urban markets and proximity sub-model (commercial farming)								
Urban market (population)	407,851.1	34.0	792.170.2	66.0	0	0	0	0
Proximity to district capitals (km)	1,200,021.3	100	0	0	0	0	0	0
Sub-overall	407,844.2	33.9	792,177.5	66.1	0	0	0	0
Final overall suitability for small-scale fish farming	32,678.9	2.7	1,166,594.5	97.2	747.8	0.1	0	0
Final overall suitability for commercial fish farming	16,768.2	1.4	1,183,253.0	98.6	0	0	0	0

According to socioeconomic and infrastructure classification, only 27.3% (327,705.3 ha) of the land were classified as being highly suitable for fish farming (Table 4). This is mainly linked to low consumption behavior, poor production experience, and long distance to travel toward the district market center. Nearly 50% of the land was unsuitable due to limited access to the district market center for the development of fish farming in the study area. Well accessible transportation facilities, including all-weather roads and well-established marketing facilities are significant to the local community for developing a network with linked industries for tilapia farming [17]. On the other hand, population density (64.5% highly suitable) and road proximity (98.8% highly suitable) were in favor of fish farming development within the study area. In general, significant parts of the basin (about 66.9%) were put

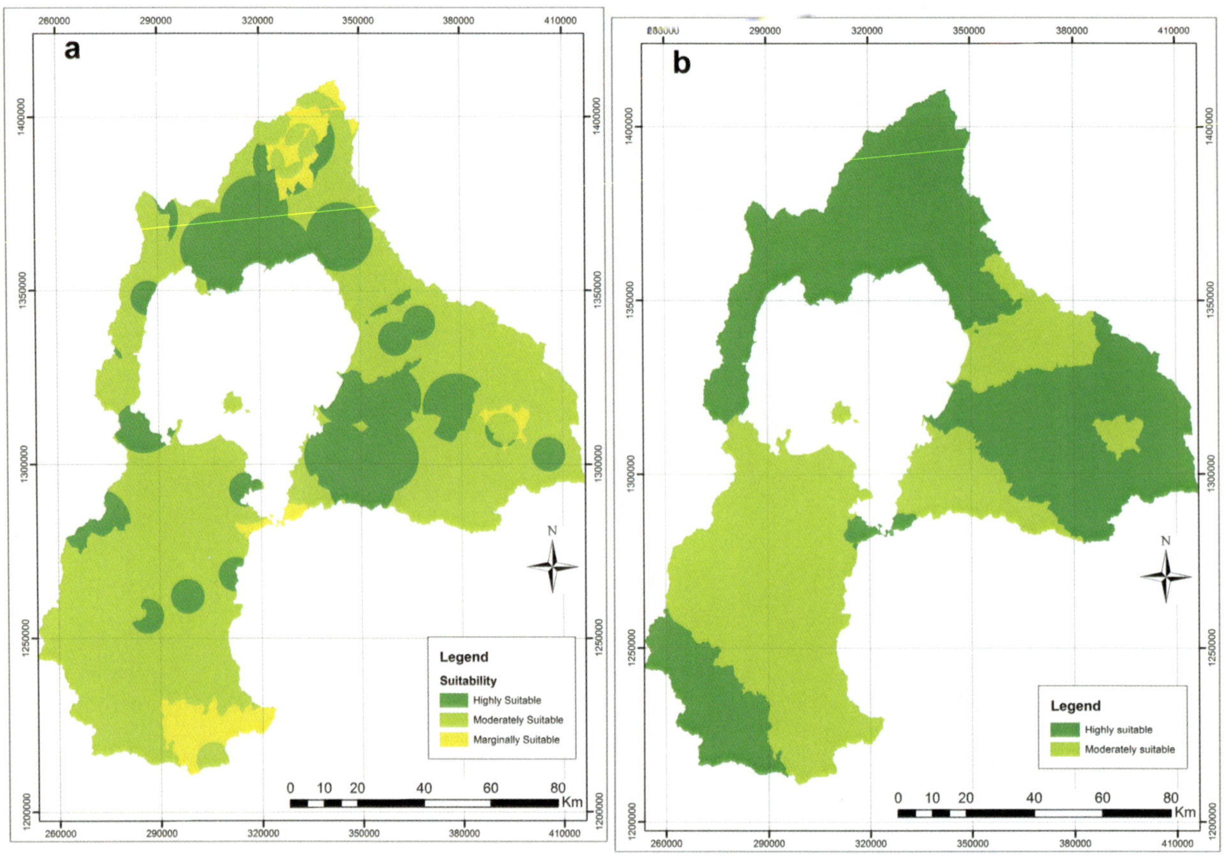

Fig. 4 Suitability map for **a** socioeconomic and infrastructure, and **b** farm inputs and land use types in the Lake Tana basin, Northwest Ethiopia

under a moderately suitable category, and no unsuitable lands were found for Nile tilapia aquaculture in terms of socioeconomic and infrastructure factors (Fig. 4a).

Based on the farm input and the land uses classification (Additional file 1:Table S1 and Additional file 2: Fig.S1), 657,983.4 ha (54.8%) and 542,037.5 ha (45.2%) of the potential areas were identified as highly suitable and moderately suitable, respectively. Manure production from livestock also showed 100% highly suitable. The contribution of the total manure from the livestock population in the basin was significant to those potential results. About 652,448.0 ha (54.4%) cultivated land, which could produce more than 80 quintals/km^2, was highly suitable, whereas 539,139.6 ha (50%) of the area was classified as being marginally suitable, and 8433.7 ha (0.7%) were classified as moderately suitable to access crop by-products for fish feed and fertilizing the pond (Fig. 4b). The livelihoods of the majority of the population in the basin primarily depend on agriculture. Therefore, agricultural by-products such as livestock waste and crop litter can be used both as sources of feed for fish and fertilizer for fish ponds [22]. Adding livestock waste to fish ponds induces the blooming of phytoplankton that serves as a primary food for the fish.

Adequate water source is also an essential element for identifying areas to establish fish farming. The water availability of the Lake Tana basin is highly dependent on rainfall availability as compared to perennial rivers, lakes and reservoirs. Significant parts of the study area in the south and east were classified as moderately suitable, whereas the northern and western parts were categorized into marginally suitable areas. On the basis of annual rainfall, approximately 70% of the area was highly suitable and there was no area found unsuitable. This confirms that a critical moisture deficit in the basin is not a problem (Table 4). On the other hand, based on perennial rivers, 1,067,649.74 ha (89%) of the area were classified under marginally suitable condition. The northern and northeastern parts of the study area, namely, Gondar Zuria, Wegera, Alefa and Lay Armachiho districts had very few perennial rivers and received marginally suitable ailment. Likewise, 587,127.3 ha (50%) of the area was found unsuitable because of long distance to the rivers. Only 22% of the land was classified as being moderately

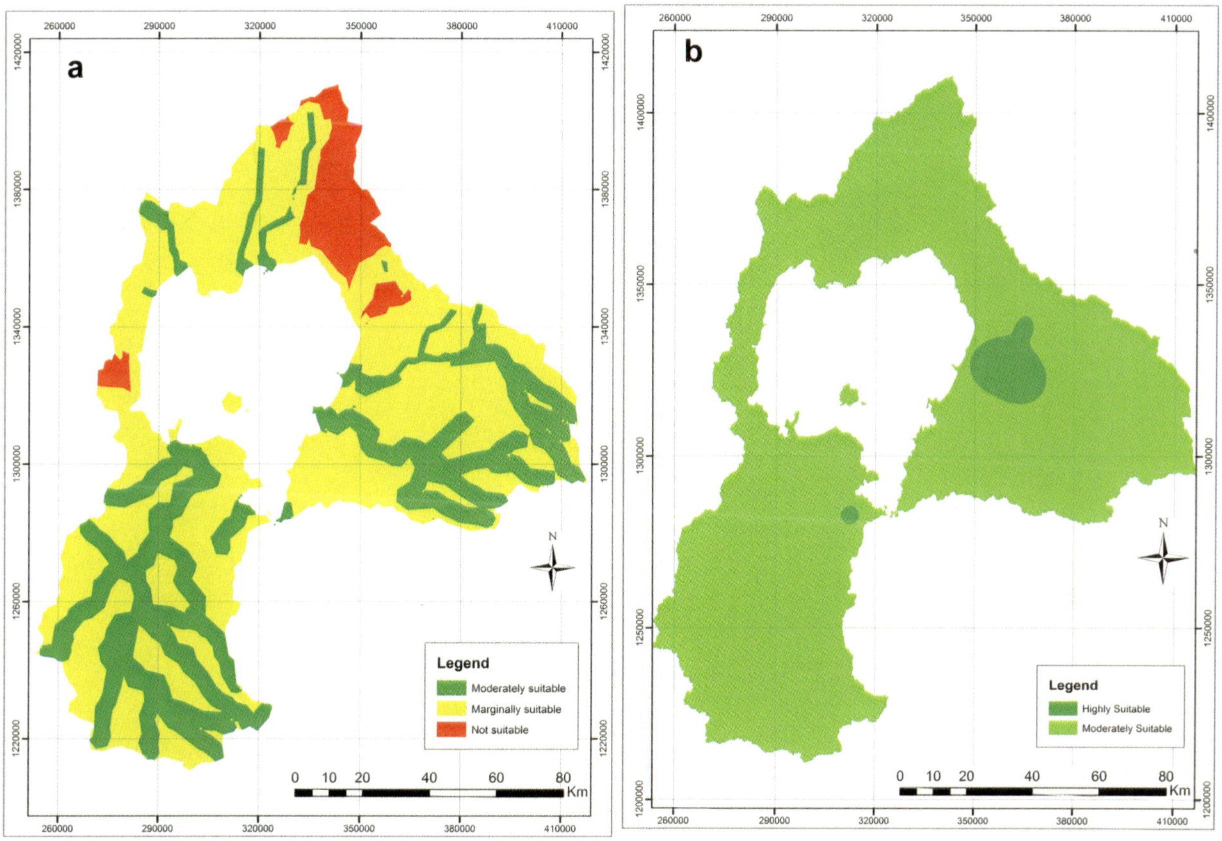

Fig. 5 Suitability map for **a** water availability, and **b** water quality in the Lake Tana basin, Northwest Ethiopia

suitable, whereas 16% and 13% classified as highly suitable and marginally suitable, respectively. Accesses to lake and reservoir waters were also difficult in the study area, as the majority parts of the land (87%) were classified under unsuitable condition. In the basin, over 60,000 ha land is planned for irrigation development and of this 6000 ha for "Koga" irrigation project is already completed [1, 24]. This can provide an enormous opportunity for the development of sustainable land-based aquaculture. The integration of rice cultivation with the existing cropping systems in the northeastern parts of the basin, particularly in "Fogera" flood plains will also create opportunity in introducing small-scale fish farming. Overall, highly suitable area for water availability accounted for 33.5% of the basin's land area, whereas 58.6% and 7.9% were belonged to moderately suitable and marginally suitable classes, respectively, and there were no unsuitable areas.

Good water quality, which is characterized by well-oxygenated, pollution free, favorable temperature and pH, provides an optimum condition for the growth of fish in a pond. The pH value within the range of 6.5–9.0 is suitable for fish farming, if it is below 4 or greater than 11, the fish will not alive (Boyd and Tucker 1998). Based on the water temperature parameter, 95.4% (1,434,139.9 ha) was moderately suitable and 76.4% (1,149,109.1 ha) of areas within the basin is classified as moderate conditions for fish growth in terms of dissolved oxygen parameter. Water temperature and dissolved oxygen critically limit the growth of fish and should be seriously considered during the establishment of fish farming [40, 47]. No areas were found unsuitable in terms of both water temperature and dissolved oxygen. With regard to temperature, highly suitable areas are located in some relatively flat areas, such as Fogera plain. In general, water quality for fish farming in the basin is classified as moderately suitable.

Soil characteristics such as, texture, organic matter content, pH, clay content, are also another important factor to be considered while selecting sites for establishing an aquaculture venture. Soil suitability analysis in terms of texture indicated that 64% (763,364.8 ha) of the basin is highly suitable; 26% (310,541.1 ha) is moderately suitable; 8% (99,842.2 ha) marginally suitable and the rest 2% (26,273.3 ha) is unsuitable. The north

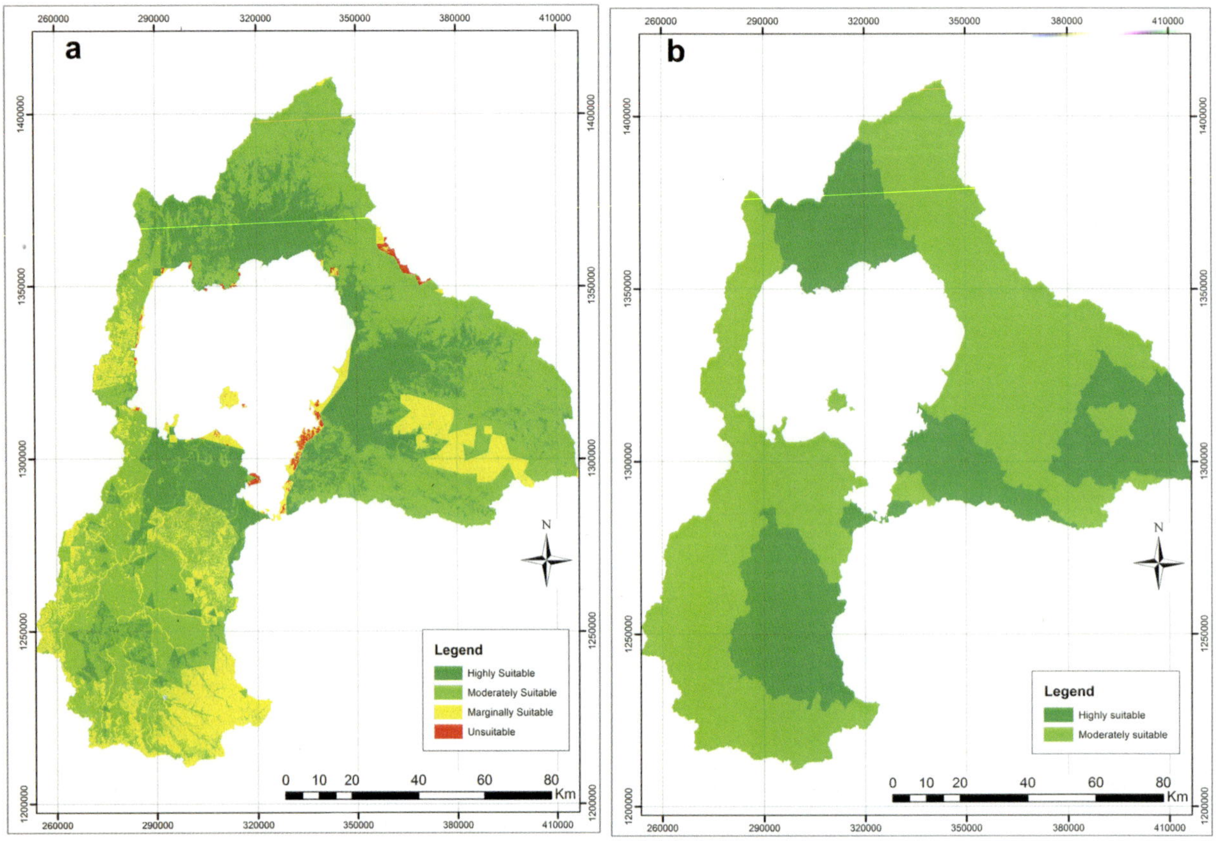

Fig. 6 Suitability map for **a** soil with regard to suitability for pond construction, and **b** urban market potential in the Lake Tana basin, Northwest Ethiopia

and northeastern parts of the basin are highly suitable, whereas greater parts of the southern and eastern of the basin are moderately and marginally suitable (Fig. 6a). According to previous study [25], Vertisols and Fluvisols are the most common soil types available in the land of flat plains and along the river banks of the basin. In terms of texture, these soils are heavy clay and sandy loam, which are suitable for the use of pond aquaculture. The majority (73%) part within the area was also classified as moderately suitable with regard to soil pH, but most of the northern parts of the basin were highly suitable (Fig. 6a).

In the study area, soil suitability for pond construction was limited by topography (slope) of the land. Evidently, slope classification shows that nearly 41% of the land was categorized as unsuitable conditions. Topography highly determines pond design and farm layout. A terrain with a slope between 2 and 4% is ideal as this feature allows the filling and draining of ponds by simple means of gravity [33]. Accordingly, the building of a fish pond on a steeper land is a very expensive business in terms of construction and water pumping cost. The areas surrounding Lake Tana were found highly suitable and the up-streams of the basin were unsuitable according to slope classification. Weighted overlay of the final overall soil suitability (Fig. 6a) revealed that more than 661,935.4 ha (50%) of the area was moderately suitable with the slope being the most limiting land characteristics.

Critical evaluation of potential fish market availability is the key issue for the viability of aquaculture operation. Thus, the urban market potential is one of the most important factors that should be considered for the development of commercial farming in Africa [22]. In this study, market potential was inferred from the proximity of cities in each district and the capital town besides to city resident population densities. On the basis of market proximity, 100% of the Tana basin was highly suitable since all the sites are within 165 km radii of the district towns that are considered to be important markets for commercial fish. On the other hand, the GIS classification of the population data showed that of the area 34%

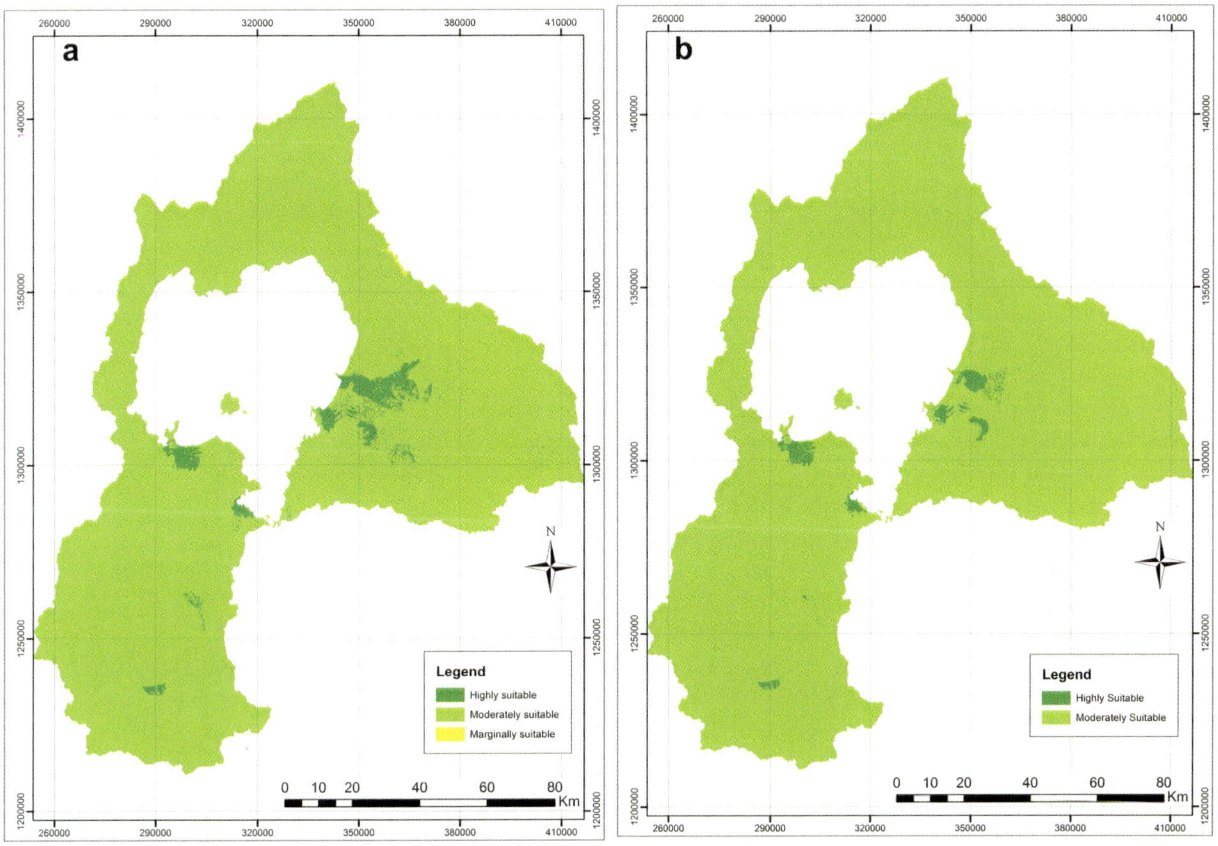

Fig. 7 Final overall suitability map for **a** small-scale fish farming, and **b** commercial fish farming in the Lake Tana basin, Northwest Ethiopia

(407,851.1 ha) and 66% (792,170.2 ha) of the area were moderately suitable and marginally suitable, respectively.

Integrated models

The derived weights were used to integrate the sub-models, and their maps are shown in Fig. 7a, b. Accordingly, the overall evaluation indicated that 97% (1,166,594.5 ha) of the area of the basin was found to be moderately suitable for small-scale fish farming. The remaining 3% (32,678.9 ha) of land was highly suitable. Negligible land in the area (0.1%) was marginally suitable. No land was identified as unsuitable for subsistence aquaculture. As shown in Fig. 7a, the land in the category of highly suitable for small-scale aquaculture was found in the Northeast and Southern parts of the lake, where most of the requirements overlapped each other in a conducive manner for the category they are assigned. The area identified as highly suitable for commercial farming (Fig. 7b) is dropped by 50% as compared to small-scale fish farming. The difference may be mainly related to marketing indicators. The majority of the area (1,183,253.0 ha) in the basin was found to be moderately suitable for commercial fish farming, which is a little larger than small-scale farming. Poor infrastructure development and low population density in some parts of the basin may influence biophysical and socioeconomic suitability for commencing commercial fish farming. Therefore, implementing commercial farming in these areas requires more investment in infrastructure development. By introducing mitigation measures for some of the land quality limitations and designing appropriate management strategies, it is possible to use unsuitable areas for the development of aquaculture [48].

Conclusions

In this study, the potential of small-scale and commercial aquaculture in the Lake Tana basin based on identified 21 biophysical and socioeconomic determinant factors was evaluated using GIS modeling. The result confirms that the prevailing situation in the basin supports promising opportunity for the establishment and development of aquaculture. The GIS evaluation predicts that from the total area of the basin, 32,678.9 ha (2.7%) and

1,166,594.5 ha (97.2%) were highly suitable and moderately suitable, respectively, for small-scale pond aquaculture system. The area delineated as marginally suitable category was negligible. Similar to small-scale aquaculture, a considerable area of (98.6% or 1,183,253.0 ha) the basin was moderately suitable and very small area (1.4% or 16,768.2 ha) was highly suitable for commercial aquaculture. Therefore, it can be concluded that the Lake Tana basin has a sizeable potential for aquaculture at small-scale and commercial production levels.

The most limiting factor for the establishment and development of aquaculture at the basin level are limited fish production experience, inadequate water access to rivers, and sloppiness of the land. Wise use of water from rainfall runoff can be a sufficient source of water for small-scale fish farming. The groundwater should also be considered as an alternative source for commercial fish farming. Commercial fish farming in the region is also limited by population densities, which is the proxy of market potential. This problem can be ameliorated selling aquaculture products to Addis Ababa and emerging cities as the demand fish is ever increasing in the peri-urban and urban markets despite the supply from the capture fishery is decreasing. Production experience related to fish farming is also found to be an obstacle to harness the potential of aquaculture in the basin. Therefore, there is a need to provide guidelines and trainings for small-scale community-based aquacultures. Models used to analyze the suitability in this study can be used to scale up the aquaculture development at the household level as well as commercial level. Based on the result of the analysis, the huge potential of the Lake Tana basin for fish farming has to be changed into reality; thereby, it can contribute more toward improving food insecurity in the region. Finally, the GIS modeling that was used in this study was a powerful tool for identifying the suitability of potential lands for the development of aquaculture ventures at the small-scale and commercial level in the Lake Tana basin. It provides useful data for policy makers and decision-makers to frame strategies for promoting aquaculture.

Authors' contributions

WWA conceived and designed the study, organized the field work for data collection, organized soil laboratory analysis, data analysis and writing of the manuscript; WBA was involved in data collection, analysis and GIS mapping. Both authors reviewed and edited the final manuscript for submission. Both authors read and approved the final manuscript.

Author details

[1] Department of Biology and Blue Nile Water Institute, College of Science, Bahir Dar University, P.O.Box 79, Bahir Dar, Ethiopia. [2] Amhara Design and Supervision Works Enterprise, Bahir Dar, Ethiopia.

Acknowledgements

The authors wish to thank agricultural offices for offering secondary data and facilitating field data collection during a sampling campaign. Blue Nile Water Institute of Bahir Dar University greatly acknowledged for supplying logistic support. We also thank Dr Belachew Getnet and Zebasil Tasew for providing helpful comments on the earlier version of the manuscript.

Competing interests

The authors declare that they have no competing interests.

Funding

This work was financially supported by the Blue Nile Water Institute through Bahir Dar University, Ethiopia. The funder had no role in the study design, data collection and analysis, decision to publish, or preparation of the manuscript.

References

1. Awulachew SB, Erkossa T, Smakhtin V, Fernando A. Improved water and land management in the Ethiopian highlands: its impact on downstream stakeholders dependent on the Blue Nile. In: Intermediate results dissemination workshop held at the international livestock research institute (ILRI), Addis Ababa, Ethiopia, 5–6 February 2009. International Water Management Institute. https://doi.org/10.3910/2009.201.
2. Dejen E, Anteneh W, Vijverberg J. The decline of The Lake Tana (Ethiopia) fisheries: causes and possible solutions. Land Degrad Dev. 2017;28:1842–51.
3. Mengistu AA, Aragaw C, Mengist M, Goraw G. The fish and the fisheries of Lake Tana. In: Stave S, Goraw G, Shimeles A, editors. Social and ecological system dynamics. Berlin: Springer; 2017. p. 157–77.
4. El-Sayed A-FM. Tilapia culture. Wallingford: CABI Publishing; 2006.
5. LRDPA. Annual report of livestock resources development promotion agency (LRDPA). Bahir Dar: Amhara Region Bureau of Agriculture; 2016.
6. Adugna BT, Goshu G. Integrating aquaculture with traditional farming system: socio-economic assessment in the Amhara region, Ethiopia. Ecohydrol Hydrobiol. 2010;10(2–4):223–30.
7. Kapetsky JM, McGregor L, Nanne LH. A geographical information system and satellite remote sensing to plan for aquaculture development: a FAO-UNESCO/GRID cooperative study in Costa Rica. FAO fisheries technical paper (287); 1987, p 51.
8. Burrough PA. Principles of geographic information system for land resources assessment. New York: Oxford University Press; 1986.
9. Aguilar-Manjarrez J, Nath SS. A strategic reassessment of fish farming potential in Africa. CIFA technical paper 32, FAO, Rome; 1998.
10. Aguilar-Manjarrez J, Ross GL. Geographical Information Systems (GIS), environmental models for aquaculture development in Sinaloa state, Mexico. Aquac Int. 1995;3(2):103–15.
11. Arnold WS, White MW, Norris HA, Berrigan ME. Hard clam (*Mercenaria* spp) aquaculture in Florida, USA: geographic information system applications to lease site selection. Aquac Eng. 2000;23:203–31.
12. Mcleod I, Pantus F, Preston N. The use of a geographical information system for land-based aquaculture planning. Aquac Res. 2002;33:241–50.
13. Giap DH, Yi Y, Yakupitiyage A. GIS for land evaluation for shrimp farming in Haiphong of Vietnam. Ocean Coast Manag. 2005;48:51–63.
14. Karthik M, Suri J, Saharan N, Biradar RS. Brackish water aquaculture site selection in Palghar Taluk, Thane district of Maharashtra, India, using the techniques of remote sensing and geographical information system. Aquac Eng. 2005;32:285–302.
15. Perez OM, Telfer TC, Ross LG. Geographical information system-based models for offshore floating marine fish cage aquaculture site selection in Tenerife, Canary Islands. Aquac Res. 2005;36:946–61.
16. Buitrago J, Rada M, Hernandez H, Buitrago E. A single-use site selection technique, using GIS, for aquaculture planning, choosing locations for mangrove oyster raft culture in Margarita Island, Venezuela. Environ Manag. 2005;35:544–56.

17. Hossain MS, Chowdhury SR, Das NG, Rahaman MM. Multi-criteria evaluation approach to GIS-based land suitability classification for tilapia farming in Bangladesh. Aquac Int. 2007;15:425–43.
18. Radiarta IN, Saitoh SI. Biophysical models for Japanese scallop, *Mizuhopecten yessoensis*, aquaculture site selection in Funka Bay, Hokkaido, Japan, using remotely sensed data and geographic information system. Aquac Int. 2009;17:403–19.
19. Hossain MS, Chowdhury SR, Das NG, Sharifuzzaman SM, Sultana A. Integration of GIS and multicriteria decision analysis for urban aquaculture development in Bangladesh. Landsc Urban Plan. 2009;90:119–33.
20. Hossain MS, Das NG. GIS-based multi-criteria evaluation to land suitability modelling for giant prawn (*Macrobrachium rosenbergii*) farming in Companigonj Upazila of Noakhali, Bangladesh. Comput Electron Agric. 2010;70:172–86.
21. Hadipour A, Vafaie F, Hadipour V. Land suitability evaluation for brackish water aquaculture development in the coastal area of Hormozgan, Iran. Aquac Int. 2014. https://doi.org/10.1007/s10499-014-9818-y.
22. Kapetsky JM. A strategic assessment of warm-water fish farming potential in Africa, CIFA Technical Paper, No. 27 edn. FAO, Rome; 1994
23. Dejen E, Mintesnot Z. A generic GIS-based site suitability analysis for pond production of Nile Tilapia (*Oreochromis niloticus*) in Ethiopia. In: Lemma B, Getahun A, Mengistu S, editors. The role of aquatic resources for food security in Ethiopia, proceedings of the fourth annual conference of the Ethiopian fisheries and aquatic sciences association (EFASA), AAU Printing Press, Addis Ababa; 2012. pp. 28–57.
24. Alemayehu T, McCartney M, Kebede S. Water resource implications of planned development in the Lake Tana catchment, Ethiopia. Ecohydrol Hydrobiol. 2010;2–4(10):211–22. https://doi.org/10.2478/v10104-011-0023-6.
25. Getahun M, Selassie YG. Characterization, Classification and Mapping of Soils of Agricultural Landscape in Tana basin, Amhara National Regional State, Ethiopia. In: Stave S, Goraw G, Shimeles A, editors. Social and ecological system dynamics. Berlin: Springer; 2017. p. 93–115.
26. National Meteorogy Agency of Ethiopia (NMA). Bahir Dar Meteorological Branch, Bahir Dar, Ethiopia; 2014. http://www.ethiomet.gov.et/.
27. Bureau of Agriculture (BoA). Amhara National Regional State, Bureau of Agriculture, Bahir Dar, Ethiopia; 2014. http://www.amhboard.gov.et/.
28. Central Statistical Authority (CSA). National statistics of the year 2014. Addis Ababa: Central Statistical Authority of Ethiopia; 2014.
29. Assefa WW, Goshu G. Incidence of fish off-flavor for the first time from Lake Tana, Ethiopia: causes and possible management strategies. IJSRR. 2014;3(4):30–47.
30. Dejen E, Anteneh W, Vijverberg J. The decline of The Lake Tana (Ethiopia) fisheries: causes and possible solutions. Land Degrad Dev. 2017;28(6):1842–51.
31. Stiassny MLJ, Getahun A. An overview of labeonin relationships and the phylogenetic placement of the Afro-Asian genus Garra Hamilton, 1922 (Teleostei: cyprinidae). Zool J Linn Soc. 2007;150(1):41–83.
32. Kalmbach E. Establishment of the Lake Tana biosphere reserve within the UNESCO world network of biosphere reserves. In: Stave S, Goraw G, Shimeles A, editors. Social and ecological system dynamics. Switzerland: Springer; 2017. p. 523–44.
33. FAO. Digital soil map of the world and derived soil properties. Version 3.5.CD-ROM. Rome: FAO; 1995.
34. Black CA. Methods of soil analysis. Part 1: physical and mineralogical properties. Madison: American Society of Agronomy; 1965.
35. Baruah TC, Barthakur HP. A text-book of soil analyses. New Delhi: Vikas Publishing House Pvt. Ltd.; 1997.
36. Tefera B, Wale M, Assefa WW, editors. Proceedings of the second national workshop on challenges and opportunities of water resources management in Tana Basin, Upper Blue Nile Basin, Ethiopia, 26–27 March 2012. Blue Nile Water Institute—Bahir Dar University, Bahir Dar, Ethiopia.
37. FAO. A framework for land evaluation. Rome: Food and Agriculture Organization of the United Nations; 1976.
38. Malczewski J. GIS-based land-use suitability analysis: a critical overview. Prog Plann. 2004;62:3–65.
39. Kapetsky JM, Nath SS. A strategic assessment of the potential for freshwater fish farming in Latin America. FAO COPESCAL Technical Paper (10); 1997.
40. Boateng E, Ayamga T, Chidley T, Elgy J, Savory D, van Velthuizen H. Geospatial modelling for agro-ecological assessments and decision support for development planning in Ghana Land suitability analysis. Accra: Council for Scientific and Industrial Research—Soil Research Institute; 2001.
41. Salam AM. The potential of geographical information system-based modelling for aquaculture development and management in Southwestern Bangladesh. PhD dissertation. Institute of Aquaculture, University of Stirling. Scotland, UK; 2000.
42. Coche AG, Laughlin T. Soil and freshwater fish culture. Simple methods for aquaculture. FAO training series 6. Rome: FAO; 1985.
43. Meaden GJ, Aguilar-Manjarrez J, editors. Advances in geographic information systems and remote sensing for fisheries and aquaculture. CD–ROM version. FAO Fisheries and Aquaculture Technical Paper No. 552. Rome, FAO; 2013.
44. Saaty TL. The analytic hierarchy process: planning, priority setting, resource allocation. New York: McGraw-Hill; 1980.
45. Saaty TL. How to make a decision: the analytic hierarchy process. Interfaces. 1994;24(6):19–43.
46. Malczewski J. GIS and multicriteria decision analysis. New York: Wiley; 1999.
47. Boyd CE, Tucker CS. Pond aquaculture water quality management. Boston: Kluwer Academic Publishers; 1998.
48. Ssegane H, Tollner EW, Veverica K. Geospatial modeling of site suitability for pond-based Tilapia and Clarias farming in Uganda. J. Appl. Aquac. 2012;24(2):147–69.

Evaluation of morphological characteristics, yield and nutritive value of *Brachiaria* grass ecotypes in northwestern Ethiopia

Wubetie Adnew Wassie[1*], Berhanu Abraha Tsegay[1], Asaminew Tassew Wolde[2] and Bimrew Asmare Limeneh[2]

Abstract

Background: A study was conducted to evaluate the effect of altitude and harvesting dates on productivity and nutritive value of three *Brachiaria brizantha* grass ecotypes (Eth. 13726, Eth. 13809 and Eth. 1377) in northwestern Ethiopia. A factorial arrangement of treatments was employed with a combination of three altitudes and three harvesting dates. The data collected consisted of plant height (PH), number of tillers, number and length of leaves and fresh yield. The yield and chemical analysis of forage samples were analyzed for dry matter yield, ash, crude protein (CP), neutral detergent fiber (NDF), acid detergent fiber (ADF) and acid detergent lignin. All data were subjected to analysis of variance procedures, with significance tested at $P < 0.05$.

Results: Results indicated that the highest plant height (PH) was recorded in all sites (low, mid- and high altitudes) by Eth. 1377 ecotype followed by Eth. 13726 ecotype (low and mid-altitudes) and in high altitude by Eth. 13809 ecotype. The highest DM yield was recorded by Eth. 13809 ecotype (7.26, 5.98 and 4.50) which was followed by Eth. 1377 ecotype (6.28, 4.68 and 4.30) at low, mid- and high altitudes, respectively. There was significant difference ($P > 0.05$) in DM yield with increasing harvesting dates for all ecotypes. CP content declined with increased harvesting date in all ecotypes [Eth. 13726 (16.33, 10.63 and 6.72), Eth. 13809 (13.87, 10.60 and 9.57) and Eth. 1377 (14.80, 10.15 and 7.86)] for 60, 90 and 120 harvesting dates, respectively. All ecotypes had the highest CP content at day 60 harvesting stage.

Conclusions: Although all ecotypes had potential as an alternative ruminant feed in all altitude areas in Ethiopia, highest CP and lowest NDF and ADF concentrations were recorded by Eth13809 ecotype. Its capability to grow at low rainfall maintaining high yields is an additional advantage over the other two. Thus, among the tested ecotype grasses Eth13809 showed outstanding potential as a forage plant especially in low altitude area of northwestern Ethiopia.

Keywords: Dry matter yield, Grass ecotypes, Plant height, Tiller number

Background

Livestock production is one of the most important agricultural land use systems in the world, with grasslands covering 25% of land surface and contributing to the livelihoods of more than 800 million people [52]. However, in Ethiopia livestock production has been mainly constrained by inadequate supply and poor quality of available feed resources [18, 21]). This could be further emphasized by the fact that feed accounts for 60–70% of the costs associated with livestock production.

Nowadays, the most important livestock feed resources in Ethiopia are natural pasture, crop residues and grass hay [1]. Natural pasture hay and crop residues which provide the bulk livestock feed in Ethiopia are seasonally produced during particular periods of the year (October–January) following the main rainy season. Excess forage production is experienced during the rainy season, but more often, acute shortages occur in the dry season

*Correspondence: wu1999as@yahoo.com
[1] Department of Biology, College of Science, Bahir Dar University, P. O. Box 76, Bahir Dar, Ethiopia
Full list of author information is available at the end of the article

as observed in some parts of the country [2]. According to CSA [14] report, the major feed resources in the country are green fodder (54.59%), crop residue (31.6%), hay (6.81%) and agricultural by product (1.53%) which have poor quality. Common forage crops adapted in the farming systems in Ethiopia like Napier grass (*Pennisetum* sp.) and silver leaf *Desmodium* sp. have been affected by the global effects of climate change and Napier grass is also threatened by the emergence of stunt and smut diseases [37], which has also limited its expansion to drier areas. Consequently, finding alternative feeds for livestock is an important step to sustain livestock production in the country. This constraint can be solved through finding a climate-friendly way through the introduction of improved forage varieties of the *Brachiaria* genus [50]. Moreover, climate smart agriculture is one of the means that can help to reduce greenhouse gas emissions and increase livestock productivity of the country through improved livestock feed and feeding practices [19]. *Brachiaria* grasses tend to be drought resistant and resilient in infertile soils, and produce well with relatively low levels of fertilizer inputs. They are also resistant to many diseases affecting baseline varieties in Eastern Africa, particularly Napier stunt and smut disease [22, 31]. Thus, as a new strategy termed using 'climate smart *Brachiaria* grass', which can withstand heavy grazing is best option.

In addition to adapted to drought, diseases and low-fertility soils, *Brachiaria* grass sequesters carbon through its large roots system enhance nitrogen use efficiency and subsequently minimize eutrophication and greenhouse gas emissions [6, 36]. *Brachiaria* grass plays important roles in soil erosion control and ecological restoration. *Brachiaria* grass species have been important component of sown pastures in humid low lands and savannas of tropical America with current estimated acreage of 99 million hectare in Brazil alone [25]. Moreover, the grass has a potential of production of high dry matter yield [49]. Additional advantages of using the *Brachiaria* genus in the integrated systems are that the species produce abundant roots which contribute to the collection of water, soil aggregation and aeration [27].

The genus *Brachiaria* consists of roughly 100 species which grow in the tropics and subtropics. Most of these species are native to Africa, where they constitute important components of the natural savanna landscape [22]. We see the distribution of *B. brizantha* is high in *Africa* including Ethiopia [15] and need more research to exploit maximum in the region. Recent trials indicate that adoption of *B. brizantha* cultivars has the potential to increase baseline milk production of 3–5 L/cow/d on participating farms in Kenya by 15–40% [50]. A farm trial in Rwanda reported a 30% increase in milk production and a 20% increase in meat production [13]. Despite the diversity of *Brachiaria* spp. in Eastern and Central Africa, comparatively little information is available on their agro-morphological characteristics, yield and chemical composition.

The productivity of the different grass species could be distinctly different and is also influenced by area of origin, including temperature, light intensity, total rainfall, soil type, fertilization level, and by stage of maturity [24, 26]. In East Africa, Bogdan [11] reported that *B. brizantha* is very variable and several varieties show striking differences in habit, morphology and seed setting capacity. It is possible that the different varieties will perform differently in different ecological zones, but information of this kind is lacking in Ethiopia. *B. brizantha grass* growing in Ethiopia is the wild type and ecologically different (an ecotype) from that in other countries, which is not developed/enriched like that of other countries such as Brazil. Therefore, the study was aimed to evaluate the performance of *B. brizantha* ecotypes and to select the best herbage yielding and quality among selected three ecotypes so as to use the better performing ecotype for wider distribution among livestock producer communities in the country.

Materials and methods

Description of the study areas

The study was conducted in south Gondar (Amhara Regional State) at three agro-ecologies using a rain fed system. The low altitude area was represented by a location called Futan at Tach Gayint district located at 11°22′N and 28°19′E at an altitude of 1230 m above sea level. Mean minimum and maximum annual temperature of the district ranges from 13 to 27 °C and the mean minimum and maximum annual rainfall ranges from 900 to 1000 mm per annum. The mid-altitude location was represented by a place called Woreta at Fogera district which is situated at 11°58′N and 37°41′E at an altitude range of 1774 masl and is predominantly classified as mid-altitude agro-ecology. The mean annual rainfall is 1216.3 mm and ranges from 1103 to 1336 mm. According to the district Office of Agriculture, the dominant soil type is black clay (ferric vertisols) (personal communication). The highland area was represented by Farta district, Tsegure Eyesus Kebele (Kebele is the smallest administration unit in Ethiopia) at a site called Melo located near Debre Tabor Town, at 11°11′N and 38°E and at an altitude of 2650 m above sea level. The soils of Melo site are characterized by clay and sand mixture with chemical composition of 2.26% organic matter, 0.11% total nitrogen and pH of 5.47. The mean annual rainfall is about 1570 mm and the mean maximum and minimum annual temperatures were reported to be 21.5 C and 9.6 C, respectively.

Land preparation, planting and experimental design

A total area of 341 m^2 was prepared for each location. The experimental land was plowed in May and harrowed in June 2017. The land was divided into three blocks each of which comprised three plots (3*3 m each). Planting materials *Brachiaria brizantha* ecotypes/wild (Eth. 13726, Eth. 13809 and Eth. 1377) were collect from International Livestock Research Institute (ILRI) Forage Gene Bank Addis Ababa, Ethiopia. The ecotypes were planted using vegetative root splits in rows. The experiment was laid out in a factorial arrangement of three altitudes (low, mid and high) and three harvesting stages (60 days, 90 days and 120 days) in a randomized complete block design with three replications. The spacing between rows and plants was 50 cm and 30 cm, respectively. Land preparation, planting, weeding, harvesting and related management practices were applied according to standard practice for the grass [42]. Artificial fertilizers (di-ammonium phosphate and urea) were applied at rate of 100 kg/ha and urea at 25 kg/ha during planting and after establishment based on the recommendations for the grass.

Data collection

Morphological parameters such as plant height and leaf length were measured from 10 plants that were randomly selected from middle rows of each plot at 60 days, 90 days and 120 days after planting. The number of tillers and leaves was computed as mean counts. To determine biomass yield, the forage harvesting was done by hand using a sickle leaving a stubble height of 8 cm according to recommended practice. The fresh herbage yield was measured immediately after each harvest using a portable balance with a sensitivity of 0.01 g. Representative samples were taken from each plot at each site and were dried in a draft oven at 65 °C for 72 h before chemical analysis.

Chemical analyses

The chemical composition of all samples of feeds collected during agronomic was conducted at Debre Berhan Agricultural Research Center Animal Nutrition Laboratory. The grass samples were dried at 65 °C for 72 h and ground to pass through a 1 mm sieve. Ash/organic matter (OM), dry mater (DM), crude protein and total ash were determined according to AOAC (1990). The neutral detergent fibers (NDF), acid detergent fiber (ADF), acid detergent lignin (ADL) were determined according to Van Soest and Robertson [55]. The CP was calculated as percentage of nitrogen in the sample multiplied by a factor of 6.25.

Data analysis

The collected data were analyzed using general linear model of statistical analysis system (SAS) procedures of 2002 version 9.0. Least significant difference (LSD) test was employed for variables whose F values declared a significant difference ($P<0.05$). The statistical model for data analysis was:

$$Yijk = \mu + Ai + Hj + Ck + Ai * Hj + +A * C + H * C + \varepsilon ijk$$

where Yijk is the response (plant morphological parameters, chemical composition, and yield of *B. brizanthag-rass*) at each altitude and harvesting days

μ = overall mean,

Ai = altitude (i = low, mid and high),

Hj = effect of harvesting days (j = 60, 90, and 120 days)

Ck = *Brachiaria brizantha* Ecotypes (Eth. 13726, Eth. 13809 and Eth. 1377).

Ai^*Hj = the interaction of ith altitude and jth harvesting date.

A^*C = interaction of altitude and cultivars.

H^*C = interaction of harvesting date and cultivars.

εijk = the residual error.

Results

Effect of harvesting dates and altitudes on plant morphological characteristics of *B. brizantha ecotypes*

The results of the effect of harvesting dates and altitudes on plant morphological characteristics of Eth. 13726, Eth. 13809 and Eth. 1377 grass ecotypes are presented in Table 1. The overall results obtained in the three agro-ecologies indicate that all morphological characteristics of Eth. 13726 grass studied were significantly affected ($P>0.05$) by harvesting dates and altitude. In mid-altitude, plant height (PH), number of tillers per plant (NTPP), number of leaves per plant (NLPP) and length of leaves per plant (LLPP) showed significant difference ($P<0.05$) while between low and high altitude there was significant difference ($P<0.05$) in PH only. All parameters were significantly ($P<0.05$) increased as harvesting stage increases from 60 to 120 days. Plant height showed high increment at day 90 while NLPP showed significant increment ($P>0.05$) by harvesting at day 120. Although there was no significant ($P<0.05$) difference in PH between second and third harvest, numerically there were high increment in NLPP.

The overall results obtained in the three agro-ecologies indicate that with the exception of PH by altitudes, other morphological characteristics of Eth. 13809 grass studied were significantly affected ($P>0.05$) by altitude and harvesting date. Except NLPP by altitude, all the morphological characteristics (PH, NTPP, NTPP and LLPP) of Eth. 1377 grass in all studied agro-ecology were significantly affected ($P>0.05$) by altitude and harvesting date.

Table 1 Effects of altitude, harvesting stages and their interactions on plant morphological characteristics of *brizantha* grass ecotypes

Factors	Parameters			
Cultivars	PH (cm)	NTPP (counts)	NLPP (counts)	LLPP (cm)
Eth. 13726				
Altitude				
Low	59.39ᵃ	24.71ᵃᵇ	4.64ᵇ	18.61ᵃᵇ
Mid	56.97ᵃ	33.67ᵃ	5.51ᵃ	19.38ᵃ
High	36.86ᵇ	20.67ᵇ	4.08ᵇ	13.89ᵇ
Sig	**	*	**	*
Harvesting stages				
60	26.69ᵇ	17.89ᵇ	3.97ᵇ	14.29ᵇ
90	56.83ᵃ	25.07ᵃᵇ	4.58ᵇ	17.23ᵃᵇ
120	69.69ᵃ	36.09ᵃ	5.69ᵃ	20.36ᵃ
Mean	51.07	26.35	4.74	17.29
SE	7.39	3.40	0.33	1.29
Sig	***	*	**	*
Eth. 13809				
Altitude				
Low	56.40	35.77ᵃ	4.77ᵃ	20.17ᵇ
Mid	52.99	35.07ᵃ	4.52ᵃ	23.89ᵃ
High	45.03	29.72ᵇ	3.78ᵇ	16.27ᶜ
Sig	NS	*	*	**
Harvesting stages				
60	33.14ᵇ	25.13ᵇ	3.62b	17.52ᵇ
90	53.01ᵃ	29.86ᵇ	4.29b	19.84ᵃᵇ
120	68.27ᵃ	45.56ᵃ	5.16a	22.97ᵇ
Mean	51.47	33.52	4.36	20.11
SE	5.52	3.26	0.27	1.38
Sig	**	***	**	*
Eth. 1377				
Altitude				
Low	65.72ᵃ	25.09ᵇ	5.48	18.01ᵃ
Mid	62.91ᵃ	32.14ᵃ	5.07	17.39ᵃ
High	47.57ᵇ	30.93ᵃ	4.48	13.25ᵇ
Sig	**	**	NS	**
Harvesting stages				
60	47.89ᶜ	18.54ᶜ	3.80ᵇ	12.97ᶜ
90	58.75ᵇ	26.36ᵇ	4.84ᵃᵇ	16.72ᵇ
120	69.56ᵃ	43.25ᵃ	6.39ᵃ	18.96ᵃ
Mean	58.73	29.38	5.01	16.22
SE	4.29	3.83	0.44	1.16
Sig	**	***	*	***

Treatments means with different letters in a column are significantly different ($P<0.05$) for altitudes and harvesting stages. NS for non-significant, *for $P<0.05$ (5%), **for $P<0.01$ (1%) and ***for $P<0.001$ (0.1%); *SE* standard error, *PH* plant height, *NTPP* number of tillers per plant, *LLPP* leaf length per plant

Effect of cutting ages and altitudes on chemical composition, dry matter yield of three *B. brizantha* grass ecotypes

The effect of harvesting dates and altitudes on chemical composition, dry matter yield (DMY) and crude protein yield (CPY) of Eth. 13726, Eth. 13809 and Eth. 1377 grasses is shown in Table 2. The effect of harvesting date and altitude on chemical composition, DMY and CPY of Eth. 13726 grass is shown in Table 2. DM, DMY and CP, NDF and ADF by harvesting date and only DM by altitude showed significant differences ($P<0.05$) in Eth. 13726 grass.

There were significant differences ($P<0.05$) in DM, DMY, CP and CPY both by altitude and harvesting date. However, there was no significant difference ($P<0.05$) between mid- and high altitudes. Moreover, there were significant differences ($P<0.05$) in ash by altitude and ADF by harvesting date of Eth. 13809. The overall results of the grass in low altitude show that greatest DM, DMY, ASH, CP and CPY than mid- and high altitudes. There were significant differences ($P<0.05$) in DM and DMY by altitude and harvesting date. CPY by altitude and CP, ADF and ADL by harvesting dates were also significantly different ($P<0.05$) in Eth. 1377 grass.

Chemical composition of three *Brachiaria* grass lines

The effect of altitude, harvesting date and their interaction on chemical composition of *Brachiaria* grasses is shown in Table 3. There were highly significant difference ($P<0.05$) in the average DM, DMY, CPY, ADF and ADL of the three ecotypes. Except NDF, ADF and ADL by attitudes and ash by harvesting dates, there were highly significant ($P<0.05$) effects of altitude and harvesting date between ecotypes and have also high interactions between ecotypes and altitudes. There were significant ($P<0.05$) overall interactions except for NDF.

Agronomic performance of three *B. brizantha* ecotypes grass lines

The agronomic comparison of three ecotype grasses (Eth. 13726, Eth.13809 and Eth.1377) is presented in Table 4. Except NLPP, the agronomic performances of PH, NTPP and LLPP differ significantly ($P>0.05$) between three *Brachiaria* grass lines. Eth. 1377 performs best in PH and NLPP from the other two ecotypes while in case of NTPP and LLPP Eth. 13809 showed best performance.

Discussion

The agronomic characteristics result indicated that all ecotypes tested recorded higher PH than other *Brachiaria* grass cultivars (Llanero, MG4, Marandu, Piata, Mulato

Table 2 Effect of altitude, cutting age and their interaction on chemical composition and yield of *Bracheria* grass ecotypes

Factors	Parameters							
Cultivar	DM (%)	DMY (t/ha)	Ash (%)	CP (%)	CPY (t/ha)	NDF (%)	ADF (%)	ADL (%)
Eth. 13726								
Altitude								
Low	37.60[a]	4.24	8.97	9.81	0.33	71.70	50.99	14.46
Mid	36.55[b]	4.63	9.66	12.36	0.47	67.57	51.02	16.57
High	35.99[b]	3.28	13.98	11.51	0.34	69.42	50.29	14.87
Sig	**	NS	NS	NS	NS	NS	NS	NS
Harvesting stages								
60	35.03[c]	1.69[b]	12.67	16.33[a]	0.29	63.87[b]	42.14[b]	11.84
90	36.64[b]	3.63[ab]	10.85	10.63[ab]	0.42	68.72[ab]	52.73[a]	15.37
120	38.47[a]	6.83[a]	9.10	6.72[b]	0.44	76.11[a]	57.44[a]	18.69
Mean	36.71	4.05	10.87	11.23	0.38	69.56	50.77	15.30
SE	0.55	0.83	1.06	1.55	0.05	1.99	2.38	1.18
Sig	**	*	NS	*	NS	*	**	NS
Eth. 13809								
Altitude								
Low	38.77[a]	7.26[a]	15.40[a]	14.93[a]	1.04[a]	65.26	41.85	10.64
Mid	38.13[b]	5.98[ab]	9.15[b]	10.54[b]	0.57[b]	66.85	44.96	11.99
High	36.35[c]	4.50[b]	12.48[a]	8.57[b]	0.36[b]	64.31	43.60	10.75
Sig	***	**	**	**	**	NS	NS	NS
Harvesting stages								
60	36.58[c]	2.95[c]	12.35	13.87[a]	0.43[b]	59.91	39.11[b]	9.12[b]
90	37.70[b]	6.51[b]	11.83	10.60[b]	0.72[ab]	63.89	42.18[ab]	10.54[ab]
120	38.96[a]	8.28[a]	12.85	9.57[b]	0.83[a]	72.62	49.11[a]	13.72[a]
Mean	37.75	5.91	12.34	11.35	0.66	65.47	43.47	11.13
SE	0.50	0.89	0.95	1.156	0.12	2.40	1.65	0.80
Sig	***	***	NS	**	*	NS	*	NS
Eth. 1377								
Altitude								
Low	36.96[a]	6.28[a]	12.30	13.04	0.75[a]	67.57	48.57	13.38
Mid	35.14[b]	4.68[b]	11.47	11.17	0.47[b]	63.81	43.07	12.93
High	35.04[b]	4.30[b]	10.03	8.60	0.34[b]	71.80	48.79	13.61
Sig	***	**	NS	NS	*	NS	NS	NS
Harvesting stages								
60	34.85[c]	2.65[c]	11.58	14.80[a]	0.41	61.01	40.70[b]	9.94[b]
90	35.57[b]	5.72[b]	9.75	10.15[b]	0.60	68.64	47.65[a]	14.28[a]
120	36.72[a]	6.89[a]	12.47	7.86[b]	0.54	73.52	52.08[a]	15.70[a]
Mean	35.71	5.09	11.27	10.94	0.52	67.72	46.81	13.31
SE	0.42	0.71	0.72	1.26	0.07	2.47	1.97	0.92
Sig	**	***	NS	*	NS	NS	**	*

DM dry matter, *DMY* dry matter yield, *CP* crude protein, *CPY* crude protein yield, *NDF* neutral detergent fiber, *ADF* acid detergent fiber, *ADL* acid detergent lignin, *SE* standard error, *CV* coefficient of variation

Mean values followed by a different lowercase superscript letter in the same column are statistically significant at $P < 0.05$. NS for non-significant, * for $P < 0.05$ (5%), ** for $P < 0.01$ (1%) and *** for $P < 0.001$ (0.1%)

II, Basilisk and Xaraes) in Kenya dry lands as reported by Nguku et al. [42]. However, height may not be an important estimate on the expected biomass yield, as it has been clearly demonstrated that the shorter Eth. 13809 than Eth. 1377 had the best primary DM yield. This high DM yield of Eth. 13809 might be due to high number of tillers.

Table 3 Mean chemical composition, yield of *Bracheriea* ecotypes

Factor	Parameters							
	DM (%)	DMY (t/ha)	Ash (%)	CP (%)	CPY (t/ha)	NDF (%)	ADF (%)	ADL (%)
Cultivars								
Eth. 13726	36.71^{b}	4.05^{b}	10.87	11.23	0.38^{b}	69.56	50.77^{a}	15.30^{a}
Eth. 13809	37.75^{a}	5.91^{a}	12.34	11.35	0.66^{a}	65.47	43.47^{b}	11.13^{b}
Eth. 1377	35.71^{c}	5.09^{ab}	11.27	10.94	0.52^{ab}	67.72	46.81^{b}	13.31^{ab}
Overall mean	36.72	5.02	11.49	11.17	0.52	67.59	47.02	13.24
Overall SE	0.319	0.47	0.53	0.74	0.05	1.32	1.26	0.64
CV (%)	0.60	16.63	13.73	11.27	26.77	7.03	5.97	13.14
Sig	***	**	NS	NS	**	NS	**	**
Altitudes								
Low	37.77^{a}	5.93^{a}	12.22^{a}	12.59^{a}	0.71^{a}	68.17	47.14	12.83
Mid	36.61^{b}	5.10^{ab}	10.09^{b}	11.35^{a}	0.50^{b}	66.08	46.35	13.83
High	35.79^{c}	4.02^{b}	12.17^{ab}	9.56^{b}	0.35^{b}	68.51	47.56	13.08
Sig	***	**	*	**	**	NS	NS	NS
Harvesting dates								
60	35.49^{c}	2.43^{c}	12.20	15.00^{a}	0.37^{b}	61.60^{b}	40.65^{c}	10.30^{c}
90	36.64^{b}	5.29^{b}	10.82	10.46^{b}	0.58^{a}	67.08^{b}	47.52^{b}	13.40^{b}
120	38.05^{a}	7.33^{a}	11.47	8.05^{c}	0.60^{a}	74.08^{a}	52.88^{a}	16.04^{a}
Sig	***	***	NS	***	*	**	***	***
P value								
cul*al	**	NS	**	**	*	NS	NS	NS
cul*hd	**	NS	NS	NS	NS	NS	NS	NS
al*hd	NS	NS	NS	NS	NS	NS	NS	NS
Interaction	***	***	*	***	*	NS	**	*

DM dry matter, *DMY* dry matter yield, *CP* crude protein, *CPY* crude protein yield, *NDF* neutral detergent fiber, *ADF* acid detergent fiber, *ADL* acid detergent lignin

Mean values followed by a different lowercase superscript letter in the same column are statistically significant at $P < 0.05$

More tillers were reported for Eth. 13809 at 120 days (45.56) in low (35.77) and mid (35.07)-altitudes as shown in Table 1. Tillers density is an important attribute of grasses as it increases the chances of survival and amount of available forage [28, 51]. Moreover, it is an indicator of resource use efficiency by the different grass species. The large numbers of tillers produced by some grass species allows them to attain maximum growth at an earlier age and recover faster after defoliation [28]. Tillering is also important in forage plants, because of its influence on leaf-area production and dry matter yield. Mean tiller number increased progressively for all ecotypes, and there were significant differences among the ecotypes ($P < 0.05$) as shown in Table 4. The results in this study are similar to the findings by Mganga [35] and Machogu [32], where the tiller density differed among the local range grasses in Kenya.

The DM content increased with delayed harvesting because of decreased moisture content in leaves as the plants aged and became lignified. This result is in agreement with other studies [4, 8] for other types of grasses. The studies reported that the DM content of grasses increased with an increase in growth and development of plants and longer time to harvest. The higher dry matter yields at later stages of harvesting were to be expected as plants were taller, had more tillers per plant and more leaves per plant and leaf elongation and stem development. All these characteristics would contribute to increased photosynthetic activity and hence higher DM production. The highest total DM yield observed at the last harvest stage was in agreement with the work of Hare et al. [23], Njarui and Wandera [45] and Ondiko et al. [47] on wild *Brachairia* grasses at different countries; Tessema and Alemayehu [53], Leta et al. [29]; Asmare et al. [9] on cultivated grasses and Feyissa et al. [20] for natural pasture in Ethiopia. These researchers indicated that the time of harvesting had a high influences on dry matter yield. The highest yield of forage for the longest harvesting date could also be attributed to the favorable rainfall, temperature and available nutrient in the soil over the extended growing period of the grass in the study area. But, when compared with time taken until consumption, the age of the grasses at day 60 could be used by farmers

Table 4 Overall morphological characteristics of three *Brachiaria brizantha* ecotypes on the effects of altitude and harvesting dates and their interactions

Factor	Parameters			
	PH (cm)	NTPP (counts)	NLPP (counts)	LLPP (cm)
Ecotypes				
Eth. 13726	51.07[b]	26.35[b]	4.74	17.29[b]
Eth. 13809	51.47[b]	33.52[a]	4.36	20.10[a]
Eth. 1377)	58.73[a]	29.38[b]	5.01	16.22[b]
Overall mean	53.76	29.75	4.70	17.87
Overall SE	3.33	2.03	0.20	0.78
CV (%)	8.65	8.15	11.44	5.25
Sig	*	***	NS	***
Altitudes				
Low	60.50[a]	28.52[b]	4.96[a]	18.93[b]
Mid	57.62[a]	33.62[a]	5.04[a]	20.22[a]
High	43.15[b]	27.12[b]	4.11[b]	14.47[c]
Sig	***	**	*	***
Harvesting dates				
60	35.91[c]	20.52[c]	3.80[c]	14.93[c]
90	56.20[b]	27.09[b]	4.57[b]	17.93[b]
120	69.17[a]	41.63[a]	5.74[a]	20.76[a]
Sig	***	***	***	***
P value				
cul*al	NS	**	NS	*
cul*hd	0.0263	NS	NS	NS
al*hd	NS	NS	NS	NS
Interaction	***	***	*	***

SE standard error, *CV* coefficient of variation, *PH* plant height, *NTPP* number of tillers per plant, *LLPP* leaf length per plant

Treatments means with different letters in a column are significantly different ($P<0.05$) for altitudes and cutting age. NS for non-significant, * for $P<0.05$ (5%), ** for $P<0.01$ (1%) and *** for $P<0.001$ (0.1%)

for forage because the relative growth is high, soft and palatable with reduced lignin.

Overall, Eth. 13809 had highest dry matter yield (5.91 t/ha) which is followed by Eth. 1377 (5.09 t/ha). Variations in DM production across the ecotypes can be attributed to differences in growth rate and growth habit, which are mediated through the genotypic and phenotypic differences. This is a common phenomenon in grasses [35, 46]. In this study, the high primary DM yield by Eth. 13809 can be largely attributed to its large size leaves (20.1 cm long) and tiller number (33.52) than the other two ecotypes. However, the leafy nature might be a disadvantage in dry areas where water supply is limited, as it facilitates rapid water loss through transpiration [48]. Grasses which yield the highest DM should be the most sought since they can supply the highest amount of forage to livestock. The average DMY of ecotypes (Eth. 13726 (4.05 t/ha), Eth. 13809 (5.91 t/ha and Eth. 1377 (5.09 t/ha) were very high compared to *C. ciliaris* with 1.6 t/ha reported by Mutimura and Everson [39] and natural pasture (4.4 ton/ha) in Ethiopia [1].

Crude protein (CP) value of the three *B. brizantha* ecotypes (Eth. 13726 (16.33, 10.63 and 6.72%), Eth. 13809 (13.87, 10.60 and 9.57%) and Eth. 1377 (14.80, 10.15 and 7.86%) differed significantly ($P>0.05$) among harvesting dates. As would be expected, the highest CP concentration was obtained at the earliest stage of harvesting, with values declining as harvesting was delayed. This result is in agreement to the findings by Njarui et al. [44] and Mutimura et al. [38] in *Brachiaria* grasses. Similarly, Bayble et al. [7] and Ansah et al. [5] reported a decreasing trend of CP with increase in harvesting age ($60>90>120$ days) for Napier grass. This is a growth dilution effect with increase in structural carbohydrate content of forage materials harvested at late maturity reducing the percentage of protein in the forage.

The higher CP Contents of Eth. 13809 and Eth. 1377 at low altitude and Eth. 13726 at mid-altitude were higher than one of the best *Brachiaria* hybrid cv. Mulato II reported by Mutimura and Everson [39] which had a CP of 11.1%. The overall highest CP of the ecotypes was recorded in low land (12.59%) compared to mid (11.35%) and high land (9.56%) which is contrary to those reported as high temperatures to have positive effect on quality of grasses [43]. This might be due to agro-ecological and management differences during the growth stages of ecotypes.

The CP content was generally high in all the *Brachiaria* grass ecotypes [Eth. 13726 (11.23%), Eth. 13809 (11.35%) and Eth. 1377(10.94%)] compared with mean of (5.3–7.7%) and (7–10%) different *Brachiaria* grass cultivars reported by Ondiko et al. [47] and Nguku [41] in coastal lowlands of Kenya and in the semiarid region of eastern Kenya, respectively. Natural pasture, Rhodes grass, Tef straw, Maize stover and Finger millet straw were also found have very low CP value of 5.5, 7.1, 4.2, 2.84 and 4.12%, respectively (Abebe et al. [1] and *desho* grass (8.35%) [9]. Furthermore, the CP content was higher than the most Ethiopian dry forages and roughages which have a CP content of less than 9% [10], which is the level required for adequate microbial synthesis in the rumen [3]. Of the factors considered, the CP content at low altitude was higher (12.59%) than at mid-altitude (11.35%) and high altitude (9.56%) which may have been associated with differences in temperature, precipitation and soil characteristics as reported by Daniel [17] where plant growth and quality were affected markedly by temperature and soil moisture conditions. Generally, all ecotypes in all altitudes including high altitude which is uncommon in other countries are highly palatable.

Forages with a CP content range of 9–12% are highly palatable [32]. Despite the reduction in CP percentage with time, crude protein yield (CPY) increased significantly as harvesting was delayed. Similar findings have been reported by Asmare et al. [9] for the desho grass species and by Melkie [34]. They recorded mean CPYs of Bana grass at 60, 90 and 120 days of age to have 0.47, 0.91 and 0.85 t/ha, respectively. Obviously, decisions on the optimal time to harvest *Brachiaria* grass will depend on a compromise between yield and quality of forage.

As would be expected, NDF, ADF and ADL concentrations increased as harvesting date was delayed though there was no significant ($P<0.05$) difference of NDF in Eth. 1377 and ADL in Eth. 13726 and Eth. 13809. Moreover, there was no significant ($P<0.05$) difference in all ecotypes by altitude. The NDF content of forage varies widely, depending on species, maturity, and growing environment [12, 33, 40]. Therefore, each plant species presents a unique NDF–ADF ratio in the feed. For legumes, <40% NDF content is classified as good quality forage, while >50% [54] is considered as poor quality forage. For grasses, <50% NDF is considered high quality and >60% as low-quality forage. In this study, forage materials from all the grass ecotypes had >60% NDF which may account for the low intake and digestibility. The higher fiber content in the Ecotype grasses, therefore, can be attributed to their phenological stage. Furthermore, increase in CF content is accompanied by increase in lignin content in forages. The forages, therefore, become less digestible because they are inaccessible to digestive enzymes [16]. The ash quantity of any feed is a positive indicator of the inorganic (minerals) content. Generally, most forage has ash content ranging from 3 to 12% (Linn and Martin [30]. All the three ecotypes investigated in this study were in the range shown by these authors.

Conclusion

Based on results, it can be concluded that the harvesting dates can affect the forage DM yield and nutritive values of *Brachiaria* grasses. Day 60 could be the optimal level for harvesting *Brachiaria* grass since the quality forage is high and yield is not compromised. The highest CP and lowest NDF and ADF concentrations were recorded in Eth13809. Thus, among the tested ecotype grasses Eth13809 showed great potential as a forage plant especially in low altitude since it has shown that it can grow under low rainfall maintaining high yields. Overall, *Brachiaria* grasses had a higher biomass yield and better chemical composition than the main feed resources of natural pasture and crop residues in the region. Therefore, it can be concluded that the studied *Brachiaria* grass ecotypes, especially Eth13809, have great potential as an alternative ruminant feed in all altitude areas in south Gondar. To fully utilize the potential of the studied *Brachiaria* grass ecotypes, further studies on agronomic and nutritional evaluation involving animal evaluation experiments are recommended.

Abbreviations

ADF: acid detergent fiber; ADL: acid detergent lignin; AOAC: association of official analytical chemists; CP: crude protein; CPY: crude protein yield; CSA: central statistical agency; DM: dry matter; DMY: dry matter yield; EIAR: Ethiopian Institute of Agricultural Research; Eth: Ethiopia; FAO: Food and Agricultural Organization the United Nations; GLM: general linear model; LLPP: leaf length per plant; NDF: neutral detergent fiber; NLPP: number of leaf per plant; NTPP: number of tillers per plant; PH: plant height; OM: organic matter; SAS: statistical analysis system.

Authors' contributions

WA contributed in research proposal writing, data collection, data analysis, data interpretation and article writing. BA, AT and BA contributed in research data collection, data analysis, data interpretation and article writing. All authors read and approved the final manuscript.

Authors' information

Adnew is a candidate of PhD at Bahir Dar University, College of Sciences in Biology Department. Berhanu Abraha Tsegay (PhD) is an Associate professor of Biology, Bahir Dar University, Plant Biology, currently teaching at graduate program and conducting research on plant ecophysiology and Ethnobotany. Tassew (PhD) is an Associate professor of animal production, currently teaching and conducting research in the area of animal feeds and feeding, animal production and productivity improvement in Ethiopia. Asmare (PhD) is an Assistant Professor of Animal Feeds and Nutrition, currently teaching and conducting research in the area of animal feeds and feeding, animal production and productivity improvement in Ethiopia.

Author details

[1] Department of Biology, College of Science, Bahir Dar University, P. O. Box 76, Bahir Dar, Ethiopia. [2] Department of Animal Production and Technology, College of Agriculture and Environmental Sciences, Bahir Dar University, P.O. Box 5501, Bahir Dar, Ethiopia.

Acknowledgements

The first author is grateful to Bahir Dar University for funding the research work. He also extends his thanks to Fogera National Rice Research Institute, Farta and Tach Gayint district agricultural offices for their permission to use the respective experimental sites. He also appreciates the contribution of Mr. Bamlaku Getie, Mr. Getachew Molla and Mr Yilak Hulgezie for their assistance during field data collection.

Competing interests

All authors declare that they have no competing interests.

Funding

Bahir Dar University, *College of Sciences*.

References

1. Abebe Y, Tafere M, Dagnew S, Tolla M, Gebre-Selassie Y et al. Best fit practice manual for Rhodes grass (*Chloris gayana*) production. BDUCASCAPE, 2015;working paper 10.
2. Adnew W, Tsegay BA, Tassew A, Asmare B. Assessments of farmers' perception and utilization status of Brachiaria grass in selected areas of Ethiopia. Biodiversitas. 2018;19(3):951–62. https://doi.org/10.13057/biodiv/d190326.
3. Agricultural Research Council. The nutrient requirements of ruminants livestock (Supp. 1). Farnham Royal, UK: Commonwealth Agricultural Bureaux; 1980. p. 351
4. Alemu B, Melaku S, Prasad NK. Effects of varying seed proportions and harvesting stages on biological compatibility and forage yield of oats (*Avena sativa* L.) and vetch (*Vicia villosa* R.) mixtures. Livest Res Rural Dev. 2007;19:1e9.
5. Ansah T, Osafo ELK, Hansen HH. Herbage yield and chemical composition of four varieties of Napier (*Pennisetum purpureum*) grass harvested at three different days after planting. Agric Biol J N Am. 2010;1(5):923–9. https://doi.org/10.5251/abjna.2010.1.5.923.929.
6. Arango J, Moreta D, Núñez J, Hartmann K, Domínguez M, Ishitani M, Miles J, Subbarao G, Peters M, Rao I. Developing methods to evaluate phenotypic variability in biological nitrification inhibition (BNI) capacity of *Brachiaria* grasses. Trop Grassl Forrajes Trop. 2014;2:6–8.
7. Bayble T, Melaku S, Prasad NK. Effects of cutting dates on nutritive value of Napier (*Pennisetum purpureum*) grass planted sole and in association with Desmodium (*Desmodium intortum*) or Lablab (*Lablab purpureus*). Livestock Research for Rural Development 2007-19, article #11. www.lrrd.org/lrrd19/1/bayb19011.htm. Accessed 11 May 2016.
8. Berihun M. Effect of planting patterns and harvesting days on yield and quality of Bana Grass [*Pennisetum purpureum* (L.) *Pennisetum americanum* (L.)] (Master thesis). School of Graduates of Haramya University, 2005, Harar, Ethiopia.
9. Asmare B, Solomon D, Taye T, Firew T, Aynalem H, Jane W. Effects of altitude and harvesting dates on morphological characteristics, yield and nutritive value of desho grass (*Pennisetum pedicellatum* Trin.) in Ethiopia. Agric Nat Resour. 2017;51:148–53.
10. Bediye S, Sileshi Z. The composition of Ethiopian feedstuffs, Research Report No. 6. Addis Ababa, Ethiopia: Institute of Agricultural Research (IAR); 1989. pp 33.
11. Bogdan AV. Herbage plants at the grassland research station, Kitale, Kenya. East Afr Agric For J. 1955;20:151–65.
12. Buxton DR, Fales SL. Plant environment and quality. In: Fahey Jr GC, editor. Forage quality, evaluation and utilization. Madison: ASA-CSSASSA; 2004. p. 155–99.
13. CSB (Climate-Smart *Brachiaria* Program). CSB annual review meeting. KALRO (Kenya Agricultural and Livestock Research Organization), Embu, Kenya 2016. goo.gl/VbzJ4D.
14. Central Statistical Agency (CSA), Agricultural Sample Survey. Livestock and livestock characteristics (private peasant holdings) (Addis Ababa Ethiopia). Stat Bull. 2017;585(2):33–5.
15. CIAT. Pastures for the tropical lowlands. Cali: CIAT; 1992.
16. Crowder LV, Chheda MR. Tropical grassland husbandry. London: Longman; 1982.
17. Daniel K. Effect of nitrogen application and stage of development on yield and nutritive value of Rhodes grass (*Chloris gayana*). Ethiop J Agric Sci. 1996;15:86–101.
18. EIAR. Livestock research strategies (2016–2030): feeds and nutrition, rangelands and animals health. Addis Ababa: Ethiopia Institute of Agricultural Research; 2017.
19. Jirata M, Grey S, Kilawe E. Ethiopia climate-smart agriculture scoping study. Addis Ababa, Ethiopia: FAO; 2016. http://www.fao.org/publications
20. Feyissa F, Shiv P, Getnet A, Seyoum B, Getu K, Aemiro K, Gezahenge K. Dynamics in nutritional characteristics of natural hays as affected by harvesting stage, storage method and storage duration in the cooler tropical highlands. Afr J Agric Res. 2014;9:3233–44.
21. Gelayenew B, Nurfeta A, Assefa G, Asebe G. Assessment of livestock feed resources in the farming systems of mixed and shifting cultivation, Gambella Regional State, Southwestern Ethiopia. Glob J Sci Front Res. 2016;XVI(V):1.
22. Ghimire SR, Njarui D, Mutimura M, Cardoso JA, Johnson L, Gichangi E, Teasdale S, Odokonyero K, Caradus JR, Rao IM, Djikeng A. Climate-smart *Brachiaria* for improving livestock production in East Africa: Emerging opportunities. In: Vijay D, Srivastava MK, Gupta CK, Malaviya DR, Roy MM, Mahanta SK, Singh JB, Maity A, Ghosh PK, editors. Proceedings of the XXIII international grassland congress, New Delhi, India, 2015. 20–24 November 2015. p. 361–370. hdl.handle.net/10568/69364.
23. Hare MD, Phengphet S, Songsiri T, Sutin N, Stern E. Effect of cutting interval on yield and quality of three Brachiaria hybrids in Thailand. Trop Grassl Forrajes Trop. 2013;1:84–6.
24. Huhtanen P, Nousiainen J, Rinne M. Recent developments in forage evaluation with special reference to practical applications. Agric Food Sci. 2006;15:293–323.
25. Jank L, Barrios SC, do Valle CB, Simeao RM, Alves GF. The value of improved pastures to Brazilian beef production. Crop Pasture Sci. 2014;65:1132. https://doi.org/10.1071/cp13319.
26. Jančík F, Koukolová V, Kubelková P, Čermák B. Effects of grass species on ruminal degradability of silages and prediction of dry matter effective degradability. Czech J Anim Sci. 2009;54:315–23.
27. Kluthcouski J, de Oliveira I, Yokoyama L, Dutra L, Portes TDA, da Silva A, Pinheiro BDS, Ferreira E, de Castro EDM, Guimarães CM, Gomide JDC, Balbino LC. The Barreirão system: recovering and renewing degraded pastures with annual crops. In: Guimarães EP, Sanz JI, Rao I, Amézquita MC, Amézquita E, Thomas RJ, editors. Agropastoral systems for the tropical savannas of Latin America 2004, vol 338. Centro Internacional de Agricultura Tropical (CIAT), Cali, Colombia.
28. Laidlaw A. The relationship between tiller appearance in spring and contribution of dry-matter yield in perennial ryegrass (*Loliumperenne* L.) cultivars differing in heading date. Grass Forage Sci. 2005;60:200–9.
29. Leta G, Duncan A, Abdena A. Desho grass (*Pennisetum pedicellatum*) for livestock feed, grazing land and soil and water management on small-scale farms. In: NBDC Brief 11, International Livestock Research Institute (ILRI), Nairobi, Kenya (2013).
30. Linn JG, Martin NP. Forage quality tests and interpretations; 1999 (**access online**).
31. Maass BL, Midega AO, Mutimura M, Rahetlah VB, Salgado P, Kabirizi JM, Khan ZR, Ghimire S, Rao IM, Homecoming of brachiaria: improved hybrids prove useful for African animal agriculture. East Africa Agric Fo J. 2015;81:71–8
32. Machogu C. A comparative study of the productivity of brachiaria hybrid cv. mulato II and native pasture species in semi-arid rangelands of Kenya: M.Sc., Thesis paper, Nairobi, Kenya;2013.
33. Mahyuddin P. Chemical composition of leaf and stem of tropical grasses at different stage of growth. J ProduksiTernak. 2007;9(3):153–9.
34. Melkie B. Effect of planting patterns and harvesting days on yield and quality of Bana grass [*Pennisetum purpureum* × *Pennisetum americanum*]. M.Sc. Thesis, 2005. Haramaya University, Harar, Ethiopia.
35. Mganga K. (2009) Impact of grass reseeding technology on rehabilitation of degraded rangelands: a case study of Kibwezi district, Kenya. University of Nairobi, M.Sc. Thesis, Nairobi, Kenya.
36. Moreta DE, Arango J, Sotelo M, Vergara D, Rincón A, Ishitani N, Castro A, Miles J, Peters M, Tohme J, Subbarao GV, Rao IM. Biological nitrification inhibition (BNI) in Brachiaria pastures: A novel strategy to improve eco-efficiency of crop-livestock systems and to mitigate climate change. Trop Grassl Forrajes Trop. 2014;2:88–91.
37. Mureithi JG, Djikeng A. Overview of the climate-smart Brachiaria grass programme. In: Njarui DMG, Gichangi EM, Ghimire SR, Muinga RW, editors. Climate-smart Brachiaria grasses for improving livestock production in East Africa: Kenya experience: Proceedings of a workshop, Naivasha, Kenya, 2016;14–15 September 2016. Kenya Agricultural and Livestock Research Organization, Nairobi, Kenya.
38. Mutimura M, Ebong C, Rao IM, Nsahlai IV. Effect of cutting time on agronomic and nutritional characteristics of nine commercial cultivars of *Brachiaria* grass compared with Napier grass during establishment under semi-arid conditions in Rwanda. Afr J Agric Res. 2017;12(35):2692–703.
39. Mutimura M, Everson T. On-farm evaluation of improved Brachiaria grasses in low rainfall and aluminium toxicity prone areas of Rwanda. Int J Biodivers Conserv Biol. 2012;4:137–54.

40. Nelson CJ, Moser LE. Plant factors affecting forage quality. In: Fahey Jr GC, editor. Forage quality, evaluation and utilization. Madison: ASA-CSSA-SSSA; 1994. p. 115–54.
41. Nguku S. An evaluation of brachiaria grass cultivars productivity in semi arid Kenya. MSc. Thesis, South Eastern Kenya University, Kenya; 2015
42. Nguku S, Musimba N, Njarui D, Mwobobia R, Kaindi E. Effects of acid scarification on germination of the genus *Brachiaria* grass Cultivars. Int J Sci Res Innov Technol. 2016;3:45–50.
43. Njarui DMG, Gatheru M, Mwangi DM, Keya GA. Persistence and productivity of selected Guinea grass ecotypes in semiarid tropical Kenya. Grassl Sci. 2015;61:142–52.
44. Njarui DMG, Gichangi E, Gatheru M, Nyambati EM, Ondiko CN, Njunie MN, Ndungu KW, Kiiya WW, Kute CAO, Ayako W. A comparative analysis of livestock farming in smallholder mixed crop-livestock systems in Kenya: 1. Livestock inventory and management. Livestock Research for Rural Development, 2016; 28, Article #066.
45. Njarui DMG, Wandera FP. Effects of cutting frequency on productivity of five selected herbaceous legumes and five grasses in semi-arid tropical Kenya. Trop Grassl. 2004;38:158–66.
46. Ogillo BP. Evaluating performance of range grasses under different micro catchments and financial returns from reseeding in Southern Kenya. Nairobi, Kenya: Unpublished M.Sc. Thesis; University of Nairobi; 2010.
47. Ondiko CN, Njunie MN, Njarui MG, Auma E, Ngode L. Effect of cutting frequency on forage production and nutritive value of Brachiaria grass cultivars in coastal lowlands of Kenya. In: Njarui DMG, Gichangi EM, Ghimire SR, Muinga RW, editors. Climate smart brachiaria grasses for improving livestock production in East Africa—Kenya experience. Proceedings of the workshop held in Naivasha, Kenya, 14–15 September 2016, Nairobi, Kenya; 2016. p. 271.
48. Pratt and Gwynne. Rangeland management and ecology in East Africa. London: Hodder and Stroughton; 1977.
49. Rodrigues RC, Sousa TVR, Melo MAA, Araújo JS, Lana RP, Costa CS, Oliveira MO, Parente MOM, Sampaio IBM. Agronomic, morphogenic and structural characteristics of tropical forage grasses in northeast Brazil. Trop Grassl Forrajes Trop. 2014;2:214–22.
50. Schiek B, González C, Mwendia S, Steven D. Prager. Got forages? Understanding potential returns on investment in *Brachiaria* spp. for dairy producers in Eastern Africa. Trop Grassl Forrajes Trop. 2018;6(3):117–133. https://doi.org/10.17138/tgft(6)117-133.
51. Skerman PJ, Riveros F. Tropical grasses. FAO Plant Production and Protection Series No. 23. Rome, Italy; Food and Agriculture Organisation, 1990.
52. Steinfeld H, Gerber P, Wassenaar T, Castel V, Rosales M, de Haan C. Livestock's long shadow: environmental issues and options. Food and Agriculture Organization of the United Nations, Rome, 2006. http://www.fao.org/docrep/fao/010/a0701e/a0701e00.pdf.
53. Tessema K, Alemayehu M. Management of Napier grass (*Pennisetum purpureum* (L) Schumach) for high yield and nutritional quality in Ethiopia: a review. Ethiop J Anim Prod. 2010;10:73–94.
54. Van Saun RJ. Determining forage quality: understanding feed analysis. Lamalink.com. 2006;3(8):18–9.
55. Van Soest PJ, Robertson JB. Analysis of forage and fibrous food. In: A laboratory manual for animal science, vol 613. Ithaca, NY, USA: Cornell University; 2015

Nutrient intake, digestibility, growth performance and carcass characteristics of Farta sheep fed urea-treated rice straw supplemented with graded levels of dried *Sesbania sesban* leaves

Lamrot Tekliye, Yeshambel Mekuriaw, Bimrew Asmare* and Fantahun Mehret

Abstract

Background: The experiment was conducted using twenty-five Farta sheep to investigate the response of supplementing dried *Sesbania sesban* leaves (SSL) on feed intake, digestibility, growth performance and carcass characteristics of Farta sheep fed urea-treated rice straw. Twenty-five male Farta sheep with initial body weight of 18.9 ± 1.7 kg (mean ± SD) were used in randomized complete block design with five replications. The treatments were 0 g *Sesbania sesban* leaves (SSL) (control T_1), 100 g SSL (T_2), 200 g SSL (T_3), 300 g SSL (T_4) and 400 g SSL (T_5). The experiment lasted 90 days of growth and 7 days of digestibility period. All experimental sheep were given urea-treated rice straw ad libitum as basal diet. Chemical composition of diets and refusals, feed intake, digestibility of dry matter and nutrients, carcass yield and quality were recorded. All collected data were analyzed using general linear model (GLM) procedure of SAS version 9.1.3. Means were separated using Duncan's multiple-range test. Correlations among nutrient intakes, weight gain and digestibility of nutrients were analyzed using Pearson linear correlation test.

Results: Results showed that supplementation of SSL significantly increased ($P < 0.001$) total dry matter, crude protein and organic matter intakes. Supplementation of SSL improved the digestibility of dry matter, organic matter ($P < 0.05$), crude protein ($P < 0.01$). Final body weight ($P < 0.01$), average daily gain and feed conversion efficiency ($P < 0.001$) for T_5 were higher for supplement groups, while sheep fed sole UTRS (T_1) lost body weight. Slaughter weight ($P < 0.01$), empty body weight ($P < 0.01$) and hot carcass weight ($P < 0.001$) increase with increase in the level of supplementation. Dressing percentage on slaughter BW basis and EBW basis was higher ($P < 0.05$) for T_5 than control. The partial budge analysis result indicated that sheep fed UTRS alone had the lowest net return (− 92.22 ETB), while sheep supplemented with 400 g SSL (T_5) had better net return (144.6 ETB) with marginal rate (1.7) compared to the other supplemented treatments.

Conclusion: From the study it could be concluded that SSL could serve as alternative CP supplement in urea-treated-based crop residues feeding of growing Farta sheep to improve the biological performance and economic feasibility in the study area.

Keywords: Digestibility, Farta sheep, Feed intake, *Sesbania sesban*, Urea-treated rice straw

*Correspondence: limasm2009@gmail.com
Department of Animal Production and Technology, Bahir Dar University, P. O. Box. 5501, Bahir Dar, Ethiopia

Background

Livestock production is an important component of the farming systems in all parts of Ethiopia and plays a vital role in the livelihood of many people. Sheep and goats are widely reared in a crop–livestock farming systems and are distributed across different agroecological zones of Ethiopia. They require smaller investments, have shorter production cycles and greater environmental adaptability and hence have a unique niche in smallholder agriculture. Although diverse sheep and goat resources are found in Ethiopia, their productivity is low mainly because of inadequate year-round nutrition, in terms of both quantity and quality [1]. Livestock in general and sheep in particular are getting their feed from natural pasture and crop residues. Among the crop residues, rice straw has become involved as livestock feed. Rice straw is abundantly available crop residues in most tropical and subtropical countries and commonly used as a diet for ruminants, although feeding only rice straw does not provide enough nutrients to the ruminants to maintain high production levels due to the low nutritive value of this highly lignified material. By treating rice straw with urea intake, degradability and growth can be enhanced, compared to feeding untreated rice straw alone [2].

Most investigations in animal nutrition have been conducted on supplementation of poor-quality roughages with concentrates [3]; however, concentrate feeds comprised of especially grains and oil seeds are expensive and highly valued as human food [4]. Moreover, agro-industrial by-products are inaccessible and unaffordable to smallholder farmers; hence, alternative supplementary feedstuffs produced on-farm other than commercial concentrates are very essential. In this regard, strategic supplementation of unexploited, cheap, less competitive and year-round available and easily accessible protein sources like *Sesbania sesban* could be a feasible alternative to mitigate protein deficiency in poor-quality feeds during periods of feed scarcity. Therefore, the current study was conducted to evaluate the effects of graded levels of *Sesbania sesban* leaves supplementation on feed intake, digestibility, live weight change and carcass characteristics of Farta sheep fed urea-treated rice straw as basal diet.

Methods

Description of the experimental area

The sheep feeding trial was conducted at Zenzelma Campus of Bahir Dar University which is located between latitude and longitude of 11°37N 37°27E coordinates and an elevation of 1900 masl. The average daily minimum and maximum temperatures were 7 °C and 29 °C, respectively. The average annual rainfall was 1445 mm.

Experimental animals, their management and treatments

Twenty-five male Farta sheep with average age of 1 year and initial body weight of 18.9 ± 1.7 kg (mean ± SD) were purchased from Debre Tabor town. The age of the sheep was determined by dentition and information from the owners. The experimental animals were ear-tagged for identification. The sheep quarantined for 15 days to observe incidence of any disease. During this quarantine period, each animal was treated against internal (albendazole) and external (ivermectin) parasites. Animals were continuously observed for incidence of any ill health and disorders during the experimental period.

The experiment was conducted in a randomized complete block design with five treatments and five replications based on their initial body weight. Treatment 1 was the control, where animals received only urea-treated rice straw basal diet ad libitum. The remaining treatments involved supplementation of dried *Sesbania sesban* leaves at 100 g (treatment 2), 200 g (treatment 3), 300 g (treatment 4) and 400 g (treatment 5) per day on top of urea molasses-treated rice straw basal diet.

Rice straw was chopped to about 5–10 cm size to make it easily compacted. Four kg of urea dissolved with 80 liter water was used to treat 100 kg of rice straw. The straw was treated with a solution prepared on a plastic sheet on the floor, mixed thoroughly and rubbed with hand to ensure proper penetration of the solution. The treated straw was well trampled and compacted batch by batch until filled to the bag capacity. After filling the bag was tightened carefully and loaded by straw on top. The ensiled straw was left unopened for 21 days for proper treatment to occur. At the end of 21 days, the bag was opened and a portion of the straw was taken daily and ventilated overnight to remove residual ammonia before offering to the animals.

Digestibility, feed intake and body weight change

Digestibility trial was conducted before the commencement of the actual feeding trial. Sheep were fitted (harnessed) with fecal collection bags. Feces were collected and recorded every morning; 20% of representative samples were taken and frozen at − 20 °C and pooled over the collection period for each animal. At the end of the collection period, each sample was mixed and dried at 60 °C for 72 h. Apparent digestibility of nutrients was calculated as the proportion of the difference between nutrient consumed and nutrient in feces to nutrient consumed.

The feeding trial lasted for 90 days. During feed trial, daily feed offered and refusals were collected from each treatment throughout the experimental period. Daily mean feed intake was measured as differences between offered and refused feeds. The sheep was offered ad libitum hay, at 20% refusal. Feed refusal samples were taken daily per animal during the feeding and digestion period, pooled on treatment basis, and used for chemical analysis.

Initial body weight of animals was determined by taking the mean of two consecutive weights, and subsequent body weight measurements were taken at 10-day interval. Final body weight was taken at the end of the experiment. Average daily body weight gains (ADG) were calculated as difference between final and initial weights divided by the number of feeding days. According to [5, 6], feed conversion efficiency of experimental animals was determined by dividing the average daily BW gain to the amount of feed consumed.

Carcass parameters

Carcass parameters were analyzed at the end of the experiment. Blood, heart, liver, kidney, tongue, reticulorumen, omasum abomasums, hind gut (small and large intestine) and internal fat (kidney, omental and mesenteric) were weighed and considered as edible offal components. Head (without tongue), skin, testicle, penis, feet, gut content, bladder, spleen and lung with trachea were recorded and considered as non-edible offal components. Empty body weight (EBW) was calculated as the difference between slaughter weight (SW) and gut content. Dressing percentage (DP) was calculated as proportion of HCW to SW and EBW. Both the right and left halves were cut between the 12th and 13th ribs perpendicular to the backbone to measure the cross section of the rib eye area (REA).

Chemical composition analysis

The representative samples of daily feed offers, refusals and feces were taken. Dry matter content was determined by oven-drying samples at 105 °C. The crude protein (CP) content determined as *N**6.25. Ash was determined by igniting the samples in muffle furnaces at 550 °C over night. Neutral detergent fiber (NDF), acid detergent fiber (ADF) and acid detergent lignin (ADL) were analyzed following recommendation [7].

Statistical analysis

Data on feed intake, digestibility, weight gain and carcass parameter analysis were subjected to general linear model (GLM) procedures SAS 9.3.1. Means were separated using Duncan's multiple-range test. Correlation between nutrient intakes, weight gain and digestibility of nutrients were analyzed using Pearson correlation procedure.

Model

$$Y_{ij} = \mu + T_i + B_j + E_{ij}$$

where Y_{ij} is the response variable (the observation in j_{th} block and i_{th} treatment), μ overall mean, T_i treatment effect, B_j block effect, and E_{ij} random error.

Partial budget analysis

The difference in selling and purchasing price was considered as total return (TR). A net return (NR) was calculated by subtracting TVC from the total return (TR).

Results

Chemical composition of treatment feeds

The chemical composition of experimental feeds and refusals is shown in Table 1. The CP content of *Sesbania sesban* leaf was much higher than urea-treated rice straw. But the fiber fraction (NDF, ADF and ADL) in *Sesbania sesban* was clearly lower than the basal feed.

Table 1 Chemical composition of experimental feeds and urea-treated rice straw refusals

Parameters	DM	OM	Ash	CP	NDF	ADF	ADL
Feed offer							
UTRS	88	88.30	11.7	10.2	73.3	64.4	25.0
SSL	89	84.45	15.55	26.4	27.5	20.2	10.0
UMTRS refusals							
T_1	88	89.23	10.77	8.08	75.11	67.22	26.66
T_2	88	90.23	9.77	9.92	75.33	65.44	28.88
T_3	89	88.23	11.77	8.02	76.44	65.44	28.88
T_4	88	89.34	10.66	9.0	76.11	65.22	26.66
T_5	88	89.34	10.66	9.70	75.11	66.22	27.77

ADF acid detergent fiber, *ADL* acid detergent lignin, *CP* crude protein, *DM* dry matter, *NDF* neutral detergent fiber, *OM* organic matter, *UTRS* urea-treated rice straw, *UMTRS* urea molasses-treated rice straw, *SSL Sesbania sesban* leaves

Feed and nutrient intake

The feed DM and nutrient intake of Farta sheep fed urea-treated rice straw supplemented with *Sesbania sesban* are presented in Table 2. The mean daily DM intake of UTRS for the control was higher ($P<0.001$) than the supplemented treatments, but the contrary was true for TDMI. There was difference ($P<0.01$) in TDMI among treatments, and TDMI increased as the level of SSL increased. The intake of CP and OM was increased ($P<0.001$) in the order $T_5>T_4>T_3>T_2>T_1$. Intake of the NDF ($P<0.001$) and ADF ($P<0.01$) increased with the level of *Sesbania sesban* supplementation.

Nutrient digestibility

The apparent digestibility of DM, OM, CP, NDF, ADF and ADL in Farta sheep fed urea-treated rice straw and supplemented with *Sesbania sesban* is presented in Table 3. Apparent digestibility of DM, OM and CP for sheep supplemented sheep was higher ($P<0.01$) than control group. NDF and ADF digestibility ($P<0.01$) increases with increase in the level of *Sesbania sesban.*

Body weight change and feed conversion efficiency

Body weight change parameters of Farta sheep fed urea-treated rice straw supplemented with *Sesbania sesban* are presented in Table 4. Final body weight was significant ($P<0.01$) in the supplemented groups. Average daily gain, body weight change and feed conversion efficiency ($P<0.001$) were increased with supplementation level increased.

Carcass characteristics

Carcass characteristics of Farta sheep fed urea-treated rice straw and supplemented with dried *Sesbania sesban* leaves are presented in Table 5. Supplemented sheep had higher value slaughter weight (SW), empty body weight (EBW) ($P<0.01$) and hot carcass weight (HCW) ($P<0.001$) than sheep fed sole urea-treated rice straw. Dressing percentage as slaughter weight (DPSW) and

Table 2 Daily feed and nutrient intake of Farta sheep fed urea-treated rice straw supplemented with *Sesbania sesban*

Intake (g/day)	Treatment feeds					
	T_1	T_2	T_3	T_4	T_5	SL
Basal (UTRS) DMI	696.00^a	691.33^a	640.56^b	631.41^b	615.72^b	***
SSL DMI	–	100^d	200^c	300^b	400^a	***
Total DMI	696.00^e	796.00^d	840.56^c	931.41^b	1015.72^a	***
DM intake (%BW)	3^b	3.1^{ab}	3.2^{ab}	3.4^{ab}	3.6^a	*
Total OMI	536.72^d	611.17^c	620.10^c	717.21^b	768.4^a	***
Total CPI	69.59^e	98.27^d	114.98^c	148.39^b	169.46^a	***
Total NDFI	508.70^{bc}	536.42^{ab}	496.16^c	552.84^a	564.37^a	***
Total ADFI	447.24^b	465.53^{ab}	462.06^{ab}	469.5^{ab}	474.08^a	**
Total ADLI	201.40^{ab}	208.80^a	190.48^b	212.49^a	212.87^a	**

T_1 = UMTRS, T_2 = UTRS + 100 g SS DM, T_3 = UMTRS + 200 g SS DM, T_4 = UMTRS + 300 g SS DM, T_5 = UMTRS + 400 g SS DM, *UMTRS* urea molasses-treated rice straw, *SSL Sesbania sesban* leaves, *SEM* standard error of mean, *SL* significance level

abc Means with different superscripts in a row are significantly different at ***($P<0.001$); **($P<0.01$); *($P<0.05$)

Table 3 Dry matter and nutrient digestibility of Farta sheep fed urea-treated rice straw and supplemented with dried *Sesbania sesban* leaves

Digestibility	Treatment					
	T_1	T_2	T_3	T_4	T_5	SL
DM	56.5^b	65.4^{ab}	65.8^{ab}	67^a	67^a	*
OM	59.7^b	67.7^{ab}	68.3^{ab}	69.7^{ab}	70.6^a	*
CP	63.8^b	65^b	69.02^{ab}	72.44^a	73.02^a	**
NDF	49.6^c	54.5^b	60.3^b	63.9^{ab}	78.4^a	**
ADF	53.1^b	53.4^b	59.7^b	63.5^{ab}	81.1^a	**
ADL	47.9^b	52^b	55.2^b	61^{ab}	76.3^a	*

DM dry matter, *CP* crude protein, *NDF* neutral detergent fiber, *ADF* acid detergent fiber, *ADL* acid detergent lignin, *OM* organic matter, *SL* significance level, *NS* nonsignificant, T_1 (control) = UTRS, T_2 = UTRS + 100 g SSL, T_3 = UTRS + 200 g SSL, T_4 = UTRS + 300 SSL, T_5 = UTRS + 400 g SSL

ab Means in the same row with different superscript differ significantly at **$P<0.01$; *$P<0.05$

Table 4 Body weight change of Farta sheep fed urea treatment rice straw supplemented with graded leaves of dried *Sesbania sesban leaves*

Parameters	Treatments					
	T_1	T_2	T_3	T_4	T_5	SL
Initial body weight (kg)	19.18[a]	18.56[a]	18.54[a]	19.08[a]	18.16[a]	ns
Final body weight (kg)	17.44[c]	19.46[bc]	19.54[bc]	21.40[ab]	22.22[a]	**
Body weight change (kg)	−1.74[c]	0.90[b]	1.00[b]	2.32[ab]	4.06[a]	***
ADG (g/day)	−19.33[c]	10.00[b]	11.11[b]	25.77[ab]	45.11[a]	***
FCE (g ADG/g TDMI)	−0.028[c]	0.012[b]	0.014[b]	0.027[a]	0.044[a]	***

NS nonsignificant, *ADG* average daily gain, *FCE* feed conversion efficiency

[abc] Means with different superscripts in a row are significantly different at ***($P<0.001$); **($P<0.01$)

Table 5 Carcass characteristics of Farta sheep fed urea-treated rice straw and supplemented with dried *Sesbania sesban* leaves

Carcass characteristics	Treatment					
	T_1	T_2	T_3	T_4	T_5	SL
SW (kg)	17.52[b]	19.5[ab]	19.6[ab]	21.5[a]	22.60[a]	**
EBW (kg)	12.70[c]	13.6[bc]	14.30[bc]	16.2[ab]	17.0[a]	**
HCW (kg)	6.48[c]	6.73[c]	7.15[bc]	8.4[ab]	9.0[a]	***
Dressing percentage						
SW basis (%)	36.7[b]	37.46[ab]	37.70[ab]	39.0[ab]	40.0[a]	*
EBW basis (%)	48.3[b]	50.0[ab]	51.15[a]	51.3[a]	51.8[a]	*
REA (cm[2)])	8.41[a]	8.43[a]	8.8[a]	9.7[a]	10.8[a]	NS

SW slaughter weight, *EBW* empty body weight, *HCW* hot carcass weight, *REA* rib eye muscle area, *SL* significance level, *NS* nonsignificant

[a, b, c] Means with the same letter in the same row are not significantly different, **$P<0.01$, *$P<0.05$

dressing percentage as empty body weight (DPEBW) bases were significant ($P<0.05$) and better for T_5. Rib eye muscle area (REA) was not affected by supplementation.

Edible offal components

The edible offal components of the carcass of yearling Farta sheep fed urea-treated rice straw and supplemented with *Sesbania sesban* are presented in Table 6. Liver, kidney, reticulorumen, hind gut, ($P<0.05$) and testicles ($P<0.01$) were higher in T_5 compared to T_1. Lower total edible offal (TEO) was recorded in T_1 and higher in T_5. Blood, heart, tongue, omasum abomasum, omental and mesenteric fat and kidney were not affected by supplementation. Skin, lung with trachea and bladder were higher ($P<0.05$) for T_5 compared to other treatments. Spleen was higher ($P<0.01$) for T_5 compared to T_1.

Non-edible offal components

The non-edible offal components of the carcass obtained from the slaughter of yearling Farta sheep fed a basal diet of UTRS and supplemented with graded levels of dried *Sesbania sesban* leaf are presented in Table 7. The non-edible offal components such as skin, lung with trachea and bladder were significantly higher ($P<0.05$) for T_5. Head without tongue, feet, esophagus, gull bladder with bile and gut content were not significantly ($P>0.05$) affected by supplementation. Head without tongue, penis, feet, esophagus, gull bladder with bile and gut content were not affected by supplementation.

Partial budget analysis

The partial budget analysis result of Farta sheep fed urea-treated rice straw and supplemented with dried *Sesbania sesban* leaves is presented in Table 8. Sheep fed sole urea-treated rice straw had a negative net return (−92.22 ETB/sheep). Among all treatment groups, sheep supplemented with the highest level of *Sesbania sesban* leaves (T_5) returned a higher net return as compared to the other levels of supplementation and the control treatment. The net return from the supplemented treatments was 21.36, 73.76, 135.16 and 144.58 ETH/head for T_2, T_3, T_4 and T_5.

Table 6 Edible offal components of Farta sheep fed urea-treated rice straw and supplemented with dried *Sesbania sesban* leaves

Edible offal	Treatment					
	T_1	T_2	T_3	T_4	T_5	SL
Blood	718.1^{a}	769.3^{a}	771.1^{a}	785.83^{a}	807.9^{a}	NS
Heart	71.6^{a}	73.1^{a}	84.5^{a}	95.2^{a}	100.5^{a}	NS
Liver	213.8^{b}	256.1^{ab}	278^{ab}	291.1^{a}	322.2^{a}	*
Kidney	53.5^{b}	55.5^{ab}	62.4^{ab}	62.6^{a}	63.9^{a}	*
Tongue	80.4^{a}	85.67^{a}	86.42^{a}	86.47^{a}	89.3^{a}	NS
Reticulorumen	435.5^{b}	487. 2^{b}	501.2^{b}	575.8^{ab}	660.6^{a}	*
Omasum abomasum	160.2^{a}	161.4^{a}	174.1^{a}	186.0^{a}	187.7^{a}	NS
Hind gut	420.3^{b}	522.7^{ab}	509^{ab}	591.4^{ab}	610.4^{a}	*
Testicles	101.4^{b}	119.1^{b}	182.8^{ab}	210.2^{a}	239.3^{a}	**
Kidney fat	11^{a}	12.3^{a}	12.5^{a}	13.8^{a}	16.9^{a}	NS
Omental and mesenteric fat	32.13^{a}	36.78^{a}	43.43^{a}	56.5^{a}	62.48^{a}	NS
Total edible offal (kg)	2.31^{b}	2.62^{ba}	2.72^{ba}	3.01^{a}	3.04^{a}	*
Total usable product (kg)	10.26^{c}	10.82^{bc}	11.67^{bc}	13.2^{ba}	14.17^{a}	**
Total usable product (%SW)	59.43^{a}	59.7^{a}	61.08^{a}	61.2^{a}	64.48^{a}	NS

SL significance level, *NS* nonsignificant

[a, b] Means with the same letter in the same row are not significantly different,*$P<0.05$, **$P<0.01$

Table 7 Non-edible offal components of Farta sheep fed urea-treated rice straw and supplemented with dried *Sesbania sesban* leaves

Non-edible offal	Treatment					
	T_1	T_2	T_3	T_4	T_5	SL
Head without tongue (kg)	1.21^{a}	1.24^{a}	1.43^{a}	1.4^{a}	1.58^{a}	NS
Skin (g)	1.68^{b}	1.93^{ab}	1.94^{ab}	2.1^{ab}	2.43^{a}	*
Penis (g)	41.7^{a}	49.93^{a}	52.25^{a}	55.3^{a}	56.55^{a}	NS
Feet (g)	514.13^{a}	515.10^{a}	531.03^{a}	560.10^{a}	553.25^{a}	NS
Lung with trachea (g)	262.15^{b}	305.5^{ab}	305.6^{ab}	315.4^{ab}	380.55^{a}	*
Spleen (g)	21.28^{c}	24.65^{c}	29.55^{bc}	35.85^{ab}	39.48^{a}	**
Esophagus (g)	33^{a}	33.39^{a}	35.88^{a}	39.75^{a}	39.15^{a}	NS
Gull bladder with bile (g)	5.5^{a}	5.75^{a}	10.7^{a}	8.03^{a}	10.88^{a}	NS
Bladder	21.25^{b}	26.43^{ab}	23.28^{ab}	38.6^{ab}	54.48^{a}	*
Gut content (kg)	5.35^{a}	4.68^{a}	4.85^{a}	4.53^{a}	4.44^{a}	NS
Total non-edible offal (kg)	6.278^{b}	6.279^{b}	7.01^{ab}	8.06^{ab}	8.7^{a}	*

SL significance level, *NS* nonsignificant

[a, b] Means with the same letter in the same row are not significantly different,*$P<0.05$; **$P<0.01$

Discussion

Chemical composition of treatment feeds

The observed CP (10.17%) content of urea-treated rice straw in the current study was higher than 7.1% and 5.5% [8, 9]. This difference might be due to the stage of rice harvested and the time and season differences. The CP and ash composition of the refusal UTRS was reduced; however, NDF, ADF and ADL were increased compared to UTRS offer. This confirms the ability of sheep to select better nutritive value and palatable parts. Protein supplements are feedstuffs containing more than 20% CP [10]. Based on this *Sesbania sesban* leaves were recognized as protein source and it demonstrated its greatest value when supplemented with low-quality roughage like the rice straw used in this experiment. The value of CP content observed in the current study was higher than 23.9% and 23.34 [11, 12], respectively, and comparable to 26.8%, 26.37%, 27.5% and 25.3% [13–16], respectively.

The NDF and ADF values of *Sesbania sesban* leaves in this study comparable [15] (28.7% NDF and 23.6% ADF),

Table 8 Partial budget analysis of Farta sheep fed natural urea-treated rice straw and supplemented with dried *Sesbania sesban*

Variables	Treatments				
	T1	T2	T3	T4	T5
Number of animals	5	5	5	5	5
Purchasing price of sheep (ETB/head)	650	650	650	650	650
Total basal diet consumed (Kg/head)	62.22	62.64	54.24	56.84	55.42
Total supplement consumed (Kg/head)	–	9	18	27	36
Total feed consumed (Kg/head)	62.22	71.64	72.24	83.84	91.42
Cost of basal diet (ETB/head)	62.22	62.64	54.24	56.84	55.42
Cost of supplement (ETB/head)	–	36	72	108	144
Total feed cost (ETB/head)	62.22	98.64	126.24	164.84	199.42
Gross income (selling price of sheep) (ETB)	620	770	850	950	994
Total return (ETB)	− 30	120	200	300	344
Net return (ETB)	− 92.22	21.36	73.76	135.16	144.58
ΔTVC	–	36.42	64.02	102.62	137.2
ΔNI	–	113.6	166	227.4	236.8
MRR ((Δ NI/ΔTVC))	–	3.1	2.3	2.2	1.7

higher than the values [13] (19.3% NDF and 15.5% ADF), [12] (25.8% NDF and 17.5% ADF) [16] (21.23% NDF and 18.04% ADF) and lower than the values [17] (39.9% NDF and 29.9% ADF). ADL content of *Sesbania sesban* (10%) in this study was comparable to 9.8% [16] and higher than 4%, 4.1%, 7.5%, 5.5% and indicated by [12, 13, 15, 17], respectively. Generally, all the above differences in nutritional composition of *Sesbania sesban* may be attributed to differences in accession, stage of plant growth, cutting frequency and harvesting regimen, drying process, season, environment, soil type and fertility status, parts of the plant (leafs, twigs, whole forage and green pods) included during feeding and chemical analysis and due to the environmental conditions in which the plant grows.

Feed and nutrient intake

The UTRS DM intake decreases in supplement groups as compared to the control. This might be due to the substitution effect of SSL and its high nutrient content, which might have satisfied the nutrient requirement of the experimental sheep without additional UTRS DM intake. The supplement DM intake of T_2, T_3, T_4 and T_5 was 12.6, 23.8, 32.2 and 38.4% of total dry matter intake, respectively. [18] indicated that if the level of supplementation is <30–40% of the DM intake of the animal, there is an increase in the intake of the basal diet, which is contrary to what has been noted in this study. Therefore, such contradictory results reported by different researchers might be due to the amount of supplement, nutrient content and ratio given to the animals. Total DM intakes increase with increase in the level of supplementation. This might be due to the fact that supplementation might have created a favorable rumen environment resulting in enhanced fermentation of the basal roughage and thus increased microbial protein synthesis, rate and extent of digestion which resulted in better DM and nutrient intakes. The total DM intake noted in this study was higher than values of 711 g/head/day reported for Kaffa sheep [15] and comparable to 856–979 g/head/day reported for Hararghe highland sheep [19]. Values of the CP intake in this study were higher than the values that ranged from 54.23 to 111.98 g/day reported [15] and 53.26 to 62.97 g/day reported [12, 20, 21]. The higher intake of NDF and ADF in the supplemented group could mainly be a contribution attributed to the fiber content of supplement.

Apparent digestibility of DM and nutrients

Apparent digestibility of DM and OM was better for supplement groups. This is in agreement with report of McDonald et al. [10] who noted that addition of dietary protein in the supplement has increased protein availability to rumen microorganisms to speed up the digestion process. However, the control was grouped under low digestibility which could be due to lower CP and higher fiber content of UTRS. Contrary to the present study, [22, 23] reported nonsignificant difference in the apparent digestibility of the same parameters among the supplemented and control treatments, which could be an attribute of the types of supplement used. In the current study, DM digestibility in supplement groups comparable to 66.6–68.6% observed by Worku et al. [15] and slightly lower 65–70% reported [13] for Kaffa sheep and Arsi Bale sheep, respectively. CP digestibility was better for higher supplemented (300 g and 400 g SSL) groups. Higher values than the current finding, [15] reported CP digestibility of 85.8% in Kaffa sheep, and lower result (67.9%) of CP digestibility was reported by [14] for Tigray local sheep.

Body weight change and feed conversion efficiency

In the current study, sheep in the control group lost weight (1.74 kg). Body weight loss of sheep in T_1 indicated that UTRS used in this study was unable to provide sufficient energy for maintenance requirement even it had enough CP for maintenance requirement. [24] reported that sheep fed on poor-quality roughages alone are forced to mobilize energy and protein from body tissue reserves, leading to weight loss. Similarly, [9, 25] reported loss of body weight in sheep fed urea-treated rice straw and natural pasture grass hay alone, respectively. Average daily gain attained in this study for T_5 was higher than yearling Menz sheep (33.4–35.7 g/day) fed *teff* straw supplemented with sole *Sesbania sesban* [26] and lower than 103 g/day for sheep fed diet containing 300 g/day *Sesbania sesban* foliage [27] and 75.71 g/day for Abergelle rams supplemented with 300 g/day DM SSL [12]. Generally, differences among treatments in ADG and FCE observed in this study are consistent with similar differences in intake and digestibility of DM and nutrients.

Carcass characteristics

In this study for slaughter weight, empty body weight and hot carcass weight were better for supplemented treatments than the control treatment, which is consistent with that observed for ADG. Hot carcass weight in this trail was comparable with 5.55–12.4, 9.23–9.8 6.97–9.30 and 6.4–10.1 kg for Kaffa, Washera, Arsi Bale and Farta sheep reported by different authors [9, 13, 15, 25], respectively. However, it was heavier in other studies for Wogera sheep and Farta sheep breeds [28, 29]. Dressing percentage as a proportion of slaughter BW and EBW was better for higher level (400 g SSL) of supplemented treatments than the control treatment. As opposed to the current findings, various studies reported dressing percentages calculated on empty body weight basis as having no significant difference between supplemented and non-supplemented animals [9]. The dressing percentage (on EBW bases) values of the current study were comparable with the average dressing percentages of tropical sheep (40–50%) reported by William et al. [30].

Rib eye area (REA) was not different among treatments. Similarly, Hailu et al. [9] reported that different levels of concentrate in the diet had no impact on rib eye muscle area. Contrary to this in [15, 28] report, supplementation had a significant effect on the rib eye muscle area. The value obtained for rib eye area (REA) in this study was comparable to the 8.1–11.9 cm^2 values noted [25] and lower than 7.4–12.6 cm^2 [29] for the same breed. Generally, the variations in carcass traits in this study and previous studies may be due to variations in initial weight and supplement feeds used for the experiment. In agreement with this, [10] noted that nutrition, age, sex, genetics, season and other related factors affect the growth and carcass traits of animals.

Edible and non-edible offal components

Supplemented treatments had better edible offal components compared to the control T_1. This might be due to high metabolic activities of the organs and high intake of protein from supplemented sheep. This is supported by Kirton et al. [31] that reported dietary status has an impact on the production efficiency of offal. The nonsignificant differences in most of the organs could be linked with organs being early maturing and consequently being little affected by dietary treatments [31]. The increase in liver weight with supplementation might be related to the storage of reserve substances such as glycogen with supplementation [32]. Lower total edible offal (TEOC) was recorded in the control group than supplemented groups of sheep. This was in agreement with various studies [9, 15, 19, 25, 33] that reported heavier TEOC in supplemented sheep than the non-supplemented sheep.

The difference in slaughter weight also reflected in the weights of non-edible offal. In the current study, the non-edible offal components such as skin, lung with trachea and bladder show better result for T_5. Significant differences were observed in the weight of skin linked was better growth of hair following better feeding regime as a result of supplementation. The other non-edible offal components such as head without tongue, feet, esophagus, gull bladder with bile and gut content were not affected by supplementation. Offal components rich in bone and/or having low metabolic activity (head and feet) varied slightly with diet [31]. Gall bladder with bile also has nonsignificant difference between treatments, which might be due to lower intake of energy to stimulate liver for secretion of bile to emulsify fat and neutralize nutrient coming from stomach. The decreasing trend of gut content with increasing *Sesbania sesban* level might be contributed to the expected higher rate of digestion and faster passage rate of the diet through the digestive tract due to consumption of more digestible feed. However, animals on UTRS are forced to fill their gut with less digestible roughage and have proportionally bigger gut content. This result was also in agreement with the views [7, 34]; non-supplemented animals fill their gut with less digestible roughage which would be retained in the gut for a long time to be degraded by rumen microorganisms. In this study total non-edible offal was highest for supplemented sheep than sheep fed on sole urea-treated rice straw. The heavier TEOC and TNEOC in supplemented sheep than the non-supplement ones suggest that most of the edible and non-edible offal components were not fully grown at the beginning of this study and were able to respond to supplementation.

Partial budget analysis

The difference in net return was in a similar trend with their weight gain, i.e., loss of 92.22 ETB per animal was observed for the control treatment because weight loss was recorded. The marginal rate of return in this study was 3.1, 2.3, 2.2 and 1.7 for T_2, T_3, T_4 and T_5, respectively. This indicates that to attain required BW by supplement feeding, each additional unit of 1 ETB increment per sheep to purchase supplement feed resulted in a profit of 3.1 ETB for $T_{2,\ 2.3}$ ETB for $T_{3,\ 2.2}$ ETB for T_4 and 1.7 ETB for T_5, respectively. The result achieved in the present study was higher than 1.07–1.15 [12]. This may be the variations in purchasing and selling price of sheep, variations in sheep breeds used and differences in basal diet and supplements used in different experiments. Supplementation of sheep (400 g/head/day on DM basis) had more satisfactory daily BW gain (45.11 g/day), highest NI (144.6 ETB), highest NI (236.8 ETB) but lowest MRR (1.7) compared to other treatments. In this study, supplementation with T_5 improved body weight change of sheep and correspondingly increased the net income from the sale of sheep at the end of the feeding trial. Therefore, among the supplemented treatments, T_5 was found to be more profitable considering the net return and marginal rate of return.

Conclusion

Supplementation of Farta sheep with *Sesbania sesban* leaves increased the total dry matter, crude protein and organic matter intakes. Supplementation with SSL also significantly improved apparent digestibility of nutrients, average daily gain, feed conversion efficiency and carcass characteristics. Hence, in the present study, it was concluded that supplementation of UTRS with 400 g/day SSL is biologically efficient, potentially profitable and socially acceptable level for growing Farta sheep. As SSL was higher in CP content, high performance could be expected from feeding high level of dried SSL. So, further investigation is recommended by combining *Sesbania sesban* leaves with another good energy source basal diet feed.

Abbreviations

EBW: empty body weight; BW: body weight; ETB: Ethiopian birr; HC: hot carcass; HCW: hot carcass weight; TVC: total variable cost.

Authors' contributions

LT, YM and BA contributed to research proposal writing, data collection, data analysis, data interpretation and article writing. FM contributed in research data collection, data analysis, data interpretation and article writing. All authors read and approved the final manuscript.

Authors' information

Lamrot is a Lecturer of Animal Production currently teaching and conducting research in animal production and productivity improvement in Ethiopia. Yeshambel is an Associate Professor of Animal Feeds and Nutrition, currently teaching and conducting research in the area of animal feeds and feeding, animal production and productivity improvement in Ethiopia. Bimrew is an Associate Professor of Animal Feeds and Nutrition, currently teaching and conducting research in the area of animal feeds and feeding, animal production and productivity improvement in Ethiopia. Fentahun is a Lecturer of Biotechnology in Animal Production, currently teaching and conducting research in animal production and productivity improvement in Ethiopia.

Acknowledgements

The authors would like to acknowledge Bahir Dar University for funding this research and Debre Berhan Agricultural Research Center Animal Nutrition laboratory for assisting in sample feed and fecal analysis.

Competing interests

The authors declare that they have no competing interests.

Funding

Not applicable.

References

1. Tibbo M. Productivity and health of indigenous sheep breeds and crossbreds in the Central Ethiopian highlands. Faculty of Medicine and Animal Science Department of Animal Breeding and Genetics. Ph.D. Dissertation. Swedish University of Agricultural Sciences. 2006.
2. Wanapat M, Polyrach S, Boonnop K, Mapato C, Cherdthong A. Effect of treating rice straw with urea and calcium hydroxide upon intake, digestibility, rumen fermentation and milk yield of dairy cows. Livest Sci. 2009;2009(125):238–43.
3. Mekoya A. Multipurpose fodder trees in Ethiopia; Farmers' perception, constraints to adoption and effects of long-term supplementation on sheep performance. Ph.D. thesis, Wageningen University, Wageningen, The Netherlands. 2008.
4. Ayenew A, Tamir B, Melaku S. Feed intake, digestibility and live weight change of lambs fed Finger millet (*Eleusine coracana*) straw

supplemented with Atella, noug seed (*Guizotia abyssinica*) cake and their mixtures. Agric Trop Subtrop. 2012;45(3):105–11.
5. Gülten K, Rad F, Kindir M. Growth performance and feed conversion efficiency of Siberian Sturger o juveniles (*Acipenser baeri*) reared in concentrate ways. Turk J Vet Anim Sci. 2000;2000(24):439.
6. Brown L, Hindmarsh R, Mcgregor R. Dynamic agricultural book three. 2nd ed. Sydney: McGraw-Hill Book Company; 2001.
7. Van Soest PJ. Nutritional ecology of ruminants. 2nd ed. London: Cornell University Press; 1994.
8. Derso T. On-farm evaluation of urea treated rice straw and rice bran supplementation on feed intake, milk yield and composition of Fogera cows, North Western Ethiopia. A Thesis Submitted to the Department of Animal Science and Technology School of Graduate Studies, Bahir dar Universitiy. 2009.
9. Hailu A, Melaku S, Tamir B, Tassew A. Body weight and carcass characteristics of Washera sheep fed urea treated rice straw supplemented with graded levels of concentrate mix. Livest Res Rural Dev. 2011;23(8).
10. McDonald P, Edwards RA, Greenhalgh JFD, Morgan CA, Sinclair LA, Wilkinson RG. Animal nutrition. 7th ed. England: Longman Group UK Ltd; 2010.
11. Sabra HA, Hassan SG, Mohammed MI. Effect of *Sesbania sesban* (*Sesbania egyptiaca*) Supplementation on the reproductive performances of Baladi Sheep as compared to Berseem (*Egyptian clover*). J Reprod Infertil. 2010;1(3):66–70.
12. Tekle D, Gebru G, Hagos H, Belay S. Effect of supplementation dried Sesbania sesban leaf on performance of Abergelle rams. J Biol Agric Healthc. 2016;6(11):125–129
13. Bekele B, Mekasha S, Melaku Y. Effect of substitution of concentrate mix with *Sesbania sesban* on feed intake, digestibility, body weight change, and carcass parameters of Arsi-Bale sheep fed a basal diet of native grass hay. Trop Anim Health Prod. 2013. https://doi.org/10.1007/s11250-013-0413.
14. Hadgu H. Effect of supplementation of concentrate mixture, dried local brewery byproduct (atella), *faidherbia albida* and *sesbania sesban* on the performance of local sheep fed hay basal diet. M.Sc. thesis Presented to Haromiya University. 2014.
15. Worku A, Animut G, Urge M. Supplementing rice bran, Sesbania (*Sesbania sesban*) leaf and their mixtures on digestibility and performance of Kafka sheep fed native grass hay. Int J Agric Sci Res. 2015;4(3):57–66.
16. Yirdaw M, Mengistu A, Tamir B, Berhane G. Replacing cotton seed cake by dried *Acacia saligna*, Sesbania sesbana and *Cowpea* on productivity of Begait sheep in North Ethiopia. Am Eurasian J Sci Res. 2017;12(1):29–36.
17. Etana D, Tolera A, Eik LO, Ragnar S. Nutritive value of morphological fractions of *Sesbania sesban* and *Desmodium intortum*. Trop Sub-tropical Agro Ecosyst. 2011;2011(14):793–805.
18. Topps, JH. Forage legumes as protein supplement to poor quality diets in the semi-arid tropics. In: Wallance RJ, Lahlov-kassi A, editors. Rumen ecology research planning. Proceeding of a workshop held at ILRI (International Livestock Research Institute). Ethiopia: Addis Ababa; 1997;13–18 March 1995 p. 45–50.
19. Yirga H. Supplementation of concentrate mix to Hararghe Highland sheep fed a basal diet of urea treated maize Stover; effect on feed utilization, live weight change and carcass characteristics. An M.Sc. thesis presented to the school of graduate studies of Haramaya University. 2011.
20. Gulilat L, Walelign E. Evaluation of milk production performance of lactating Fogera cows fed with urea and effective micro-organisms treated rice straw as basal diet. Int J Sci Res Publ. 2017;7:111–9.
21. Gizachew A. Comparative feedlot performance of Washera and Horro sheep fed different roughag to concentrate ratio M.Sc. Thesis submitted to Haromiya University. 2012.
22. Mekuriaw Y, Urge M, Anmut G. Intake, digestibility, live weight changes and rumen parameters of Washera sheep fed mixture of low land bamboo (*Oxytenanthera abyssinca*) leaves and natural pasture grass hay at different ratios. Pak J Nutr. 2012;11(4):322–31.
23. Bonsi MLK, Osuji PO, Tuah AK. Effect of supplementing teff straw with different levels of leucaena or sesbania leaves on the degradabilities of teff straw, sesbania, leucaena, tagasaste and vernonia and on certain rumen and blood metabolites in Ethiopian Menz sheep. Anim Feed Sci Technol. 1995;52:101–29.
24. Assefa A, Animut G. Supplementation of raw, malted and heat treated grass pea (*Lathyrus Sativus*) grain on body weight gain and carcass characteristics of Farta sheep. Int J Soil Crop Sci. 2013;1(1):1–6.
25. Melaku S, Peters KJ, Tegegne A. Feed intake, live weight gain and reproductive performance of Menz sheep ewes supplemented with *Lablab purpures*, graded levels of *Leucaena pallida* 1403 and *Sesbania sesban* 1998. Livest Prod Sci. 2004;87:131–42.
26. Manaye T, Tolera A, Zewdu T. Feed intake, digestibility and body weight gain of sheep fed Napier grass mixed with different levels of *Sesbania sesban*. Livest Sci. 2009;122(1):124–9.
27. Moges M, Tamir B, Yami A. The effects supplementation of grass hay with different level of brewery dried grain on feed intake digestibility and body weight gain in intact Wogera lambs. East Afr J Sci. 2008;2:105–10.
28. Dejen. M.Sc. thesis on effect of supplementation of hay with graded level of rapeseed cake and rice bran mixture on feed intake, digestibility, body weight change and carcass characteristics of Farta sheep. M.Sc. thesis. 2011.
29. Williams CB, Jenkins TG. A dynamic model of metabolizable energy utilization in growing and mature cattle. II. Metabolizable energy utilization for gain. J Anim Sci. 2003;81:1382–9.
30. Kirton AH, Bennet LG, Dobbie LJ, Mercer KG, Duganzich MD. Effect of sire breed, sex and growth path on carcass composition of cross-breed lambs. J Agric Res. 1995;38:105–14.
31. Hag MGEI, Shargi KMEI. Feedlot performance and carcass characteristics of local (Dhofari) and exotic (Cashmere) goats fed on a high-fibre by-products diet supplemented with fish sardine. Rumais Livestock Research Station. Ministry of Agriculture and Fisheries. Muscat, 1996;391–395.
32. Lawrence TJ, Fowler VR. Growth of farm animals. Wallingford: CABI; 1998.
33. Defar G, Urge M. Bodyweight gain and carcass characteristics of Horro sheep fed urea treated maize husk and untreated maize husk supplemented with different levels of concentrate mix at Bako, Western Ethiopia; 2017.
34. Pond WG, Church DC, Pond KR. Basic animal nutrition and feeding. 4th ed. New York: Wiley; 1995.

PERMISSIONS

All chapters in this book were first published in A&FS, by BioMed Central; hereby published with permission under the Creative Commons Attribution License or equivalent. Every chapter published in this book has been scrutinized by our experts. Their significance has been extensively debated. The topics covered herein carry significant findings which will fuel the growth of the discipline. They may even be implemented as practical applications or may be referred to as a beginning point for another development.

The contributors of this book come from diverse backgrounds, making this book a truly international effort. This book will bring forth new frontiers with its revolutionizing research information and detailed analysis of the nascent developments around the world.

We would like to thank all the contributing authors for lending their expertise to make the book truly unique. They have played a crucial role in the development of this book. Without their invaluable contributions this book wouldn't have been possible. They have made vital efforts to compile up to date information on the varied aspects of this subject to make this book a valuable addition to the collection of many professionals and students.

This book was conceptualized with the vision of imparting up-to-date information and advanced data in this field. To ensure the same, a matchless editorial board was set up. Every individual on the board went through rigorous rounds of assessment to prove their worth. After which they invested a large part of their time researching and compiling the most relevant data for our readers.

The editorial board has been involved in producing this book since its inception. They have spent rigorous hours researching and exploring the diverse topics which have resulted in the successful publishing of this book. They have passed on their knowledge of decades through this book. To expedite this challenging task, the publisher supported the team at every step. A small team of assistant editors was also appointed to further simplify the editing procedure and attain best results for the readers.

Apart from the editorial board, the designing team has also invested a significant amount of their time in understanding the subject and creating the most relevant covers. They scrutinized every image to scout for the most suitable representation of the subject and create an appropriate cover for the book.

The publishing team has been an ardent support to the editorial, designing and production team. Their endless efforts to recruit the best for this project, has resulted in the accomplishment of this book. They are a veteran in the field of academics and their pool of knowledge is as vast as their experience in printing. Their expertise and guidance has proved useful at every step. Their uncompromising quality standards have made this book an exceptional effort. Their encouragement from time to time has been an inspiration for everyone.

The publisher and the editorial board hope that this book will prove to be a valuable piece of knowledge for researchers, students, practitioners and scholars across the globe.

LIST OF CONTRIBUTORS

Alison L. Van Eenennaam and Amy E. Young
Department of Animal Science, University of California, 2113 Meyer Hall, One Shields Avenue, Davis, CA 95616, USA

Chloé Poitevin DesRivières
Department of Geography and Environmental Studies, Carleton University, Ottawa, ON, Canada

Ratana Chuenpagdee and Charles Mather
Department of Geography, Memorial University of Newfoundland, St. John's, NL, Canada

Sagana Pushparajah and Jeyalingawathani Sinniah
Department of Animal Science, Faculty of Agriculture, University of Jaffna, Ariviyal Nagar, Kilinochchi, Sri Lanka

Fsahatsion Hailemariam, Dawit Gebremicheal and Hagos Hadgu
Department of Animal Science, College of Agriculture, Aksum University Shire Campus, Tigray, Ethiopia

Jack B. Hetherington
Australian Centre for International Agricultural Research, Bruce, ACT 2612, Australia
Sydney School of Veterinary Science, The University of Sydney, Camperdown, NSW 2006, Australia

Anke K. Wiethoelter and Siobhan M. Mor
Sydney School of Veterinary Science, The University of Sydney, Camperdown, NSW 2006, Australia
Charles Perkins Centre, The University of Sydney, Camperdown, NSW 2006, Australia

Joel Negin
Sydney School of Public Health, The University of Sydney, Camperdown, NSW 2006, Australia

H. K. S. De Zoysa
Department of Bioprocess Technology, Faculty of Technology, Rajarata University of Sri Lanka, Mihintale, Sri Lanka
Department of Physical Sciences, Faculty of Applied Sciences, Rajarata University of Sri Lanka, Mihintale, Sri Lanka

E. M. R. K. B. Edirisinghe
Department of Physical Sciences, Faculty of Applied Sciences, Rajarata University of Sri Lanka, Mihintale, Sri Lanka

B. K. K. K. Jinadasa and G. D. T. M. Jayasinghe
Institute of Post-Harvest Technology (IPHT), National Aquatic Resources Research and Development Agency (NARA), Colombo, Sri Lanka

Y. Dagnew
Department of Animal and Range Sciences, College of Agriculture, Bule Hora University, Bule Hora, Ethiopia
School of Animal and Range Sciences, College of Agriculture and Environmental Science, Haramaya University, Dire Dawa, Ethiopia

M. Urge and Y. Tadesse
School of Animal and Range Sciences, College of Agriculture and Environmental Science, Haramaya University, Dire Dawa, Ethiopia

S. Gizaw
Herd Health, Livestock Expert, International Livestock Research Institute, Addis Ababa, Ethiopia

Wuletaw Mekuria
Centre for Rural Development, College of Development Studies, Addis Ababa University, Addis Ababa, Ethiopia

Kindu Mekonnen
Crop-Livestock Systems, International Livestock Research Institute, Addis Ababa, Ethiopia

Bimrew Asmare
Department of Animal Production and Technology, College of Agriculture and Environmental Sciences, Bahir Dar University, Bahir Dar, Ethiopia

Zemenu Yayeh
Department of Animal Sciences, College of Agriculture and Natural Resources, Debre Markos University, Debre Markos, Ethiopia

Robert Zougmoré, Samuel Partey and Mathieu Ouédraogo
CGIAR Research Program on Climate Change, Agriculture and Food Security (CCAFS), International Crops Research Institute for the Semi-Arid Tropics (ICRISAT), BP 320, Bamako, Mali

Bamidele Omitoyin
Department of Aquaculture and Fisheries Management, University of Ibadan, Ibadan, Nigeria

Timothy Thomas
Environment and Production Technology Division (EPTD), International Food Policy Research Institute (IFPRI), 2033 K Street NW, Washington, DC 20006, USA

Augustine Ayantunde
International Livestock Research Institute, Ouagadougou, Burkina Faso

Polly Ericksen and Mohammed Said
International Livestock Research Institute, Nairobi, Kenya

Abdulai Jalloh
West and Central African Council for Agricultural Research and Development (CORAF/WECARD), BP 48, Dakar RP, Senegal

Shewangzaw Addisu Mekuria, Aschalew Assefa Teshager, Addis Getu Endeshaw and Malede Birhan Atinaw
University of Gondar, Gondar, Ethiopia

Assemu Tesfa Sendeku
Andassa Livestock Research Center, Bahir Dar, Ethiopia

Mekete Bekele Tahir
Department of Animal Science, College of Agriculture and Natural Resource Sciences, Debre Berhan University, Debre Berhan, Ethiopia
Department of Animal Production Studies, College of Veterinary Medicine and Agriculture, Addis Ababa University, Debre Zeit, Ethiopia

Ashenai Mengistu Wossen and Berhan Tamir Mersso
Department of Animal Production Studies, College of Veterinary Medicine and Agriculture, Addis Ababa University, Debre Zeit, Ethiopia

Syma Alexi Ebbin
Department of Agricultural and Resource Economics, University of Connecticut, Avery Point Campus, 1080 Shennecossett Rd, Groton, CT 06340, USA

Yeshambel Mekuriaw and Bimrew Asmare
Department of Animal Production and Technology, Bahir Dar University, Bahir Dar, Ethiopia

David Clifford Love
Johns Hopkins Center for a Livable Future, Johns Hopkins University, 615 N. Wolfe Street, Baltimore, MD, USA
Department of Environmental Health and Engineering, Bloomberg School of Public Health, Johns Hopkins University, 615 N. Wolfe Street, Baltimore, MD, USA

Jillian Parry Fry
Johns Hopkins Center for a Livable Future, Johns Hopkins University, 615 N. Wolfe Street, Baltimore, MD, USA
Department of Environmental Health and Engineering, Bloomberg School of Public Health, Johns Hopkins University, 615 N. Wolfe Street, Baltimore, MD, USA
Department of Health, Behavior, and Society, Bloomberg School of Public Health, Johns Hopkins University, 624 N. Broadway, Baltimore, MD, USA

Patricia Pinto da Silva, Julia Olson and Patricia Mary Clay
Social Sciences Branch, Northeast Fisheries Science Center, National Marine Fisheries Service, 166 Water Street, Woods Hole, MA, USA

Workiyie Worie Assefa
Department of Biology and Blue Nile Water Institute, College of Science, Bahir Dar University, Bahir Dar, Ethiopia

Wubneh Belete Abebe
Amhara Design and Supervision Works Enterprise, Bahir Dar, Ethiopia

Wubetie Adnew Wassie and Berhanu Abraha Tsegay
Department of Biology, College of Science, Bahir Dar University, Bahir Dar, Ethiopia

Asaminew Tassew Wolde and Bimrew Asmare Limeneh
Department of Animal Production and Technology, College of Agriculture and Environmental Sciences, Bahir Dar University, Bahir Dar, Ethiopia

Lamrot Tekliye, Yeshambel Mekuriaw, Bimrew Asmare and Fantahun Mehret
Department of Animal Production and Technology, Bahir Dar University, Bahir Dar, Ethiopia

Index